Introduction to Planetary Geomorphology

Nearly all major planets and moons in our Solar System have been visited by spacecraft, and the data they have returned have revealed the incredible diversity of planetary surfaces. Featuring a wealth of images, this textbook explores the geologic evolution of the planets and moons.

Introductory chapters discuss how information gathered from spacecraft is used to unravel the geologic complexities of our Solar System. Subsequent chapters focus on current understandings of planetary systems. The textbook shows how planetary images and remote sensing data are analyzed through the application of fundamental geologic principles. It draws on results from spacecraft sent throughout the Solar System by NASA and other space agencies.

Aimed at undergraduate students in planetary geology, geoscience, astronomy, and Solar System science, it highlights the differences and similarities of planetary surfaces at a level that can be readily understood by non-specialists.

Electronic versions of figures from the book are available at www.cambridge.org/Greeley.

RONALD GREELEY (1939–2011) was a Regents' Professor in the School of Earth and Space Exploration, Arizona State University, Director of the NASA–ASU Regional Planetary Image Facility, and Principal Investigator of the Planetary Aeolian Laboratory at NASA-Ames Research Center. He co-authored several well-known books on planetary surfaces, including *The Compact NASA Atlas of the Solar System* and *Planetary Mapping* (both available from Cambridge University Press).

Introduction to Planetary Geomorphology

Ronald Greeley

Arizona State University

CAMBRIDGE
UNIVERSITY PRESS

CAMBRIDGE
UNIVERSITY PRESS

University Printing House, Cambridge CB2 8BS, United Kingdom

One Liberty Plaza, 20th Floor, New York, NY 10006, USA

477 Williamstown Road, Port Melbourne, VIC 3207, Australia

314–321, 3rd Floor, Plot 3, Splendor Forum, Jasola District Centre, New Delhi – 110025, India

103 Penang Road, #05–06/07, Visioncrest Commercial, Singapore 238467

Cambridge University Press is part of the University of Cambridge.

It furthers the University's mission by disseminating knowledge in the pursuit of
education, learning and research at the highest international levels of excellence.

www.cambridge.org
Information on this title: www.cambridge.org/9780521867115

First published 2013
7th printing 2021

Printed in the United States of America by Sheridan Books, Inc.

A catalogue record for this publication is available from the British Library

Library of Congress Cataloguing in Publication data
Greeley, Ronald.
Introduction to planetary geomorphology / Ronald Greeley.
  p. cm.
ISBN 978-0-521-86711-5 (hardback)
1. Planets – Geology – Popular works. 2. Planets – Crust – Popular
works. 3. Geomorphology. I. Title.
QB603.G46.G74   2013
551.410999'2–dc23

2011053270

ISBN 978-0-521-86711-5 Hardback

Additional resources for this publication at www.cambridge.org/greeley

For
Randall and Lidiette
Thomas, Rebecca, and Jennifer

CONTENTS

FOREWORD

Robert T. Pappalardo

Ron Greeley's *Introduction to Planetary Geomorphology* is the single most outstanding and complete compendium of the science of planetary geology that exists. It is a fully complete and up-to-date synopsis of the science of planetary geology, written in Greeley's characteristically succinct and clear style. This is the ideal primer for an upper undergraduate course, and an excellent compendium for the interested amateur or professional astronomer. The figures within are all the "right" ones – the very best for illustrating fundamental concepts and "type examples" of terrestrial and planetary processes – compiled here in one place.

Ron Greeley passed away suddenly in the fall of 2011, just a month after submitting this complete book for publication. It will remain a tribute to his life's work, encapsulating his passions both for research and for teaching.

Greeley was a scholar and a gentleman, and a pioneer in the methods of planetary geology. His Ph.D. research at the University of Missouri at Rolla included field work on the Mississippi Barrier Islands, where he studied modern living forms of organisms that he was researching in the fossil rock record. This work laid the foundation for his practical approach to deciphering the processes that have shaped the surfaces of other planets by studying their modern Earth analogues. In the laboratory and the field, Greeley would effectively visit other worlds and other times.

Greeley's career in planetary geology began in 1967, when he was called to active military service just a year after receiving his Ph.D. Fortunately, he was assigned to NASA's Ames Research Center to work on *Apollo*-related problems. He occasionally mused about whether this interesting assignment came about by someone's misunderstanding of his dissertation topic of "lunulitiform bryozoans" as somehow related to the geology of the Moon!

Greeley trailblazed the burgeoning field of planetary geomorphology at Ames. While the *Apollo* missions explored the Moon, Greeley conducted detailed comparisons of lunar sinuous rilles with terrestrial volcanic landforms in Hawaii and the Snake River Plain of Idaho, making important contributions to understanding lunar processes. Then, as early 1970s *Mariner 9* photos began to reveal Mars, Greeley used wind tunnels at Ames to simulate how aeolian processes might operate on the Red Planet. His seminal work on terrestrial and planetary aeolian processes is being applied anew today to explanation of dunes on Saturn's moon Titan, which were recently discovered by the *Cassini* spacecraft.

Greeley was involved in nearly every major spacecraft mission flown in the Solar System since *Apollo*. This includes the *Magellan* mission to Venus and the *Galileo* mission to Jupiter. He contributed to a panoply of missions to Mars: *Mariners 6, 7,* and *9, Viking, Mars Pathfinder, Mars Global Surveyor,* the two *Mars Exploration Rovers,* and the European Space Agency's (ESA's) *Mars Express.* He chaired many NASA and National Research Council (NRC) panels, including NASA's Mars Exploration Program Analysis Group, the *Mars Reconnaissance Orbiter* Science Definition Team, the NASA–ESA Joint Jupiter Science Definition Team, the NRC's Committee on Planetary and Lunar Exploration, and most recently the Planetary Science Subcommittee of the NASA Advisory Council.

For those fortunate enough to have known Ron Greeley first-hand, through his teaching, research, committees, or friendship, this book will serve as a lasting tribute. For those learning of him, and from him, for the first time, welcome to this man and his work.

As Greeley would say: a journey of a thousand miles begins with a single step. In your introduction to the planets and moons of our Solar System, the journey of 4.5 billion kilometers begins with a turn of the page. An adventure awaits.

PREFACE

Planetary geoscience had its inception with the birth of the Space Age in the early 1960s. In the ensuing decades, it has evolved into a discipline that is recognized by sections of professional organizations such as the Geological Society of America and the American Geophysical Union, as well as being taught at the university level. Much of our understanding of the geology of extra-terrestrial objects is derived from remote sensing data – primarily images that portray planetary surfaces. In fact, discoveries such as the dry river beds on Mars, the tectonic deformation of Venus, and the actively erupting volcanoes on Jupiter's moon Io all came from pictures taken from spacecraft. Thus, the focus of this book is on the geo-morphology of solid-surface objects in our Solar System and the interpretations of the processes that led to the diverse landforms observed. Geomorphology, however, must be analyzed in the context of broader geoscience; consequently, in the chapters on the individual planetary systems, the geophysics and interior characteristics are reviewed along with our current understanding of surface compositions and the general geologic histories. Of course, our knowledge of the Solar System is far from uniform from one planet to another, dependent upon the numbers and types of spacecraft that have returned data. Thus, the chapters on the Moon and Mars are more detailed than those on the outermost planet systems, Uranus and Neptune, because dozens of successful space-craft have visited our nearest planetary neighbors, in con-trast to the limited data returned from "flybys" of the *Voyager* spacecraft to the planets beyond Saturn.

Our journey to explore the geomorphology of the Solar System begins with introductory chapters that introduce the planets and other objects of planetary geoscience interest, discuss the methods used in studying extraterres-trial objects, and review the fundamental geomorphic processes on Earth that can be compared with what we see on other planets and satellites.

Key references are given in the text and listed at the back of the book. The end of this book includes additional reading for those who wish to delve into the chapter topic in more detail. Because images form the basis of much of planetary geomorphology, figure cap-tions generally include the basic NASA or other space agency data for the frames shown, to enable the use of various electronic search engines for obtaining additional information.

I hope that you find the exploration of the Solar System a rewarding experience. While many planets and satellites show landforms that are quite familiar to geologists, others hold surprises that have not yet been explained or understood. Have fun, and maybe you can solve some of these mysteries!

ACKNOWLEDGMENTS

I thank the countless graduate and undergraduate students from planetary geoscience classes who have provided the stimulus for this book. Students have the marvelous capacity to ask thought-provoking questions that remind planetary scientists that there is still a great deal to learn about the Solar System. I also thank my colleagues who are at the forefront of planetary exploration for their keen insight into the complexities of geologic surface evolution. While they have helped me tremendously in understanding these complexities, any errors in fact or interpretation contained herein are solely my responsibility.

The preparation of this book was facilitated by the talents of Sue Selkirk for illustrations, Amy Zink for preparation of the final versions of the images, and Dan Ball and the NASA Space Photography Laboratory for access to the planetary images; I am grateful for their substantial help. I thank Stephanie Holaday for her tireless word-processing of many draft iterations as well as tracking down permissions for previously published figures. Her assistance has been invaluable and is much appreciated. Finally, I thank Cindy Greeley for her editing skills and corrections to my flawed writing!

CHAPTER 1

Introduction

The early part of the twenty-first century saw the completion of the reconnaissance of the Solar System by spacecraft. With the launch of the *New Horizons* spacecraft to Pluto in early 2006 and its expected arrival in 2015, spacecraft will have been sent to every planet, major moon, and representative asteroid and comet in our Solar System. With the return of data taken by spacecraft of these objects, the study of planetary surfaces passed mostly from the astronomer to the geologist and led to the establishment of the field of **planetary geology**.[1] The term geology is used in the broadest sense to include the study of the solid parts of planetary objects and includes aspects of geophysics, geochemistry, and cartography. Much of our knowledge of the geologic evolution of planetary surfaces is derived from remote sensing, *in situ* surface measurements, geophysical data, and the analysis of landforms, or their geomorphology, the primary subject of this book.

In this chapter, an overview of Solar System objects is given, the objectives of Solar System exploration are outlined, and the strategy for exploration by spacecraft is discussed. In the following chapters, the approach used in understanding the geomorphology of planets is presented, including the types and attributes of various data sets. The principal geologic processes operating on planets are then introduced, and the geology and geomorphology of each planetary system is described in subsequent chapters. The book ends with a discussion of future missions and trends in Solar System exploration.

1.1 Solar System overview

Our Solar System consists of a fascinating array of objects, including the Sun, planets and their satellites,

comets and asteroids, and tiny bits of dust. Most of the mass of the Solar System is found within the Sun, a rather ordinary star that generates energy through nuclear fusion with the conversion of hydrogen to helium. Coupled with astrophysical models, analyses of meteorites suggest that the Solar System began to form at about 4.6 Ga (Ga is the abbreviation for giga or 10^9 *annum*, or years).

Planets are relatively large objects that are in orbit around the Sun. As we learned at a very young age, the planets are Mercury, Venus, Earth, Mars, Jupiter, Saturn, Uranus, and Neptune. And then there is Pluto! The year 2006 saw an interesting controversy emerge when the International Astronomical Union (the scientific group responsible for formal naming of objects in the heavens) declared that Pluto was no longer a "planet" and demoted it to a new class of objects called "dwarf planets." This issue will be discussed later.

All of the planets originally formed through the accretion of dust and smaller objects, making **protoplanets**. As the protoplanets grew in size, still more dust and accreted materials were swept up, a process that continues even today. For example, it is estimated that more than 10,000 tons of materials are added to Earth each day. Although this addition is impressive, it is insignificant in comparison with the orders-of-magnitude larger rates of accretion in the early stages of planetary formation. In the first 0.5 Ga, so much material was amassed that the heat generated by impacts probably melted the planets completely, leading to their **differentiation**, in which the heavier elements, such as iron, sank to their interiors to form planetary cores while the lighter elements floated toward the surface.

1.1.1 The terrestrial planets

Mercury, Venus, Earth, and Mars are called the **terrestrial planets** because they share similar attributes to Earth (which in Latin is *terra*). As shown in **Fig. 1.1**, these planets are small in comparison with the other planets

---

[1] Terms when first used are in bold and defined. These terms are given in the index, where the page number in bold indicates where the term is defined.

Figure 1.1. A family portrait of the planets imaged by spacecraft. The inner planets (Mercury, Venus, Earth and Moon, and Mars) are shown to scale with each other and are enlarged relative to the giant planets (Jupiter, Saturn, Uranus, and Neptune), which are shown to scale with each other. Earth is somewhat smaller than the Giant Red Spot (indicated by the arrow) of Jupiter (NASA PIA01341).

and are found closest to the Sun, leading to their alternative description as the **inner planets**. They are composed primarily of rocky material and have solid surfaces. In planetary geology, Earth's Moon is typically included with the terrestrial planets because of its large size and similar characteristics.

As the terrestrial planets began to cool and form crusts, elements combined and crystallized into rocks and minerals. For the most part, these elements are silicon, oxygen, iron, magnesium, sodium, calcium, potassium, and aluminum in various combinations that collectively make up the **silicate minerals**. The most important silicate minerals fall into two groups. Light-colored silicate minerals are common in continental rocks on Earth and include quartz, orthoclase feldspar, plagioclase feldspar, and muscovite mica. Dark-colored silicate minerals are common on Earth's sea floor and are rich in iron and magnesium; they include olivine, pyroxene, hornblende, and biotite mica. Silicate minerals are the basic building blocks of most rocks in the crusts of Earth and the Moon, and they are thought to make up most of the rocks on Mercury, Venus, and Mars.

Venus, Earth, and Mars all have significant atmospheres composed of gasses that are gravitationally bound to the planets (**Table 1.1**). Mercury and the Moon are too small to retain anything but the most tenuous atmospheres, measurable only by very sensitive instruments. Although some gasses were accumulated by all of the terrestrial protoplanets during their initial formation, these **primary atmospheres** were lost to space. **Secondary atmospheres** were later released as gasses escaped from the interior and interacted with the surface. Earth's atmosphere may be termed a **tertiary atmosphere** because it has been greatly modified by biologic processes.

Table 1.1. **Basic data for planets**

Name	Orbit semi-major axis (10⁶ km)	(AU)[a]	Revolution period (yr)	Diameter (km)	Rotation (days)	Mass (10²⁴ kg)	Density (g/cm³)	Escape velocity (km/s)	Surface	Atmosphere
Mercury	57.9	0.39	0.24	4,879	58.65	0.33	5.4	4.3	Silicates	Trace Na
Venus	108	0.72	0.62	12,104	243.0 (R[b])	4.87	5.2	10	Basalt, granite?	90 bar: 97% CO_2
Earth	150	1.00	1.00	12,756	1.00	5.97	5.5	11	Basalt, granite, water	1 bar: 78% N_2, 21% O_2
Mars	228	1.52	1.88	6,794	1.03	0.64	3.9	5.0	Basalt, clays, ice	0.07 bar: 95% CO_2
Jupiter	778	5.20	11.86	142,984	0.41	1,899	1.3	60	None	H_2, He, CH_4, NH_3, etc.
Saturn	1427	9.54	29.46	120,536	0.44	569	0.7	35	None	H_2, He, CH_4, NH_3, etc.
Uranus	2871	19.19	84.02	51,118	0.72 (R[b])	86.8	1.3	21	None (?)	H_2, He, CH_4, NH_3, etc.
Neptune	4498	30.07	164.79	49,528	0.67	102	1.8	24	None (?)	H_2, He, CH_4, NH_3, etc.
Pluto	5906	39.48	247.9	2,302	6.39 (R)	0.013	2	1.3	CH_4, ice	Trace CH_4

[a] 1 AU (astronomical unit) = Earth–Sun distance, or ~149.6 × 10⁶ km.
[b] R = retrograde.

Figure 1.2. The heavily cratered surface of the Moon, shown in this view obtained by the *Apollo 13* astronauts, represents the final stages of planetary accretion in the first 0.5 Ga of the Solar System. The dark, smooth area is Mare Moscoviense on the lunar far side (NASA 70–H–700).

The presence of large impact craters on the terrestrial planets (**Fig. 1.2**) shows that their crusts had cooled and solidified in the first 0.5 Ga of Solar System history before all of the miscellaneous debris had been swept up.

1.1.2 The giant planets

Jupiter, Saturn, Uranus, and Neptune are referred to as **giant planets**. Relative to the terrestrial planets, these planets are enormous and contain most of the mass in the Solar System outside the Sun. Jupiter and Saturn are composed mostly of hydrogen and helium, while Uranus and Neptune are composed mostly of water, ices, and other volatile materials. Collectively, the giant planets and Pluto are called the **outer planets**, referring to their location in the Solar System.

The early history of the giant planets is similar to that of the terrestrial planets. The giant planets also formed by the accretion of smaller bodies, with each forming a nucleus large enough to capture the lighter elements that had escaped from the inner Solar System to the outer frigid parts of the Solar System. As this process continued, the giant planets grew to their large sizes, with heavier elements sinking to their interior. Most models of the giant planets suggest that each contains a rock-like core, some of which are larger than Mars.

Each of the giant planets resembles the Sun in composition, but not even the largest, Jupiter, was destined to grow to a size sufficient to initiate nuclear fusion. However, giant planets do resemble the Sun in one important way – each grew and evolved to have a family of smaller bodies in orbit about them so that each resembles the Solar System in miniature.

Although the giant planets have no "geology" because they lack solid surfaces, their satellites are of great interest for planetary geomorphology (**Table 1.2**). Collectively,

Table 1.2. **Basic data for selected satellites**

Planet	Satellite name	Discovery	Period (days)	Diameter (km)	Mass (10^{20} kg)	Density (g/cm³)	Surface material
Earth	Moon	–	27.32	3,476	735	3.3	Silicates
Mars	Phobos	Hall (1877)	0.32	27	1×10^{-4}	2.2	Carbonaceous
	Deimos	Hall (1877)	1.26	13	2×10^{-5}	1.7	Carbonaceous
Jupiter	Io	Galileo (1610)	1.77	3,660	893	3.6	Sulfur, SO_2
	Europa	Galileo (1610)	3.55	3,130	480	3.0	Ice
	Ganymede	Galileo (1610)	7.15	5,268	1,482	1.9	Dirty ice
	Callisto	Galileo (1610)	16.69	4,806	1,076	1.8	Dirty ice
Saturn	Mimas	Herschel (1789)	0.94	396	0.376	1.2	Ice
	Enceladus	Herschel (1789)	1.37	504	0.74	1.10	Pure ice
	Tethys	Cassini (1684)	1.89	1,048	6.27	1.0	Ice
	Dione	Cassini (1684)	2.74	1,120	11	1.4	Ice
	Rhea	Cassini (1672)	4.52	1,528	23	1.3	Ice
	Titan	Huygens (1655)	15.95	5,150	1,346	1.9	Methane ice
	Hyperion	Bond, Lassell (1848)	21:3	360	8×10^{-3}?	?	Dirty ice
	Iapetus	Cassini (1671)	79.3	1,436	16	1.1	Ice/carbonaceous
	Phoebe	Pickering (1898)	550 (R^a)	220	0.004	?	Carbonaceous?
Uranus	Miranda	Kuiper (1948)	1.41	474	0.7	1.3	Dirty ice
	Ariel	Lassell (1851)	2.52	1,159	14	1.6	Dirty ice
	Umbriel	Lassell (1851)	4.14	1,170	12	1.4	Dirty ice
	Titania	Herschel (1787)	8.71	1,578	35	1.6	Dirty ice
	Oberon	Herschel (1787)	13.5	1,522	30	1.5	Dirty ice
Neptune	Triton	Lassell (1846)	5.88 (R^a)	2,704	214	2.0	Methane ice
Pluto	Charon	Christy (1978)	6.39	1,186	16.2	?	Ice

^{a} R = retrograde.

these moons represent a myriad of objects of different sizes, compositions, and geologic histories. They are classified as **regular satellites** (orbiting in the same direction as the parent planet's spin direction) or **irregular satellites** (orbiting in the opposite direction) that are probably captured objects. Jupiter's moons Ganymede and Callisto are about the size of the planet Mercury. At least three moons, Jupiter's satellite Io, Saturn's Enceladus, and Neptune's Triton, are currently volcanically active – in fact, Io is the most geologically active object in the Solar System (**Fig. 1.3**). Other outer planet satellites appear to have remained relatively unaltered since their initial formation. Many of the geologic processes that operate on terrestrial planets are also seen on outer planet satellites; however, because of their different compositions (mostly ices, plus some silicates) and extremely cold environments, the outer planet satellites also display features representing processes unique to the outer Solar System.

1.1.3 Small bodies, Pluto, and "dwarf planets"

Asteroids, comets, and the smaller moons of the outer planets are often called **small bodies**, even though the largest asteroids are hundreds of kilometers in diameter. Comets consist of primordial material left over from the early stages of Solar System formation. Most comets are found in the **Oort cloud** and the **Kuiper belt**, both beyond the orbit of Pluto. The Oort cloud forms a spherical zone some 3×10^{12} km from the Sun and is the apparent source of long-period comets (those that take more than 200 years to complete an orbit around the Sun), while the Kuiper belt is a disk-shaped region extending from Neptune's orbit to ~8×10^9 km from the Sun and is the source for short-period comets (those that orbit the Sun in less than 200 years). Just to make things a little more complex, objects that are in orbit in this belt are referred to as **Kuiper belt objects**, or KBOs. Some of the outer planet satellites, such as Neptune's Triton, could have been captured from the Kuiper belt.

Often described as "dirty snowballs," comets are composed of dust grains and carbonaceous (carbon-rich) materials embedded in a matrix of water-ice (**Fig. 1.4**). Study of cometary material collected from Comet Wild 2 by NASA's *Stardust* mission (Brownlee *et al.*, 2006) and returned to Earth suggests that at least some comets are composed of grains that were heated in the inner Solar

Figure 1.3. One of the moons of Jupiter, Io, is the most volcanically active object in the Solar System. This image was taken by the *New Horizons* spacecraft in 2007 during its flyby of the Jupiter system on the way to Pluto. The huge umbrella-shaped plume at the top of the image is pyroclastic material rising 290 km from the active volcano Tvashtar. Also visible (left side) is a smaller (60 km high) plume erupted from the volcano, Prometheus (NASA PIA09248).

System, were driven outward from the Sun, and then coalesced to form some comets. This led to the reference to some comets as "icy dirt balls," a concept that was supported in 2005 when the *Deep Impact* spacecraft launched a roughly half-ton metal ball into Tempel 1, a comet measuring 7.6 km by 4.9 km. The impact explosion released a plume of icy dust, suggesting the properties of freshly fallen, fluffy snow with dust. Images taken by *Deep Impact* and those taken by the NASA *NExT* space-craft in 2011 after the impact show that Tempel 1's surface has smooth terrains and areas that have been eroded.

Most asteroids are found in the zone between the orbits of Mars and Jupiter, known as the **main asteroid belt**. However, asteroids are also found in orbits of larger planets and are called **Trojan asteroids**, while those in orbits that come close to the Earth are called **near-Earth objects**, or NEOs. Asteroids can be further classified in terms of their spectral properties and comparisons with meteorites, many of which were derived from asteroids. Historically, asteroids were thought to be either remnants of a former planet that broke apart or objects that never accreted to form a planet early in Solar System history. As with many ideas in planetary science, this was an over-simplification. It is now fairly clear that some asteroids

Figure 1.4. This view of the 5 km in diameter Comet Wild 2 was taken by the *Stardust* spacecraft in January 2004 (NASA *Stardust* Project).

(and the corresponding meteorites) represent fragments of a larger body that had been differentiated. Thus, "metal-lic" objects are thought to represent the core of a planet, while those having signatures of the mineral olivine would represent a planetary mantle, and "stony" objects would represent the crust. Other asteroids have the signa-tures of carbon-rich materials and are considered to rep-resent "unprocessed," or primitive, planetary material. In this regard, many planetologists suggest that some of these types of asteroids are actually the rocky material left over from comets that have lost most or all of their volatile materials.

Numerous missions have flown past, orbited, and even landed on asteroids, with one mission returning samples to Earth. The first images of asteroids up-close were taken in

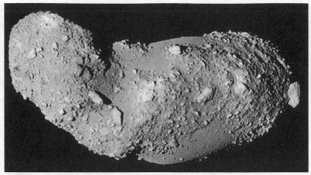

Figure 1.6. An image of the asteroid Itokawa, taken by the JAXA *Hayabusa* spacecraft, which touched down on the surface, collected samples, and returned them to Earth in the fall of 2010. This asteroid measures 535 m by 294 m by 209 m and appears to be a "rubble-pile" of boulders, the biggest of which is about 50 m across. More than 1,500 small grains were collected, and initial analyses show the presence of silicate minerals, such as olivine.

Figure 1.5. The first close-up view of an asteroid was obtained by the *Galileo* spacecraft in October 1991, shown in this view of asteroid Gaspra, which is of dimensions 19 km by 12 km by 11 km. Gaspra's irregular shape suggests that it might be a piece of a larger object that fragmented from one or more collisions. More than 600 impact craters ranging in size from 100 to 500 m are visible on Gaspra's surface (NASA JPL P-40449).

1991 and 1993 by NASA's *Galileo* spacecraft in the main asteroid belt (**Fig. 1.5**) and included the discovery that asteroids could even have their own small moons. In 2003, Japan launched the *Hayabusa* spacecraft, which rendezvoused with the NEO Itokawa (**Fig. 1.6**) in 2005; the spacecraft touched down on the asteroid and collected samples that were returned to Earth in 2010 for analyses.

Pluto was discovered telescopically in 1930 and for 76 years was classified as a planet. But it does not fit neatly into either the terrestrial planet or the giant planet classification; it is relatively small and has an orbit that is substantially inclined to the general ecliptic plane and at times is inside the orbit of Neptune. In the past few decades, many more large objects have been discovered in orbit around the Sun, including Eris, which is slightly larger than Pluto. It is estimated that, as a minimum, some several dozen large objects reside within the zone of Pluto's orbit, and many hundreds could well be found in the Kuiper belt. These factors led the International Astronomical Union to "demote" Pluto as a main planet in 2006 and to define a new category, the so-called "dwarf planets," currently consisting of Pluto, Ceres (formerly classified as an asteroid), Haumea, Makemake, and Eris, some of which have one or more moons. None of these objects has been visited by spacecraft, but the

Dawn spacecraft will visit Ceres in 2014, and the *New Horizons* spacecraft is slated to fly past Pluto in 2015.

1.2 Objectives of Solar System exploration

October 1957 saw the launch of the Soviet orbiter *Sputnik* around Earth and the beginning of the "Space Age." About the size of a basketball, *Sputnik* did little more than send a "beep–beep" radio signal, but it was the starting gun for the space race. The United States responded with President Kennedy's decision to send men to the Moon before the end of the 1960s and the formation of the National Aeronautics and Space Administration, or NASA, in October 1958. Although the decision was motivated by politics and military considerations (an orbiting spacecraft has the ability to deliver warheads to any place on Earth), the National Academy of Sciences was asked to define the scientific goals for Solar System exploration. After careful consideration by a group of distinguished scientists, the principal goals were defined as determining: (1) the origin and evolution of the Solar System, (2) the origin and evolution of life, and (3) the processes that shape humankind's terrestrial environment. Although these goals have evolved over the years, the basic concepts remain the foundation for Solar System exploration.

1.2.1 Planetary geology objectives

Geologic sciences figure prominently in the goals for Solar System exploration. Basic geologic questions include the following. (a) What is the present state of the planet? (b) What was the past state of the planet? (c) How do the present and past states compare with those of other objects in the Solar System?

The question dealing with the present state seeks to determine the composition, distribution, and ages of rocks on the surface, identify active geologic processes, and characterize the interior.

Determining the past state of a planet, including Earth, is a fundamental aspect of geology and involves determining its geologic history. For example, is the present state representative of previous conditions on the planet, or has there been a change or evolution in the surface or interior? Answering these questions is typically accomplished through geologic mapping, coupled with the derivation of a stratigraphic framework and geologic time scale.

Comparative planetology addresses the third aspect in the geologic study of planets. Once the present and past states have been assessed, the results are then compared among all of the planets to determine their differences and similarities. This comparision enables a more complete understanding of geologic processes in general and of the evolution of all solid-surface objects in the Solar System.

1.2.2 Astrobiology

Are we alone? That fundamental question has been posed in various forms throughout humankind's history and constitutes one of the key motivations in the exploration of space. The term **astrobiology** was coined to encompass all aspects of the search for present and past life, including research on the conditions for the origin of life and study of the environments conducive for biological processes. The NASA Astrobiology Institute (NAI), which was formed in 1998 and is headquartered at the NASA-Ames Research Center in California, consists of an international consortium of universities and institutions conducting a wide variety of research projects in astrobiology. The NAI organizes annual spring meetings to review the latest results in astrobiology (http://nai.nasa.gov); these meetings are well attended and open to the public.

The *Viking* mission to Mars in the mid 1970s was the first project to search for life beyond Earth. Experiments for the two *Viking* landers (**Fig. 1.7**) were developed to search specifically for life-forms and to assess possible biological processes. The results from these experiments were negative, and the general search for life was out of vogue for some 20 years. However, during this period, careful considerations were given as to how astrobiology

questions should be addressed. For example, when targeting specific planetary objects for astrobiology exploration, at least three factors should be considered: the presence of water (preferably in the liquid state), a source of sufficient energy to support biological processes, and the availability of organic chemistry and other elements essential for life processes (primarily carbon, nitrogen, hydrogen, oxygen, phosphorus, and sulfur). With current data, the search narrows to Mars, Jupiter's moon Europa, and possibly Saturn's moons Enceladus and Titan. If the search is expanded to include potential *past* environments, objects such as Jupiter's moon Ganymede might be included.

In 1996 a meteorite (designated ALH84001) found in Antarctica was thought to have been ejected from Mars and was suggested to show evidence for biology. Although much of this evidence has been rejected, interest in astrobiology increased substantially, especially as related to the exploration of Mars. The current search strategy focuses on identifying the present and past environments conducive for biology and is a "win–win" approach. Obviously, if life or the signs of life (e.g., fossils) are found, the result would be truly profound (**Fig. 1.8**). However, a negative result is equally intriguing; if present or past environments are found that are amenable for life, but life is not found, then one must ask why not, and what is it about Earth that would make our planet unique for life if indeed we are truly "alone?"

As the field of astrobiology has moved forward, life has been found to be much more pervasive on Earth than had previously been suspected. In recent years life-forms have been found in extreme conditions of temperature, pressure, pH, and other environmental parameters, showing that biology can occur in a much greater range of settings than previously suspected, thus widening the search for life throughout the Solar System.

1.3 Strategy for Solar System exploration

Determining the present and past states of planets and comparative planetology requires observations and measurements from orbit, placement of instruments on planetary surfaces, and the return of samples to Earth. Thus, the general exploration of the Solar System by spacecraft follows a strategy involving a series of missions of increasing capabilities. However, even before spacecraft are launched, Earth-based telescopic observations are made to determine the fundamental characteristics of

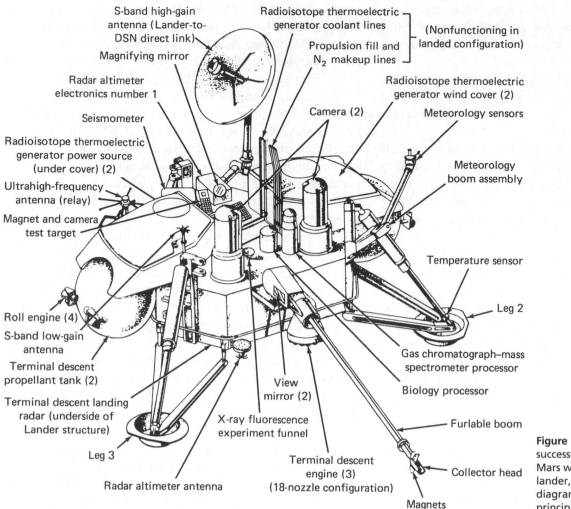

S-band high-gain antenna (Lander-to-DSN direct link)

Magnifying mirror

Radar altimeter electronics number 1

Seismometer

Radioisotope thermoelectric generator power source (under cover) (2)

Ultrahigh-frequency antenna (relay)

Magnet and camera test target

Roll engine (4)

S-band low-gain antenna

Terminal descent propellant tank (2)

Terminal descent landing radar (underside of Lander structure)

Leg 3

Radar altimeter antenna

Radioisotope thermoelectric generator coolant lines

Propulsion fill and N_2 makeup lines

(Nonfunctioning in landed configuration)

Camera (2)

Radioisotope thermoelectric generator wind cover (2)

Meteorology sensors

Meteorology boom assembly

Temperature sensor

Leg 2

Gas chromatograph–mass spectrometer processor

Biology processor

View mirror (2)

X-ray fluorescence experiment funnel

Terminal descent engine (3) (18-nozzle configuration)

Furlable boom

Collector head

Magnets

Figure 1.7. The first successful landing on Mars was the *Viking 1* lander, shown in this diagram with its principal components.

planetary objects, such as their size and density, and the presence or absence of atmospheres.

The first exploratory missions are usually "flybys," in which spacecraft zoom past planetary objects and, over a period of only hours or a few days, collect data. Although limited, these data provide the first glimpses of the object up-close and are far better than those obtained from Earth-based telescopes. For example, in 1979 and the 1980s the spectacular *Voyager 1* and *2* spacecraft (**Fig. 1.9**) revealed the complexities of the moons of Jupiter, Saturn, Uranus, and Neptune during brief flybys of those planetary systems.

Next in exploration comes the use of orbiting space-craft. Remaining in orbit for days, months, or even years, orbiters provide the opportunity for more complete map-ping and observations of potential seasonal changes.

Spacecraft in polar or near-polar orbits can obtain remote sensing data for the entire planet, enabling assessments of the surface complexity, collection of geophysical data, and measurements of topography. Thus, one of the primary advantages of orbiters is the collection of global data.

Once a planet has been surveyed from orbit, the missions that follow can include landed spacecraft. Landers enable "ground-truthing" of the remote sensing data obtained from orbit. Such data include *in situ* measurements of surface chemistry and mineralogy, determinations of the physical properties of the surface enviroment, and geophysical measurements, including seismometry. Landed missions are significantly enhanced by surface mobility as afforded by robotic systems, such as the *Mars Exploration Rovers* (**Fig. 1.10**). The advantage of

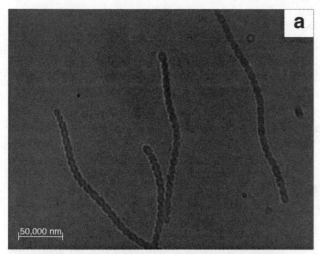

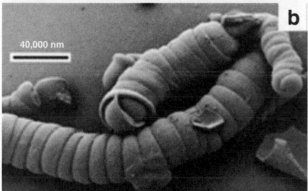

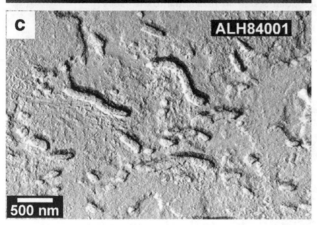

Figure 1.8. What are the signs of life that might be sought in the search for present or past life beyond Earth? From our "Earth bias," we might think that we know what fossils look like. But even on Earth, some cases are not so clear: (a) living cyanobacteria (courtesy of Jennifer Glass, Arizona State University), (b) synthetic non-biological filaments containing silica and the mineral witherite (from Garcia-Ruiz, J. M., Hyde, S. T., Carnerup, A. M. *et al.* (2003), Self-assembled silica–carbonate structures and detection of ancient microfossils, *Science*, **302**, 1,194–1,197. Reprinted with permission from the AAAS), and (c) an image of martian meteorite ALH84001 (courtesy of NASA Astrobiology Institute).

landers and rovers is the ability to obtain data directly from planetary surfaces and near-surface materials, as from drill cores, which was first done robotically by the Soviets on the Moon. The disadvantage is the relatively limited number of sites that can be visited; can you imagine characterizing the complex geology of the Earth from only a handful of stations on the surface?

Samples returned from planetary objects represent the next stage in exploration. These enable sophisticated laboratory analyses of compositions, measurements of physical properties, and searches for signs of past or present life. Although significant advances in instruments that can be applied on robotic missions have been made in recent years, none can approach the accuracy and precision afforded by full laboratory facilities on Earth. Particularly critical for geology are the ages of rocks determined on returned samples using techniques based on the decay of radioactive materials (see **Section 2.4**). While some measurements can be made from robotic spacecraft, the complexities of obtaining and properly handling samples in order to make the measurements have not been solved satisfactorily for determining ages.

The ultimate in planetary science is human exploration. Humans have the ability to analyze and synthesize data quickly, make decisions on the spot, and respond to the results. No machine can match these attributes. But, of course, sending humans into space is both risky and costly. Currently, it is far more cost-effective to send robotic spacecraft throughout the Solar System. However, the time will come when humans will be required for the ultimate step in exploration.

Figure 1.11 shows the "score-board" for the different stages of Solar System exploration. Nearby objects, such as our Moon, have been explored extensively, while most of the outer Solar System has been viewed only by flyby missions. Despite this uneven coverage, we are now well poised to address many of the fundamental aspects of the origin and evolution of the major planetary objects.

1.4 Flight projects

Getting a NASA spacecraft "off the ground" is a long process that involves many constituencies, including NASA, Congress (which appropriates the money), the aerospace industry (which builds much of the hardware),

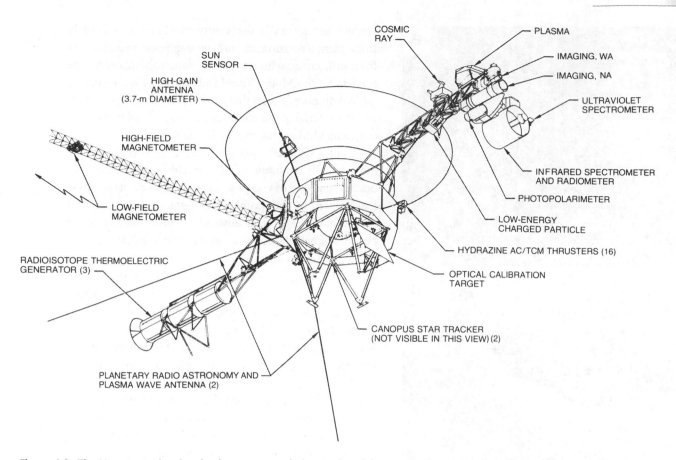

Figure 1.9. The *Voyager* project involved two spacecraft that explored the outer Solar System in 1979 and into the 1980s with flybys of Jupiter and Saturn (*Voyagers 1* and *2*) and Uranus and Neptune (*Voyager 2*), providing the first clear images of their major moons.

Figure 1.10. The Mars Exploration Rovers, *Spirit* and *Opportunity*, landed in early 2004. Shown here is *Spirit* before launch, compared with the flight-spare of the *Mars Pathfinder* rover on the right (NASA PIA04421).

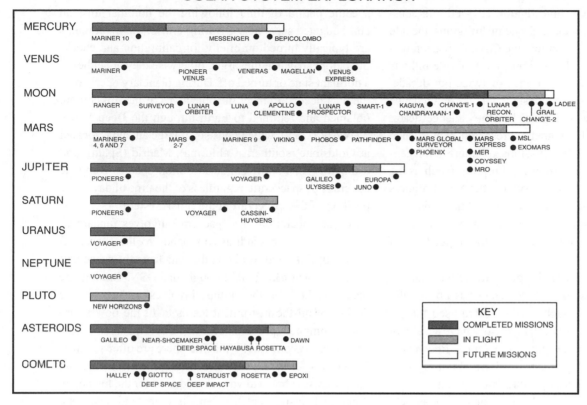

Figure 1.11. Status of Solar System exploration by spacecraft important for planetary geomorphology.

the science community, and the public. NASA is an independent federal agency, meaning that there is no "Department of Space" and its Administrator is appointed by the President. Direct science input is through NASA committees, with members appointed from universities, NASA centers, and other research organizations. The National Academy of Sciences (through its working organization, the National Research Council, or NRC) provides science guidance. This is accomplished by means of formal reports prepared by scientists from the planetary community that recommend missions and research activities covering a ten-year period, known informally as **decadal surveys** (NRC, 2011). Individuals influence Solar System exploration by making their opinions known through communication with NASA, Congress, and the Administration. Coordinated input is often conducted through organizations such as The Planetary Society (http://www.planetary.org/home/), the National Space Society (http://www.nss.org/), and the Mars Society (http://www.marssociety.org/).

In principle, the various constituencies work together to derive the specific goals for a mission, the means to

achieve those goals (e.g., the kind of spacecraft), and the budget to make it all happen. In practice, the process is often more haphazard, yet most of the constituencies are still involved to varying degrees. The time from initial mission concept to the return of data is usually years, or even decades.

Once a project has been approved, the mission is assigned to a NASA research center or run through the Jet Propulsion Laboratory (JPL) in California or the Applied Physics Laboratory (APL) in Maryland, both of which are NASA contract research centers. Flight projects go through various phases from design and development through mission operations. Early in the process, a Science Definition Team (SDT) is appointed from the scientific community, which has the responsibility for determining the specific objectives for the mission. After this has been completed, an Announcement of Opportunity is released by NASA, enabling proposals to be submitted for building the spacecraft and providing the science payload or instruments. Individuals and organizations can then compete for selection, which is made through peer-reviews of the proposals.

NASA missions can be categorized as (1) strategic missions, (2) Principal Investigator (PI)-led missions, and (3) supporting missions. Strategic missions include the *Mars Science Laboratory* and the *Cassini* spacecraft in orbit around Saturn. Such missions cost multiple billions of dollars and might be flown once or twice per decade. PI-led missions are proposed, designed, and executed by a planetary scientist, who assembles the science team, industry partners, and a planetary research center, such as the JPL or APL. PI-led missions of interest for geoscience are found in the Discovery Program (such as the *MESSENGER* mission to Mercury) and the New Frontiers Program (such as the *New Horizons* Pluto mission). Specific missions within these programs are "cost-capped," with New Frontiers being at the upper level of $1 billion.

Supporting missions are designed to collect data to enable follow-on missions. While science is not usually the primary motivation, such data are often used for scientific research, and the missions typically have a cadre of scientists involved. For example, the *Lunar Reconnaissance Orbiter* has the primary goal of obtaining data necessary for the eventual return of humans to the Moon, but these data are of high value for science as well.

For strategic and supporting missions, scientists can propose to be the PI for an instrument or suite of instruments as part of the payload. The selected PI forms the science team, designs the instrument, has it built, and implements the experiments through operation of the instrument and collection of the data. In some cases, *facility instruments* are provided directly by NASA and scientists can propose to be a member or team leader of that instrument science team; the team is then responsible for carrying out the investigation.

The European Space Agency (ESA) also flys planetary missions, but operates differently from NASA. The ESA is composed of 17 member nations and is headquartered in Paris, with its primary operations center in the Netherlands. Once a mission has been selected for flight, the ESA develops and builds the spacecraft and is responsible for its operation. The scientific payload, however, is competed for among the member nations through their science communities; if selected, that nation is responsible for funding and delivering the instrument or suite of instruments to the ESA.

Operation of an active flight project is exciting and complicated! After launch, the mission goes through *cruise* (the journey from Earth to its destination), *nominal operations* (at the target for the duration approved in the budget), and, if all goes well, an *extended mission* (a specific period of time following the nominal mission and budgeted separately). During cruise, the instruments are typically turned on briefly for calibration and check-out before arrival at the target; otherwise they are either in a dormant state or turned off. During planetary operations, the data needed to meet the objectives of the mission are obtained and returned to Earth through the **Deep Space Network** (DSN), which consists of large antennas located at Goldstone (southern California), Madrid (Spain), and Canberra (Australia). This distribution enables complete coverage of spacecraft, regardless of the time of day or the position of the spacecraft in the Solar System.

Science operation of a spacecraft involves fundamentally two aspects: sending commands to the spacecraft (called **uplink**) and receiving the data from the spacecraft (called **downlink**), both through the DSN. Putting the plan together for the uplink involves integrating the desires of all the instrument teams to fit the power, onboard computer processing, and other resources of the spacecraft. As one might imagine, there are often competing wishes for these resources among the scientists, and compromises almost always are required for the final plan. After each instrument has sent its commands, data are downlinked and, again, there is often competition for downlink resources. Modern instruments generally can take far more data than can be returned, and decisions must be made to satisfy the overall mission objectives.

1.5 Planetary data

As soon as a successful mission goes into operation, the science flight team plans the acquisition of data (such as targeting areas to be imaged), collects the data, and initiates their analysis. Some of these data are posted on the website for that particular mission (go to the general NASA website http://www.nasa.gov/, or the ESA website http://www.esa.int/, and look for the specific mission by name). These data are for general public interest and often have not been calibrated or verified for accuracy. It is considered "bad form" for the science community to publish results from such data before they are officially released for scientific analysis. Such release is done on a project schedule after validation by the science team, posted in the Planetary Data System (PDS, http://pds.jpl.nasa.gov/), and publicly announced. Because of the pace of mission operations, the volume of data from modern missions, the complexity of the data, and the possibility of errors in the data stream, releases typically occur no earlier than about six months from their acquisition. Once released, the NASA data are available to everyone.

Following (or during) a mission or set of missions, NASA will organize a Scientific Data Analysis Program. These programs provide funds to support the analysis of data by the community through open competition and peer review. Such programs are usually of a limited duration, such as three years. In addition, each scientific discipline, such as the Planetary Geology and Geophysics Program at NASA, has funds for basic research, including geologic mapping, laboratory studies, and integrated data analysis. These, too, are open through competition and peer review. The NASA Research Opportunities (ROSES; http://nspires.nasaprs.com/external/) posts the procedures and schedules for proposing for these and other opportunities from NASA.

1.6 Planetary research results

Knowledge of the Solar System is expanding rapidly and is enabled primarily by data returned from spacecraft missions. Even for planetary scientists, it is often difficult to keep up with the advances in exploration. Typically, the first results from flight projects are announced through press releases from NASA or the space agency responsible for the mission. While the releases are generally prepared by the project science team, many of the ideas presented are not very mature. The next stage is the oral presentation of results at scientific meetings. By this time, the results and the ideas have been more widely discussed within the science teams and have been somewhat refined. Although abstracts (short summaries of the content) of the presentations are published for the meeting, the abstracts are usually submitted months before the actual meeting; with active flight projects, the abstracts that are submitted are often simply placeholders and might not have much real content, unlike the oral presentation itself.

Key scientific meetings for planetary science are the Lunar and Planetary Science Conference (LPSC), held every March in Houston, the American Geophysical Union (AGU) meeting held each fall in San Francisco, the Division of Planetary Science (DPS) meeting of the American Astronomical Society held each fall, the European Geosciences Union (EGU) meeting and the Europlanet meeting held in Europe, the Geological Society of America (GSA) fall meeting, and the Meteoritical Society meeting held each fall. These meetings all publish abstracts of the presentations, which are usually available on-line from the sponsoring scientific organizations. The AGU, GSA, and EGU meetings are very large and include a wide variety of subjects in addition to planetary science.

Most large scientific meetings are attended by professional science writers who are very skilled in extracting new and exciting results. Their articles are then published in venues such as *Science News*, *Space News*, and *The Planetary Report*.

Traditionally, the first papers from flight projects are published in *Science* or *Nature*, often as special sections or editions of the journal. These papers are "peer reviewed," meaning that scientists not involved with the project but who are knowledgeable of the field have reviewed and evaluated the results.

The first full papers from planetary missions are typically published a year or two after data acquisition. By this time, the ideas have matured and the manuscripts have been rigorously peer-reviewed. Key journals include *Icarus*, the *Journal of Geophysical Research – Planets* (an AGU publication), and *Planetary and Space Science*.

Additional sources of planetary information are specialized topical meetings. These range in size from small workshops involving a dozen or so people to international conferences attended by hundreds of participants. Topics can range from the latest results from a large flight project to highly specialized research topics. In most cases, abstracts of papers are available at the meeting and full peer-reviewed papers are published in journals or as a special conference book.

Planetary science series of books published by organizations such as the University of Arizona Press and Cambridge University Press contain collections of review papers, with most individual volumes focusing on specific planetary objects. These books typically follow international meetings that are organized to synthesize new, as well as mature, results from spacecraft missions and general investigations.

While this outline has focused on results from new planetary flight projects, the venues listed are also where results from active planetary research projects can be found. As noted throughout the text, various key websites are identified for sources of information. These and related websites relevant for planetary exploration and data are listed in **Appendix 1.1** at the end of this book, and can also serve as "spring boards" for additional websites. An example is the Java Mission-planning and Analysis for Remote Sensing (JMARS) website (http://jmars.asu.edu) for a geospatial information system (GIS) that enables

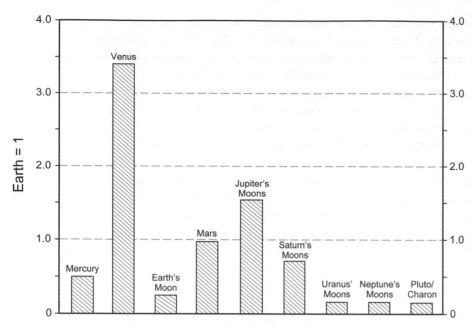

Figure 1.12. The surface areas of the rocky planets, the Moon, and the larger outer planet satellites as a function of Earth's land surface area (Earth = 1); note that the surface area of Mars is just about equal to the surface of Earth not covered by water, while the surface area of Venus is nearly 3.5 times that of Earth's land surface.

rapid searches and provides analytical tools for planetary data. For example, a user can construct maps that combine images, topographic information, and multispectral data for areas and scales of the user's choice. Currently, JMARS principally covers Mars and is being adapted for the Moon and Earth. Public downloads are readily accessible through the website. In addition, NASA maintains a network of Regional Planetary Image Facilities that have sets of images available for viewing and staff who can answer specific questions; **Appendix 1.2** lists these facilities and their locations.

In summary, the exploration of the Solar System affords a great opportunity to study geology and geomorphology in a wide variety of settings and over time scales from the earliest formation of planetary crusts to geologically active planets. As shown in **Fig. 1.12**, there is an enormous potential for such studies when considering the total surface areas of planets and satellites amenable for geology.

Assignments

1. Briefly explain how the analysis of geologic features on planetary surfaces is relevant to the search for life beyond Earth.

2. Go to the website for *Science News* and summarize a scientific result from a currently operating spacecraft that is relevant to planetary geology.

3. Compare and contrast the types of planetary data returned from an orbiting spacecraft and a landed spacecraft.

4. Go to the websites for planetary missions currently being conducted by spacefaring nations and agencies, such as the ESA, and identify one spacecraft each for an inner planet, a gas giant planet, and a small body (such as an asteroid). List the launch dates, dates of operations at the "target" planetary body, and one or two key results relevant to planetary geology for each of the three spacecraft identified.

5. Examine the tables for the primary characteristics of the planets and satellites. Give one example of how the environment for a terrestrial planet of your choice would influence the geology in comparison with an icy satellite of your choice.

6. Go to the website for *Space News* and summarize one budget issue for the current year that has an impact on planetary exploration.

CHAPTER 2

Planetary geomorphology methods

2.1 Introduction

For many years, the study of the geomorphology of the Earth was primarily descriptive. In the middle of the twentieth century, the emphasis shifted to a more process-oriented approach, with the goal of understanding the reasons behind a landform's appearance. The analysis of planetary surfaces has gone through a similar history. When the first close-up images of the Moon and planets were obtained, their surfaces were described, and some attempts were made to interpret their origin and evolution. Unfortunately, some of these attempts were rather immature. Planetary scientists with a geology background drew on their experiences with Earth, taking a simplified "analog" approach; i.e., if it looks like a volcanic crater, it must be of volcanic origin. Scaling the sizes of features and considerations of planetary environments took a back seat to the simple "look alike" answer.

As the *Apollo* program drew to a close in the early 1970s and the exploration of the full Solar System emerged, planetary geomorphology became more process-oriented, with attempts to take differences in planetary environments into account, while maintaining fundamental geologic principles.

In this chapter, the following question will be addressed: how can one study the geology of a planet or satellite without actually going there? This will include the approaches used in planetary geomorphology and the types of data that are commonly available for the study of planetary surfaces.

2.2 Approach

The general approach in planetary geomorphology involves three elements: (a) analysis of spacecraft data, (b) laboratory and computer simulations of key geologic processes in different planetary environments, and (c) the study of terrestrial analogs. Each element has its advantages and disadvantages but collectively provides a powerful means to decipher present and past planetary surface histories.

The starting point is the analysis of planetary data, typically in the form of images. From these studies, the overall terrains and varieties of landforms are identified and characterized. Various hypotheses are proposed to explain the possible formation and evolution of the landforms observed. With further study and new data, the number of hypotheses can be reduced, or new ideas emerge. The history of the study of craters on the Moon is a good case to review. Beginning with telescopic views, the origin of lunar craters was debated for centuries, leading to the time of the Space Age. Even with the return of data from spacecraft sent to the Moon in preparation for the *Apollo* program, there were two primary competing ideas for craters, impact versus volcanic origins. Images of lunar craters showed features that were used to support both ideas. While the characteristics of volcanic craters on Earth were fairly well understood, little thought had been given to extrapolation of volcanic processes to the low-gravity, airless environment of the Moon. In the early days of the Space Age, impact cratering as a process was little appreciated in the geologic context, and there was no understanding of the physics of the process. At about the same time as robotic missions were returning new, close-up data for the Moon, experiments to study the physics of impact events were initiated. Although similar work had been conducted for decades by the military to understand how projectiles could penetrate armor, much of this work was classified; moreover, the work was more applicable to man-made targets than to natural, rocky material. It is interesting to note that in the late 1880s the American geologist G. K. Gilbert dropped small cannon balls into mud targets (**Fig. 2.1**) to see what might happen. Gilbert was very interested in the origin of lunar craters, and proposed that the Imbrium feature on the Moon was the result of an impact, as discussed in **Chapter 4**.

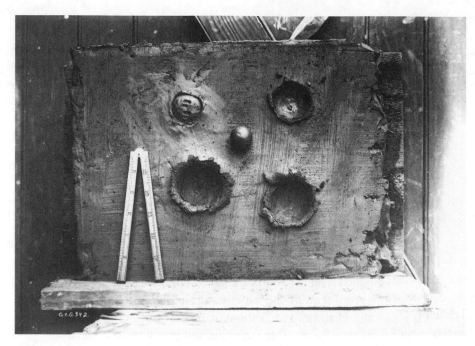

Figure 2.1. Geologist G. K. Gilbert performed experiments in the late 1880s to simulate impact processes, shown in this photograph of small cannon balls that were dropped into a target of stiff mud (courtesy of the US Geological Survey).

Building on the work of Gilbert, Don Gault at NASA-Ames Research Center constructed in the early 1960s a facility for conducting sophisticated impact experiments using a hydrogen gas-gun (**Fig. 2.2**). This gun can fire projectiles at velocities as high as 7.5 km/s into a target contained in a vacuum chamber to simulate the Moon. In some experiments, the target is placed on a platform that can be dropped at the time of the impact to simulate reduced gravity conditions (much like the "weightless" feeling when an elevator descends rapidly). Results from this facility enabled the fundamental physics of impacts to be derived and provided critical insight into the geologic aspects of impact cratering (**Fig. 2.3**).

At about the same time as impact experiments were being conducted, Gene Shoemaker of the US Geological Survey (USGS) was synthesizing results of his field studies of Meteor Crater in Arizona (**Fig. 2.4**), which included assessments of rock structure and deformation, including the presence of overturned stratigraphy in the crater rim, which was so well seen in the Gault experiments (**Fig. 2.3**). Subsequently, field sampling at Meteor Crater led to the discovery of **coesite** and **stishovite**, the high-pressure forms of quartz that are formed by impact processes. Concurrently, other geologists were scouring remote sensing data to identify possible impact structures on Earth (**Fig. 2.5**), followed by field investigations.

The mid 1960s also saw investigations of nuclear explosion craters at the Nevada test site and their study by Hank Moore of the USGS for comparisons with

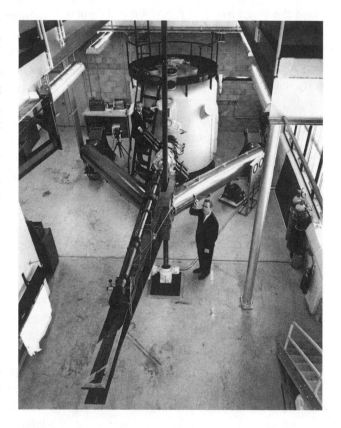

Figure 2.2. The Vertical Gun at NASA-Ames Research Center consists of an "A" frame mounted with a gun barrel that uses compressed gasses to launch small projectiles at velocities as high as 7.5 km/s into a vacuum chamber tank to simulate impact cratering processes. The "A" frame can be rotated so that the gun can fire projectiles at different impact angles from near-horizontal (seen here) to vertical.

Figure 2.3. Cross-section of a target produced in the Ames Vertical Gun Range (**Fig. 2.2**). The target consisted of layers of loose sand grains dyed different colors mixed with epoxy resin. After the "shot," the target was baked to fuse the grains and epoxy resin, and then sawn into a cross-section. Shown here is the inverted stratigraphy in the crater rim (the "overturned flap" of ejecta), which is characteristic of impact craters. The crater is about 0.4 m across.

Figure 2.4. An aerial view of Meteor Crater in northern Arizona, the best-preserved impact structure on Earth. This crater 1.2 km in diameter was formed by a 30 m iron meteoroid some 50,000 years ago (courtesy of Mike Malin).

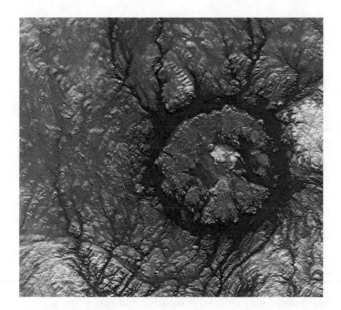

Figure 2.5. A radar image (C-band) taken on NASA's *Shuttle Radar Topography Mission* of the Manicouagan crater, Quebec, Canada; this impact structure of diameter 100 km formed about 214 million years ago. Erosion (mostly by glaciation) has removed about 1 km of rock from its original surface, exposing the deep structure, including the "root" of the central uplift, which is now surrounded by Manicouagan reservoir, seen here in dark gray (NASA PIA03385).

2.3 Planetary geologic maps

Geologic maps represent a fundamental tool for characterizing the geology and geomorphology of an area and deciphering its history. The British planetary geologist John Guest once said "a geological map is (to a geologist) like a graph to a physicist; it allows an understanding of many observations in a comprehensive form that would be otherwise difficult." The basic elements of a geologic map show the distribution of three-dimensional rock units (**Fig. 2.6**), the configuration of the rock units exposed on the surface of the area mapped, structural features, such as faults, and the ages of the rocks and structural features. As is true for all maps, geologic maps include a scale, orientation (e.g., a north arrow), a legend explaining the symbols on the map, and the location of the map area (typically indicated by geographic coordinates).

The **formation** is the basic rock unit in mapping. Formations consist of material of similar rocks, all formed at the same time, in the same place, and by the same process. For example, a lava flow resulting from a single eruption in Hawaii could be treated as a formation that would be different from a lava flow erupted from Mount Etna in Sicily at the same time, even though both might be of the same type of rock.

Some formations can be subdivided into **members**. For example, during a given eruption sequence, a lava flow might be covered by ash from an explosion; the lava flow and the ash could be called members of the same formation. Two or more formations that share common

features seen on the Moon. Computer codes were also being developed for large explosions, which would be applied later to planetary impact processes.

Through this combination of laboratory experiments, analysis of remote sensing data, and field studies, a general model of impact processes emerged that could be applied successfully to the interpretation of planetary data. The study of impact craters set the stage for the approach used in investigations of other geomorphic processes, such as aeolian activity and volcanism.

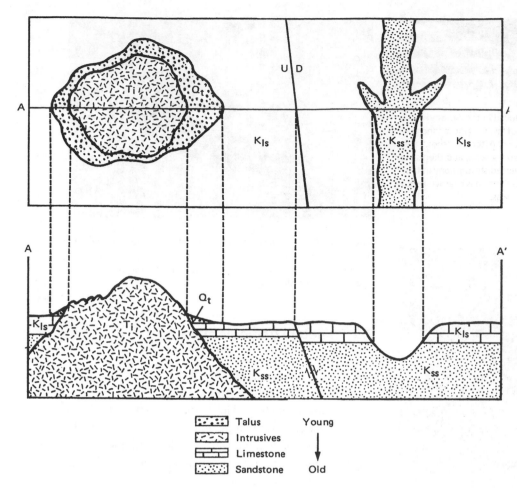

Figure 2.6. Geologic maps (top) show three-dimensional rock units (bottom cross-section), such as limestone, sandstone, and intrusive rocks, and structural features such as faults. The relative ages of the units and structural features are commonly indicated in a stratigraphic column organized from youngest (top of column) to oldest (bottom of column).

attributes in a given area are sometimes combined into a **group**. For example, a series of lava eruptions and ash flows mapped as formations and members on a volcano could be defined as a stratigraphic group.

Formations can be more complicated than the simple case of the lava flow. Imagine a lake basin that receives run-off and sediments from the surrounding mountains. As the streams empty into the lake, there is a decrease in the speed of the water flow and, hence, a decrease in their ability to carry sediments. Coarser materials, such as boulders, would be deposited close to the shore, with progressively smaller rocks and sediments being deposited outward from the shore. Over time, the sediments in this sequence continue to accumulate and can eventually be **lithified** (turned into rock). If one were to see only the part formed close to shore, the rock would be a **conglomerate** (a rock composed of large, rounded rock fragments); farther from shore, the rock would be **sandstone** (rock composed of sand-size particles); still farther from the shore, the rock would be **shale** (rock composed of very

fine grains, such as clay). Thus, three different rocks are found, and each could be treated as a different formation; however, the boundaries between these different rocks would be gradational (smaller-size materials away from shore) and would represent the local environment of formation (i.e., progressively less energy to carry the sediments away from the shoreline). In cases such as this, the conglomerate, sandstone, and shale would be called **facies** (parts) of the same basic formation.

Structural attributes of rock units are indicated by various map symbols to show faults, folds, and the "attitude" of the rock units. Attitude refers to orientation, such as horizontal, vertical, or tilted.

Geologic maps also indicate the ages of the rock units and the timing of deformation. These relations are portrayed as a **stratigraphic column**, in which the oldest materials are at the bottom of the sequence and the youngest are at the top. In geology, time is usually indicated from the bottom to the top, to reflect the **principle of superposition**. This principle states that in any sequence of undisturbed rocks those on the bottom must be the oldest because they had to be present before the subsequent rocks could be put on them.

Geologic mapping on Earth began in the eighteenth century. In the ensuing years, stratigraphic columns reflecting local sequences of rocks have been defined for most regions of our planet and have been **correlated** (connected) into regional and global associations. From this synthesis, a generalized **geologic time scale** that divides the history of the Earth into formal **eras, periods**, and **epochs** was defined. Early versions of the geologic time scale indicated only the relative ages of rocks, and it was not until dating methods based primarily on radioactive decay were developed that absolute ages could be assigned, as will be discussed below.

As for the Earth, planetary geologic maps are critical for understanding surface histories and for providing a framework for other observations. This understanding was recognized very early in Solar System exploration, and the first planetary geologic maps were compiled by the US Geological Survey (USGS) for the Moon by Robert Hackman and Gene Shoemaker, from telescopic observations (**Fig. 2.7**). Their techniques were later codified and standardized for planetary mapping by Don Wilhelms and Ken Tanaka, also of the USGS.

Geologic maps of Earth are commonly assembled from a combination of remote sensing data and field work, all combined on standard maps, which usually include topography. The identification of formations and other rock units on planets is based primarily on remote sensing data using photogeological techniques that are commonly employed for Earth. Images enable obvious features, such as lava flows, to be identified, while the general appearance of terrains is used to infer different rock units. Compositional mapping on the basis of infrared and other data is also used to distinguish units, when such data are available. Further insight is provided by quantitative studies of **albedo** (a measure of the reflectivity of the surface) and surface textures at the sub-meter scale derived from radar signatures. Unfortunately, a full suite of remote sensing data is seldom available for planetary surfaces.

After the rock units and structures have been identified and their distributions mapped, the next step is to place them in a chronological sequence. In addition to superposition, **embayment** and **cross-cutting relations** are used to determine the relative ages among units. For example, embayment refers to the "flooding" aspect of some units, as seen on the Moon (**Fig. 2.8**), in which the "flooding" unit is the younger. For the application of

Figure 2.7. Part of the geologic map of the Kepler region on the Moon, published in *Geotimes* in 1962. This was one of the early prototype maps to show that geologic maps could be produced for planetary surfaces (reprinted with permission from *Geotimes* magazine, a publication of the American Geological Institute).

cross-cutting relations, a rock unit, fault, or other structure that cuts across another must be the younger, as shown in **Fig. 2.9**.

Planetary geologic mapping is hampered by a lack of field observations except for those at the local *Apollo* sites and a handful of "ground-truth" sites gained from robotic landers. Consequently, planetary geologic maps are formatted a little differently in comparison with those for Earth. On planetary maps the rock unit descriptions are divided into two parts, *observations* and *interpretations*. The observation part describes the characteristics of the unit in objective terms on the basis of the available data, while the interpretation part explains the possible origin and evolution of the unit according to the opinion of the author. In principle, the observation part should remain

Figure 2.8. View of the Taurus–Littrow region of the Moon, showing flooding by dark mare lavas into the more heavily cratered highlands on the right. The arrow points to a lava-"embayed" crater. The paucity of craters in the mare (left side) indicates the relative youth of the lava in comparison with the highlands. The star indicates the location of the *Apollo 17* landing site. The darker part of the mare surface marks the presence of *dark mantle deposits* of volcanic origin (NASA AS17–0939).

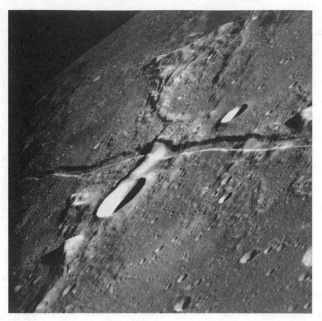

Figure 2.9. A view of Rima Ariadaeus, taken by the *Apollo 10* astronauts. This linear rille, formed by faulting, is about 2 km wide and cuts across older terrain (NASA AS10–4646).

valid (at least until new data are available), but the interpretation part could change or be different, depending on the analyst.

The first planetary geologic maps by Hackman and Shoemaker set the stage for a series of mapping programs that extended to Mars, Mercury, Venus, and outer planet satellites and that continues today through the USGS. In these programs, each planet is divided into map quadrangles of systematic scales, which serve as cartographic bases for geologic mapping.

2.4 Geologic time

Geologic time can be considered from two perspectives, **absolute time** and **relative time**. In absolute time, rocks and geologic events, such as faulting, are determined as being of a specific age expressed in years. In relative geologic time, rocks and events are simply stated as being older or younger than other rocks or events, without expressing their age differences in years. Determining the absolute ages of rocks is accomplished primarily by using radiogenic "clocks" that are based on the principle that certain unstable radioactive elements (i.e., isotopes) contained in some

rocks decay or convert to more stable isotopes at a known rate. If we know this rate and can measure the amounts of unstable and stable isotopes in a sample, it is possible to determine the age of the rock. Of course, radioactive isotopes of the right type must be available for measurement in the rock sample and, unfortunately, not all samples contain these isotopes. Consequently, only some rocks can be dated. Typically, radiogenic dating provides the age of a rock in years from its formation.

How is geologic time assessed on other planets and satellites? Practical limitations require that rock samples be analyzed in laboratories on Earth to determine radiometric ages because automated dating systems for robotic spacecraft have not yet been developed. Consequently, planetary absolute dates have been obtained only for rock samples from the Moon and for meteorites, some of which are from Mars.

Because no direct samples are available from other planets, only relative ages can be assigned with confidence. The principles of superposition, embayment, and cross-cutting relations are routinely applied to planet and satellite surfaces for this purpose.

An additional method for establishing the relative ages of planetary surfaces is based on the size–frequency distribution of impact craters. Old surfaces have been exposed to the impact environment for longer than have younger surfaces and statistically should contain more impact craters (**Fig. 2.8**). By counting the number of craters superposed on planetary surfaces, their age relative

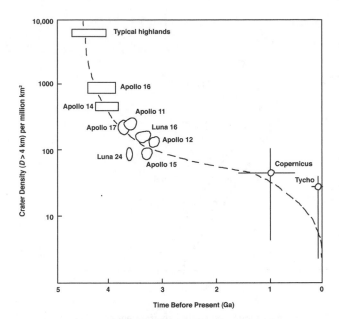

to other surfaces can be determined. This concept was developed for the Moon and was verified when rock samples were returned to Earth for analysis and radiogenic dating (**Fig. 2.10**). Research by NASA planetologist Don Gault showed that, with time, cratered surfaces reach a stage, called **equilibrium**, in which craters of a given size are obliterated by impact erosion at the same rate as they are formed, as shown in **Figs. 2.11** and **2.12**. Thus, only surfaces that have not yet reached equilibrium for the crater sizes being considered can be dated (**Fig. 2.13**).

In practice, difficulties can arise in using crater statistics for age determinations. For example, non-impact craters, such as those formed by volcanic processes, might be indistinguishable from impact craters, and, if non-impact craters were present, the surface would appear anomalously old. In addition, **secondary craters** are formed by the impact of rocks ejected from primary impact craters. Their presence adds to the total crater population and must be taken into account by various models that predict how many secondary craters would form as a function of the primary crater size. Unfortunately, such models are imperfect, and it is difficult to determine the presence and

Figure 2.10. Number of craters larger than 4 km per unit surface area versus the age in gigayears (Ga), calibrated against absolute dates obtained from lunar samples. This curve enables extrapolation of ages to surfaces lacking samples on the basis of crater counts (from Spudis, 1996, after Heiken et al., 1991; reprinted with permission from Smithsonian Institution Press).

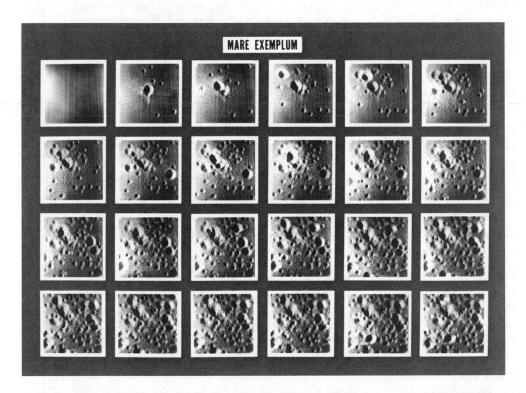

Figure 2.11. Photographs showing a NASA experiment (called *Mare Exemplum*) to simulate the evolution of a cratered surface. In this experiment, a box 3.2 m by 3.2 m was filled with loose sand, smoothed (upper left), and then impacted with bullets of different sizes (ranging from birdshot to high-powered rifle). Placement of each shot followed a grid system and a random-number generator; the ratio of differently sized impacts was based on the size–frequency distribution of craters seen on lunar mare surfaces. The end of the series is in the lower right. The series illustrates how crater counts can be used to date surfaces; surfaces in the top photographs represent younger surfaces in comparison with those in subsequent photographs. Note, however, that in the last row of photographs the crater size frequencies are essentially the same, representing cratering *equilibrium*, in which craters are being destroyed at the same rate as that of crater formation; thus, it is not possible to date surfaces within the last row of photographs (courtesy of Don Gault).

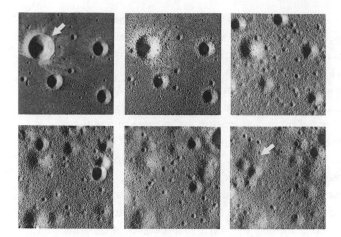

Figure 2.12. An experiment similar to that shown in **Fig. 2.11** to illustrate stages in crater modification by impact degradation. The arrows in the first and last images point to a crater that is subjected to small impacts, which gradually wear down the crater rim and fill in the crater floor until it is nearly "erased" from the surface (NASA-Ames photograph AAA481–8, courtesy Don Gault).

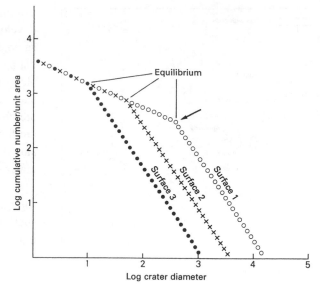

Figure 2.13. A diagram of idealized crater size–frequency distributions (number of craters per unit area versus crater diameter) for three surfaces; surface 1 is the oldest, reflected by its having the greatest number of craters and craters of largest sizes. The steep parts of all three curves represent crater *production*, while the less steep parts represent crater *equilibrium*. Relative age-dates can be determined either by the "break point" between production and equilibrium (in which progressively older surfaces have their break point at larger crater size), or by the position of the production curve, which shifts to the right with increasing age.

number of secondary craters. Other considerations include target properties that could cause variations in crater sizes. For example, experiments show that craters formed in targets containing fluids are larger than craters formed in dry targets. This difference could cause the size–frequency distribution for the volatile-containing target to be interpreted as representing an older surface.

Despite these difficulties and uncertainties, impact crater statistics are commonly used as a means for obtaining relative ages for different planetary surfaces. Ages derived from crater counts have been compared with (and calibrated against) radiometric dates obtained from lunar samples and demonstrate the validity of the technique, at least on the Moon, where surface-modifying processes are minimal. This result suggests that crater counts can be used to obtain dates for surfaces on other "airless" bodies, such as Mercury. Great caution must be exercised, however, in using crater counts on planets where differences in erosion might occur as a function of location, or where significant differences in target properties may alter the crater morphology.

In principle, crater counts can also be used to derive absolute ages for planetary surfaces, as discussed by Michael and Neukum (2010). This has been done with some confidence on the Moon where cratered surfaces have been sampled and radiometric ages determined (Neukum *et al.*, 2001); extrapolation of the calibrated crater curve (**Fig. 2.10**) to surfaces that have not been sampled enables estimates of their ages. The same can be done for surfaces on other planets by extrapolation of the calibrated

lunar crater counts. This requires, however, that correct adjustments can be made for gravity (which influences the sizes of craters), impact flux as a function of location in the Solar System (proximity to the asteroid belt, as with Mars, which experiences a higher impact rate than that on the Moon), and the potential for degradation in the presence of an atmosphere, as on Venus. These and other factors result in complex algorithms for extrapolation, and potentially large error bars on the results, depending on the assumptions and uncertainties in the age calculations.

2.5 Remote sensing data

Most of our knowledge of the geomorphology of Solar System objects is derived from remote sensing, defined as the collection of information without coming into physical contact with the object of study. This is accomplished by designing instruments that can be carried on some "platform" to collect useful information. Typically, platforms on Earth include airplanes and spacecraft, but can also include balloons, helicopters, or robots operating on the surface. To the extent possible, similar platforms are used

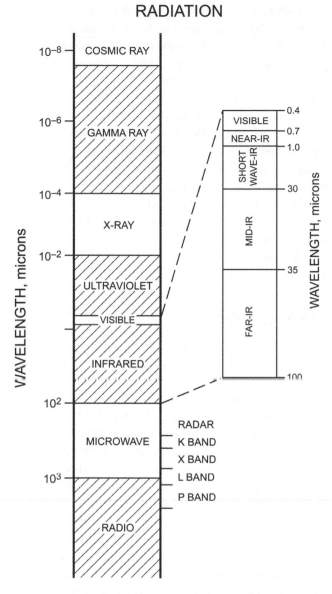

Figure 2.14. The electromagnetic (EM) spectrum, showing important wavelengths used in remote sensing. Radar systems are subdivided and given letter designations that were arbitrarily assigned (see **Table 2.1**).

detect specific parts of the EM spectrum. The energy detected depends on the interaction of the energy with the surface being analyzed, generating spectra that are distinctive for various properties of the surface. For example, sunlight shining on a planetary surface can be reflected, absorbed, and/or transmitted, depending on the EM wavelength, the temperature, and characteristics of the surface materials, such as composition and grain size.

Remote sensing systems are classified as either passive or active. Passive system use natural radiation, such as sunlight, whereas active systems illuminate the surface with an artificial energy source. For example, radar imaging systems beam energy toward a surface, some of which is reflected and then recorded by a radar sensor. Active systems also include non-imaging systems, such as laser and radar altimeters, which measure the distance from the instrument to the surface.

2.5.1 Visible imaging data

Nearly every spacecraft sent in the exploration of the Solar System has carried a camera or imaging system as part of its scientific payload. Currently, most imaging systems use **charge-coupled devices** (CCDs) as the detector, rather than film. However, in planetary geomorphology, images from previous missions are still useful, and it is important to be familiar with the imaging systems used in these missions, as described in **Appendix 2.1**.

CCDs were invented in 1969 by Bell Laboratories and are used in a variety of solid-state imaging devices. Today, modern digital cameras all use CCD technology, including simple cell-phone cameras and sophisticated video systems. A CCD "chip" consists of a layer of metallic electrodes and a layer of silicon crystals, separated by an insulating layer of silicon dioxide. When used as an imaging system, the CCD chip is structured as an array of picture elements, or **pixels**. Light focused onto the chip by a lens causes a pattern of electrical charges to be created. The charge on each pixel is proportional to the amount of light received and provides an accurate representation of the scene. Each charge can be transmitted separately and then reconstructed using conventional image-processing techniques.

CCD imaging systems can be either line arrays or two-dimensional arrays. In line arrays, a single line of CCDs sweeps across the scene as the spacecraft (or aircraft) moves over the terrain, building up the image. Two-dimensional arrays consist of a chip with CCDs on an X–Y coordinate system and are used as framing cameras in which the CCDs record the scene as a "snapshot."

in Solar System exploration, with most information coming from spacecraft located well above the surface.

Remote sensing instruments make use of **electromagnetic (EM) radiation (Fig. 2.14)**, which is generated whenever there is a change in the size or direction of an electrical or magnetic field. For example, electrons shifting from one orbit to another orbit around an atomic nucleus results in X-rays and visible (light) radiation, while fluctuations in the electrical/magnetic field generate microwaves and radio waves of the sort used in radar systems. Remote sensing instruments are designed to

In simple systems, images are produced in a single wavelength range as a "black and white" (and shades of gray) picture. The wavelength range can be either narrow (as in the near-infrared) or "broadband" to produce a panchromatic picture. Color images are produced by obtaining data for more than one wavelength over the same scene. For example, three frames could be exposed, one each through red, blue, and green filters. Each filter is sensitive to a given wavelength and the data corresponding to that wavelength are recorded. The three frames are then combined to produce a color image.

2.5.2 Multispectral data

Mapping surface compositions using remote sensing techniques is critical in planetary science. The crystal structure of minerals behaves in characteristic ways when exposed to EM energy. In the visible–near-infrared (**Fig. 2.14**), or VNIR, crystals absorb energy, resulting in **absorption bands** that are diagnostic for specific minerals or groups of minerals. To take advantage of this phenomenon, **multispectral spectrometers** are designed to measure the reflected energy as a function of wavelength using filters and detectors that are responsive to the absorption bands of interest. Generally, the narrower the band widths covering the VNIR spectrum, the more precise the analysis. For example, the Near-Infrared Mapping Spectrometer (NIMS) which flew on the *Galileo* mission to Jupiter could develop a 408-wavelength spectrum for each pixel over the range from 0.7 to 5.2 microns.

Multispectral spectrometers can be either "point" systems or mapping systems. In point systems, a single line is traced across a surface, measuring the spectra as a compositional profile over terrains. The advantage of this approach is that the instrument is relatively simple and the total data volume is small. The disadvantage is that sampling of the surface is very limited and important areas might be missed. Mapping spectrometers have two-dimensional arrays of detectors that collect data over an area rather than as a line-trace. Mapping spectrometers are much more useful for geologic studies.

2.5.3 Thermal data

Surface materials radiate, or emit, energy ("heat") in the 0.5–300 micron range of the EM spectrum, which can be recorded as digital files or transformed into images. Detectors that can measure this energy (sensitive thermometers known as **bolometers**) were developed by the military.

For example, such a detector flown on an aircraft over terrain at night can pick up the heat generated by vehicle engines or even body heat from individuals. In the mid 1960s, some thermal detectors were declassified for civilian use, leading to remote sensing systems for use on Earth and in planetary exploration. The Thermal Emission Spectrometer (TES), developed by Phil Christensen of Arizona State University and flown on the *Mars Global Surveyor* spacecraft, revolutionized our understanding of the surface of Mars. The energy recorded by a thermal detector, such as the TES, is a function of many complex variables, including surface composition and texture, the atmosphere between the surface and the detector, and the detector sensitivity. Understanding the physics of the transfer of energy from the surface to the recorder enables the determinations of factors such as the mineralogy and grain sizes of surface materials.

2.5.4 Radar imaging data

Dense clouds obscure two objects of planetary geologic interest, Venus and Titan, a moon of Saturn. Because of its long wavelength (**Fig. 2.14**), radar can "see through" clouds and has been used to map both Venus and Titan using **synthetic aperture radar (SAR)** imaging systems. Radar is an active remote sensing technique in that the energy is generated artificially, in effect illuminating the scene to be imaged. SAR systems are mounted on a moving platform (an airplane or spacecraft) and send short pulses of radio energy obliquely toward the side, striking the surface at an angle. Some of the energy is reflected back to the spacecraft, where it is received as an echo, by which time the motion of the spacecraft has carried it to a new position. Several thousand pulses are sent per second, at the speed of light, resulting in an enormous data set that must be highly processed. To construct an image, three factors must be integrated: (1) the round-trip time from the instrument to the surface, (2) the Doppler shift due to the motion of the spacecraft, and (3) the radar reflectivity of the surface, which is a function of composition, surface roughness, and other factors.

Radar images (**Fig. 2.5**) can be confused with images in the visible part of the EM spectrum, and there are significant differences that must be understood for proper interpretation. First, the brightness and apparent shadows in a radar image do not result from sunlight, but from "illumination" by the radar beam. Thus, the geometry of the bright and dark terrains depends on the position of the spacecraft and characteristics of the surface such as topography and composition.

Table 2.1. **Radar bands**

Designation	Wavelength	Frequency	Examples
Ku, K, Ka	0.8–2.4 cm	37.5–12.5 GHz	Aircraft–Earth
X	2.4–3.8 cm	12.5–8 GHz	SIR–Earth
C	3.8–7.5 cm	8–4 GHz	SIR–Earth
S	7.5–15 cm	4–2 GHz	*Magellan*–Venus
L	15–30 cm	2–1 GHz	*Seasat*–Earth
P	30–100 cm	1 GHz–300 MHz	MARSIS–Mars

Different parts of the radio wavelength EM spectrum are used in SAR imaging, designated with letters of the alphabet (**Table 2.1**). These were defined mostly during the Second World War as classified information, and the letters were arbitrarily assigned. In addition, radar data are typically polarized, by which means the waveform energies both in the "send" and in the "receive" phases are filtered into either a horizontal plane or a vertical plane. Thus, the data can occur in one of four combinations: (a) HH for horizontal send, horizontal receive; (b) VV for vertical send, vertical receive; (c) HV for horizontal send, vertical receive, or (d) VH for vertical send, horizontal receive. Each mode has advantages and disadvantages depending on the application and the nature of the surface.

The Soviet *Venera 15* and *16* and the US *Magellan* missions carried radar imaging systems to Venus to obtain the first detailed views of the surface from orbit. NASA's *Cassini* orbiter, sent to Saturn, carried a radar system to obtain images of the moon Titan.

Very-long-wavelength–low-frequency radar systems are capable of penetrating into the subsurface, depending on the nature of the surface materials. In general, some of the radar energy is reflected from subsurface boundaries, such as contacts between rock units, and is recorded; from the geometry of the received signal, the depth to the boundary can be calculated. Instruments using this principle were used on the Moon during *Apollo* and were flown on the European Space Agency *Mars Express* mission and on NASA's *Mars Reconnaissance Orbiter*. Penetrating radar is likely to fly on missions to the outer Solar System as a means of investigating the ice structure of some satellites, such as Jupiter's moon Europa.

2.5.5 Ultraviolet, X-ray, and gamma-ray data

At very short wavelengths (**Fig. 2.14**), EM energy is strongly influenced by gasses, and the ultraviolet (UV) part of the EM spectrum is used to study planetary atmospheres and surfaces that lack atmospheres. UV spectra provide useful information on some of the physical properties of the surface, such as grain size and the presence of frost. X-ray spectrometers detect energy generated by the Sun, in which the X-ray spectra are diagnostic for elements such as aluminum, silicon, and magnesium. Gamma rays are produced from radioactive decay and from the bombardment of surfaces by cosmic rays from deep space. Gamma-ray spectrometers measure this energy and can be used to map the distributions of some elements, such as titanium. For example, X-ray and gamma-ray spectrometers were flown on *Apollos 15* and *16* to map parts of the Moon.

2.6 Geophysical data

Various techniques are employed to obtain geophysical data relevant for planetary geomorphology, including the use of instruments to measure altimetry, gravity, and magnetic fields. Topographic data are fundamental for most studies and are derived from images (see **Section 2.10**) or from altimeters that measure the distance from the spacecraft to the ground. For example, topographic data from the Mars Orbiter Laser Altimeter (MOLA) are commonly used to make images that have the appearance of photographs in which the illumination direction and angle, as well as the vertical exaggeration, can be controlled.

Both radar systems and lasers have been used in planetary exploration for altimetry, in which the time between energy emission from the spacecraft and energy receipt from the surface yields the altitude. The spacecraft position referenced to the overall shape of the planet then enables the surface topography to be derived for comparison with the geomorphology.

Mapping variations in local gravity is a common technique in geophysical studies on Earth and for mineral prospecting. Areas containing high-density materials will have slightly higher gravitational accelerations than areas of low-density materials, when differences in local topography are taken into account. In planetary exploration, mapping gravity variations, as was done for the Moon, provides insight into subsurface structure and the possible relation to surface features.

Magnetic fields vary in space and time on planets. On Earth, measuring the orientations of the fields "locked" in rocks helped build the case for the theory of plate tectonics and showed that the magnetic poles have shifted with respect to the spin axis. Not all planets have magnetic fields today, but magnetometers flown on spacecraft have revealed the presence of **remnant magnetic fields**. For example, the *Mars Global Surveyor* spacecraft recorded a remnant field in the oldest rocks exposed on the surface. Its absence in the younger terrains suggests that there was a fundamental change in the interior of Mars early in its history, thus demonstrating that geophysical measurements combined with surface geology provide powerful tools for deciphering planetary histories.

2.7 Image processing

Digital "snapshot" cameras and computers typically have built-in routines to manipulate images, such as "red-eye" removal. Such digital image processing has become so commonplace that many people do not understand what is involved. In this section, some of the principles and practices of planetary image processing are outlined.

The basic units of digital images are **pixels** (an abbreviation for "picture elements"), which can be arrayed in horizontal lines and vertical rows to make a picture. The amount of light received by the detector for each pixel is encoded by its brightness level. In 8-bit encodement (2 raised to the 8th power), 256 shades of gray can be assigned, with each level referring to a specific **"DN" (digital number)**, in which a DN of 0 is black (no light received) and a DN of 255 is a perfectly white level. These are typically shown on an image by a DN histogram that gives the distribution of the various levels of gray in an image (**Fig. 2.15**). Because the human eye can discriminate only about 30 shades of gray, this means that digital images potentially contain much more information than can be visually detected. Various algorithms can be applied to extract this information after calibration of the image.

Calibrations. Digital image detectors are not uniform in their response to light. Within any given array, some detector pixels will be more sensitive than others. Consequently,

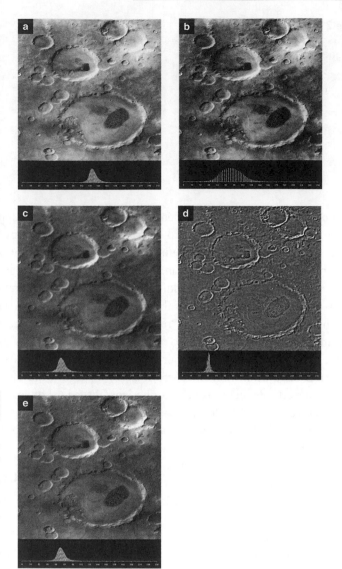

Figure 2.15. Images of Mars to illustrate some common image processing techniques: (a) calibrated image showing a histogram (bottom of image) of DN levels within a relatively narrow range centered at about 134; (b) a "stretched" image in which the DN levels are spread over a wider range than in (a) and shifted toward a darker (lower DN) level; (c) a *low-pass filter* image resulting in a smoother appearance; (d) a *high-pass filter* image in which differences in brightness are enhanced; and (e) a *sharpened* image in which boundaries among DN changes are emphasized (part of Mars Orbiter Camera image mc27–256).

after the camera has been assembled, but before it goes into use, it must be calibrated to take these differences into account. In its simplest form, an image of a "flat field," which consists of a perfectly uniform, perfectly illuminated surface is taken by the camera. Each pixel is then adjusted by adding or subtracting DNs until all the pixels have the same value. This mapping of DNs is then standardized for

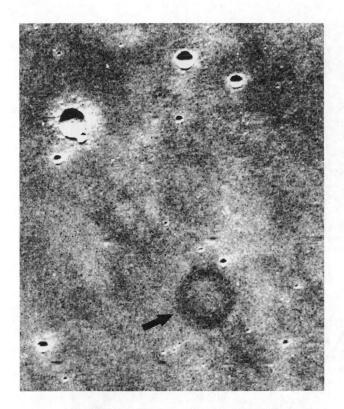

Figure 2.16. "Donuts" are image blemishes (arrow) caused by dust grains in camera systems. These and other artifacts can be removed in image processing by mapping the pixels that are affected and then assigning DN values to them on the basis of the values of the surrounding unaffected pixels. This represents cosmetic processing and users need to be aware that the assigned pixel values are artificial (NASA *Viking Orbiter* 826A68).

that particular camera so that, when images are subsequently taken, they can be calibrated or adjusted pixel-by-pixel.

Despite the best efforts to maintain cleanliness, dust grains find their way into imaging systems and can produce artifacts, such as "donuts" (**Fig. 2.16**). As part of the calibration routine, these and other artifacts are mapped so that they can be taken into account and cosmetically corrected, as noted below.

Calibrations are also typically performed during flight because detectors can change with time. Such calibrations are accomplished by taking images of known surfaces, such as the Moon or star fields, with individual pixels adjusted, just as is done in pre-flight calibrations. New artifacts can also occur, as when cosmic rays "zap" the detector and degrade or knock out one or more pixels. These artifacts are also mapped and can be corrected cosmetically.

Stretches. Stretching digital images involves shifting the distribution of DN levels (**Figs. 2.15(a)** and **(b)**), or bringing about a simple increase in brightness by moving all the DNs to a higher level without changing the shape of the histograms, or redistributing the DNs following some specific function, such as a Gaussian distribution. More complicated stretches involve giving one part of the distribution more weight than other parts in order to emphasize detail.

Filters. Filtering involves manipulating multiple pixels as sets within the image. For example, **boxcar filters** give a weighted value to each pixel as a function of the value of its neighbors (the "box"), which is then slid across the image, adjusting each pixel one-by-one. In a **low-pass filter** the value of the central pixel is the average value of the neighboring pixels, and the image tends to be smoothed, enhancing broad changes in the scene (**Fig. 2.15(c)**); the larger the boxcar, the smoother the result.

In **high-pass filters** the DN values from the low-pass filter are subtracted from the image, leaving only the smaller variations in the scene and producing a somewhat sharper image (**Fig. 2.15(d)**). **Edge-enhancement filters** decrease the contrast where pixels have similar values and enhance the contrast where pixels change, in order to emphasize the boundaries in the scene (**Fig. 2.15(e)**).

Common "snapshot" digital cameras automatically apply some form of stretching and filtering to produce pleasing images. Once in the computer, stretching, filtering, and various color-enhancement techniques are applied with user-friendly "black box" programs that are based on the processes outlined above.

Geometric projections. In most spacecraft images, the position of each pixel is referenced to some system, such as geographic coordinates by latitude and longitude. This allows the image to be re-cast into standard projections. For example, an image might be taken that is **oblique**, or viewed looking at the terrain at an angle similar to the view from an airplane window. Because the geometric position of each pixel is known, they can be shifted so that the image is portrayed **orthographically** as though it were taken as viewed looking straight down on the terrain (**Fig. 2.17**). Alternatively, the image can be re-projected into a standard cartographic product, such as a Mercator projection, depending on the intended use.

Mosaics. Multiple frames can be put together as **mosaics** (**Fig. 2.18**), in which the boundaries between individual frames are seamless. This begins with the identification of individual **tie points** that consist of specific features, such as small craters, that are visible on more than one frame. The pixels in the frames are then re-projected geometrically so that all of the features match. Various filters are then applied so that the pixel DN values along the frame boundaries are averaged to reduce the

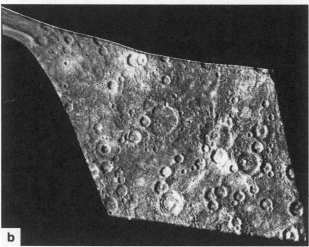

Figure 2.17. Geometric projections involve shifting individual pixels into new positions; for example, this image of Mercury was taken at an oblique angle from the *Mariner 10* spacecraft, viewed off to the side (a), which was then orthographically reprojected (b) so that the scale is near-uniform over the entire scene (as though viewed "straight-down" from the spacecraft; NASA *Mariner 10* FDS 27321).

Figure 2.18. Mosaics combine more than one image to matching pixels along frame boundaries and assigning average values to the pixels, giving the appearance of a single image. (a) This set of *Mariner 10* images of the Caloris basin on Mercury was assembled by hand; (b) the same scene as a computer-generated mosaic (courtesy of the Jet Propulsion Laboratory).

contrast. This overall process can be done automatically or by hand.

Cosmetic processing. A wide variety of processes can be applied to generate images that are more pleasing to the eye. For example, individual pixels or blocks of pixels might have DN "0" levels because parts of the detector have been damaged, or because data were lost during transmission from the spacecraft to Earth. The simplest method of filling in the missing data is to use some average value of the neighboring pixels for the missing pixel. Similarly, blemishes, such as those caused by dust grains, can be removed by reference to the calibration files

and the missing data can be applied. One must remember that cosmetically improved images include DN values that are not real (**Figs. 2.16**, **2.19**, and **2.20**).

2.8 Resolution

The scale of individual pixels in a digital image and the resolution of that image are somewhat related, but are distinctive parameters that are often confused, even in the planetary science community. Put simply, the scale of a pixel is related to the dimension of the terrain projected onto the detector through the imaging system lens. Thus, it is dependent primarily on the optics, the distance of the camera from the terrain, and the size of each pixel of the detector. The **pixel scale** then can be stated as some length per pixel, such as 10 m per pixel, meaning that 10 m

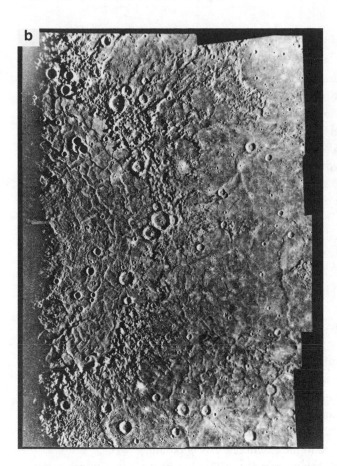

Figure 2.18. (*cont.*)

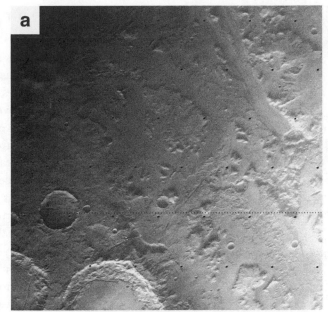

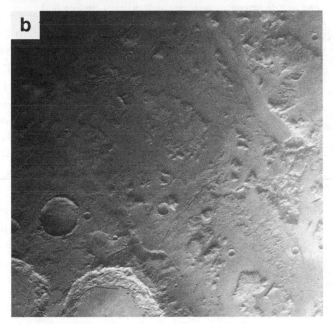

Figure 2.19. Various cosmetic image processing techniques remove artifacts such as "donuts" (**Fig. 2.16**), filling-in pixels or lines of pixels "dropped" during electronic data transmission (a) by assigning average values of surrounding pixels to the dropped pixels, and by removing reseau ("r," registration marks built into the imaging system used for precise location of surface features) to produce a final image (b).

on the ground is registered as a single DN value by the pixel detector.

Resolution is a much more complicated parameter than pixel scale, since it refers to the smallest object that can be identified in the image. Thus, the contrast of the object in relation to its background, the shape of the object, the responsitivity of the particular detector to the composition and surface texture, and other factors all play a role in resolution, as well as the size of the object in relation to the pixel scale. For example, a stark white object ten times larger than the pixel scale might not be seen on an image if the object were placed on a pristine snow field, while some linear features, such as fault traces, might be detected even though the width of the fault might be smaller than the pixel scale, simply because of its shape and the contrast with the surrounding terrain.

Because of the complications in defining resolution, digital images are often (erroneously) stated to have a "resolution" of x meters per pixel, when this phrase actually refers to the pixel scale or angular resolution. Obviously, it takes more than one pixel to discern an object, depending on the object's shape. To further complicate the issue, some images will be stated to have a resolution of x meters per *line pair of pixels*, which is an attempt to take into account the need for more than one pixel for object detection.

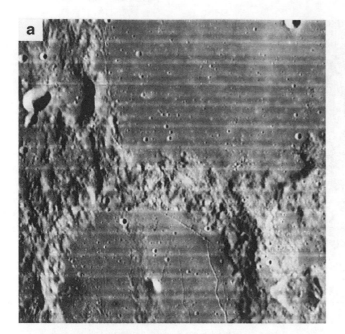

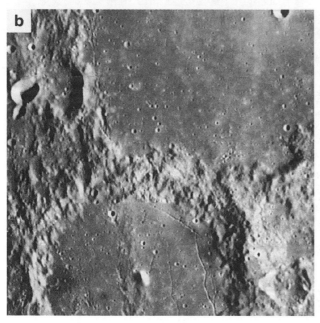

Figure 2.20. *Lunar Orbiter* photographs were transmitted and reconstructed by assembling strips of film, resulting in the distinctive pattern seen in (a); mosaicing and image processing results in a smooth, continuous image as shown in (b).

Pixel scale can also be erroneously set with the image processing tools commonly available on computers. Some of these tools resize the image, in effect adding pixels to fit some format, and leading to an imaginary pixel scale that is smaller than the actual pixel scale of the original data. When using such tools, it is best to use a "sanity check," keeping in mind that the pixel scale cannot be better than that of the original image.

2.9 Electronic data records (EDRs)

EDRs are the files from the mission (this refers to all data, not just images). For images, there are different levels of processing, as outlined below.

Level 0 refers to the **raw data** (no processing) as received from the spacecraft. This version is generally preferred by scientists who need to conduct quantitative studies, such as **photometry** (precise measurements of surface brightness), for which they have their own algorithms for customized processing. Level 0 is often the preferred archival method to preserve the original files in order that algorithms developed later can be applied.

Level 1 data are **decompressed**. To make data acquisition and transmission more efficient, various data compression techniques are applied; these can be either hardware techniques built into the imaging system or software techniques that can be updated on board the spacecraft. For example, all pixels that have a DN of zero (black) might be automatically eliminated from the data transmission. Compression is either **lossless** (preserving all the data) or **lossy**, in which some data are lost. In level 1, the compression is "reversed" to restore the original image to the extent that this is possible.

Level 2 data have the calibration files applied to produce **radiometrically corrected** images in which the brightness levels are correctly given, taking into account illumination, etc. On some missions, radiometric calibration is done on level 1 files.

Level 3 data have been custom processed for specific uses; for example, images might now be in some uniform map projection.

Level 4 data are further processed and might be merged with other data; for example, level 4 images might have topographic information incorporated.

2.10 Cartography

Maps are essential for exploration. No doubt, early humans scratched simple maps in the dirt to describe hunting grounds, clan boundaries, and other geographic locations critical for their survival. Planetary maps are generated from image mosaics using conventional cartographic projections. The USGS is supported by NASA to produce maps with a variety of scales, projections, and portrayal methods. Most of the terrestrial planets and the satellites that have been adequately imaged have been mapped in standard cartographic series, or quadrangles. Indexes and availability of cartographic materials are maintained on the USGS website (http://pdsmaps.wr. usgs.gov) and include geologic maps and a gazetteer of officially named planetary surface features.

Deriving a coordinate reference system for planets is challenging. Once the spin axis for an object has been determined, the equator can be set with latitude running north and south from the equator. The prime meridian is arbitrarily defined, using some recognizable feature, such as a small crater. Now come some potential problems! Longitude can run either east or west from the prime meridian, and both have been used in planetary science, even on the same planet. Thus, some instrument data sets give coordinates in east longitude, while others use west longitude. In principle, so long as east or west is designated, there should be no problems, but in practice, especially with some digital files where E or W is not designated directly with the values, errors can arise. For example, both systems have been used on Venus, and in the early planning stages for a mission that would have sent small probes through the atmosphere toward the surface, the scientists used one system and the engineers used the other system; had the mission flown (it did not) and the error not been caught, the probes would have descended through a part of the venusian atmosphere totally different from that targeted.

Today, the generally accepted use on maps follows the **planetographic system**, in which west longitudes are used for objects that spin in the same direction as that in which the object orbits. For example, looking down on a planet's north pole, the planet would spin in a counter-clockwise direction and the planet would orbit the Sun in a counter-clockwise direction. When viewing the planet toward the equator from a fixed position in space, the longitudes would increase in value toward the west as the planet rotated. For an object that is in retrograde motion (spinning in the opposite direction to its orbit, such as Venus), longitudes increase toward the east.

The "zero" reference elevation also poses problems; what does one use on a planet not having a sea level? Different systems have been used on different planets. For example, on the Moon and Venus, the mean radius of the planet is taken as the "zero" contour, while on Mars, the original reference was based on the triple point of carbon dioxide, the main component of the atmosphere. The **triple point** of a substance is the pressure–temperature condition at which all three phases (solid, gas, liquid) exist. As discussed in **Chapter 7**, a different system is used on Mars currently. Thus, depending on data

availability and the specific planet, different systems are used throughout the Solar System, and, as with longitude, one must be familiar with the conventions used for the specific planetary object of interest.

Topographic maps are generated using a variety of techniques, including photogrammetry, photoclinometry, and altimetry (see **Section 2.6**). Photogrammetry has long been used in making topographic maps of Earth from aerial photographs. This technique is based on **stereoscopic models** in which two or more images are taken of the same terrain but from different viewing angles. From knowledge of the geometry of the camera system optics and the altitude from which the images are taken, the relief of the terrain can be derived. **Photoclinometry**, also known as "shape from shading," uses the amount of light reflected from the surface to determine the slope of that surface; this technique requires that the illumination and viewing geometry be known and that the surfaces are homogeneous with regard to texture, composition, and other variables that influence their reflectivity. In these cases, the reflectivity of each pixel is measured, from

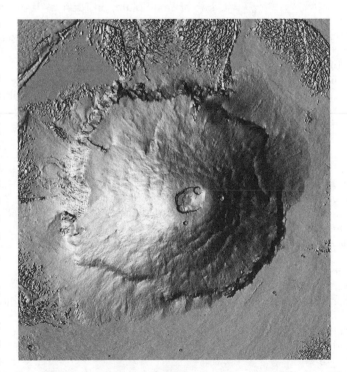

Figure 2.21. Digital elevation models (DEMs) enable topography to be portrayed by a variety of techniques, including as shaded-relief maps, shown here for the Olympus Mons shield volcano on Mars. Manipulation of DEMs can provide oblique views of terrain, as might be seen from an airplane window, and the illumination direction and angle can be changed to include early-morning to mid-day views (MOLA topographic DEM courtesy of the US Geological Survey).

Table 2.2.   **Terms used for features in planetary nomenclature (US Geological Survey)**

Feature[a]	Description
Albedo feature	Geographic area distinguished by amount of reflected light
Arcus, arcūs	Arc-shaped feature
Astrum, astra	Radial-patterned features on Venus
Catena, catenae	Chain of craters
Cavus, cavi	Hollows, irregular steep-sided depressions, usually in arrays or clusters
Chaos, chaoses	Distinctive area of broken terrain
Chasma, chasmata	A deep, elongated, steep-sided depression
Collis, colles	Small hills or knobs
Corona, coronae	Ovoid-shaped feature
Crater, craters	A circular depression
Dorsum, dorsa	Ridge
Eruptive center	Active volcanic centers on Io
Facula, faculae	Bright spot
Farrum, farra	Pancake-like structure, or a row of such structures
Flexus, flexūs	A very low curvilinear ridge with a scalloped pattern
Fluctus, fluctūs	Flow terrain
Flumen, flumina	Channel on Titan that might carry liquid
Fossa, fossae	Long, narrow depression
Insula, insulae	Island (islands), an isolated land area (or group of such areas) surrounded by, or nearly surrounded by, a liquid area (sea or lake)
Labes, labēs	Landslide
Labyrinthus, labyrinthi	Complex of intersecting valleys or ridges
Lacus, lacūs	"Lake" or small plain; on Titan, a "lake" or small, dark plain with discrete, sharp boundaries
Landing site name	Lunar features at or near Apollo landing sites
Large ringed feature	Cryptic ringed features
Lenticula, lenticulae	Small dark spots on Europa
Linea, lineae	A dark or bright elongate marking, may be curved or straight
Lingula, lingulae	Extension of plateau having rounded lobate or tongue-like boundaries
Macula, maculae	Dark spot, may be irregular
Mare, maria	"Sea;" large circular plain; on Titan, large expanses of dark materials thought to be liquid hydrocarbons
Mensa, mensae	A flat-topped prominence with cliff-like edges
Mons, montes	Mountain
Oceanus, oceani	A very large dark area on the Moon
Palus, paludes	"Swamp;" small plain
Patera, paterae	An irregular crater, or a complex one with scalloped edges
Planitia, planitiae	Low plain
Planum, plana	Plateau or high plain
Plume, plumes	Cryo-volcanic features on Triton
Promontorium, promontoria	"Cape;" headland promontoria
Regio, regiones	A large area marked by reflectivity or color distinctions from adjacent areas, or a broad geographic region
Reticulum, reticula	Reticular (netlike) pattern on Venus
Rima, rimae	Fissure
Rupes, rupēs	Scarp
Satellite Feature	A feature that shares the name of an associated feature. For example, on the Moon the craters referred to as "Lettered Craters" are classified in the gazetteer as "Satellite Features."
Scopulus, scopuli	Lobate or irregular scarp
Sinus, sinūs	"Bay;" small plain
Sulcus, sulci	Subparallel furrows and ridges
Terra, terrae	Extensive land mass

Table 2.2.   (cont.)

Feature[a]	Description
Tessera, tesserae	Tile-like, polygonal terrain
Tholus, tholi	Small domical mountain or hill
Unda, undae	Dune
Vallis, valles	Valley
Vastitas, vastitates	Extensive plain
Virga, virgae	A streak or stripe of color

[a] Singular, followed by plural form.

which the slope of the terrain covered by that pixel is determined. For example, a pixel covering ground that is flat and horizontal, with the Sun directly overhead, would appear brighter than a pixel of the same type of surface, but tilted with respect to the horizontal. An overall topographic map is then generated by integrating the slopes for the areas covered by each pixel.

Once the topography has been derived, whether from images or from other techniques, **digital elevation models (DEMs) can be constructed**. Topography then can be shown by contour lines, by colors, or as **shaded-relief maps (Fig. 2.21)** to portray the terrain as it might appear to a viewer from above.

Names for planetary objects and surface features are determined by the International Astronomical Union (IAU) through a committee and various subcommittees

(typically, one for each planet). Names on planetary surfaces derive from a variety of sources, including historic telescopic usage on maps of the Moon made centuries ago. In the Space Age, there has been an attempt to set specific themes for naming surface features. For example, small craters on Mars are named for Earth villages or towns of less than 100,000 population, while volcanoes on Io are named for ancient gods dealing with fire, such as Prometheus.

Classes of surface features typically are Latinized, as given in **Table 2.2**. With features named for people (mostly craters) it is required that the individual be deceased for at least five years before the name is applied. The USGS maintains a gazetteer of named features, which can be accessed at their website http://planetarynames.wr.usgs.gov/.

Assignments

1. Discuss the fundamental differences between images produced from the visible part of the electromagnetic spectrum and images produced from radar systems.

2. Discuss the fundamental concept of using impact crater counts for age-dating planetary surfaces and explain the difference between *equilibrium* and *production* distributions.

3. Outline the advantages and disadvantages in the use of laboratory simulations, computer modeling, and terrestrial field analog studies to understand geologic processes on other planets.

4. Explain the difference between *pixel scale* and *resolution* for images.

5. Visit the USGS website for planetary maps and review how names are assigned to surface features and summarize the process.

6. Discuss how an astrobiologist might use a geologic map of Mars to plan a future landed mission to search for evidence of past or present life beyond Earth.

CHAPTER 3

Planetary morphologic processes

3.1 Introduction

Earth is a dynamic planet. That simple statement can be supported by our own direct observations. Earthquakes, river banks collapsing during flooding, erupting volcanoes – all are experienced or documented on the news every year and show that our planet is everchanging. These examples represent three of the four fundamental processes that shape Earth's surface: **tectonism**, **gradation**, and **volcanism**.

The fourth fundamental process, **impact**, which is generally less often observed, is also documented, sometimes in quite newsworthy events as when a meteoroid plunges through the roof of a house. As the geologic record shows, the history of Earth can be profoundly altered by impacts, such as the well-known Chicxulub structure in the Yucatan peninsula of Mexico. This structure, now buried beneath a kilometer of sediments, has been mapped by geophysical methods and drill-holes to be more than 80 km in diameter and is estimated to have formed from an impact that released the energy equivalent of some 10 billion tons of TNT. The resulting fireball ignited world-wide fires, generated enormous amounts of CO_2 from the vaporization of limestone present at the impact site, and triggered tsunamis throughout the Gulf of Mexico and adjacent waters. As is now widely accepted, these catastrophic events led to mass extinctions, including that of the dinosaurs, and marked the boundary between the Cretaceous Period and the Tertiary Period 65 million years ago. It was not so much the direct impact that led to extinctions, but the effects on the surface environment, including firestorms, enhanced greenhouse processes, and disruption of the food chain.

Geologic exploration of the Solar System shows that the surfaces of the terrestrial planets, satellites, and small bodies, such as asteroids, have been subjected to one or more of the four fundamental processes. In some cases, the processes are currently active; in other cases, the geomorphology of the surface reflects events that happened in the past but are no longer taking place. Learning to recognize the landform "signatures" left by tectonism, gradation, volcanism, and impacts is one of the main goals of planetary geomorphology.

The relative importance of the surface-modifying processes among the planets is a function of many factors, including the history of the object and the local environment. For example, gradation on Earth is dominated by water, but in the current cold, dry martian environment, wind dominates. Thus, we must understand how gravity, surface temperature, the presence or absence of an atmosphere, and other variables influence the manner in which the processes operate and lead to specific landforms.

In the following sections, tectonism, volcanism, gradation, and impact processes are described and illustrated, using examples taken primarily from Earth to serve as a basis for planetary comparisons.

3.2 Tectonism

Road-cuts along many highways show ample evidence for rock deformation, or tectonic processes. As can be seen in **Figs. 3.1** and **3.2**, Earth's crust can be broken along **faults** or bent into distinctive **folds**. These and other features can be related, in part, to the style of tectonic deformation of the crust, such as tension or compression. However, knowledge of global-scale crustal deformation on Earth was not gained until the unifying concept of global plate tectonics was formulated in the 1960s. This insight was important for understanding the evolution of Earth's crust and is critical in the interpretation of other planets. Determining the styles of tectonic deformation on the planets provides clues to their general evolution and the configuration of their interiors.

Seismic and other geophysical data show that the interior of Earth consists of distinctive zones. The outer zone,

Figure 3.1. Road-cut near Death Valley, California, showing rocks that have been sheared by a reverse fault as a result of compression; the fault plane extends toward the upper right from the figure standing by the road; note that the light gray rocks (A) have been shoved upward some 5 m vertically (arrows indicate direction of relative motion).

Figure 3.2. These rocks in southern Israel were originally deposited in flat-lying, horizontal beds, but subsequently have been deformed into folds by compression.

or the crust, consists of chemically differentiated materials termed **sima** (referring to the elements Si, silicon, and Mg, magnesium), which floors the ocean basins, and a lower-density **sial** (Si and Al, aluminum) that forms the continental crust. Analyses of the physical properties of the Earth's interior zones from seismic data led to the recognition of the lithosphere and asthenosphere. The **lithosphere** includes the crust and the upper part of the mantle, which together behave as a relatively rigid, solid shell. The lithosphere rests on the **asthenosphere**, which is the mechanically plastic (or "slippery") layer of the mantle. The mantle is volumetrically the largest of Earth's interior zones, beneath which is the core, consisting of an outer liquid zone and an inner solid zone.

Heat sources within Earth's interior generate **convection** cells within the mantle and plastic flow that can upwarp zones in the lithosphere and concentrate heat. These zones can fracture, pull apart, and become sites of volcanism, which introduces new rock to the surface. Lateral flow of the "slippery" asthenosphere can, in turn, drag segments of the lithosphere outward from the fractures, forming rift zones. However, heat sources are not evenly distributed within the mantle; nor are they of equal magnitude, resulting in convection cells of different sizes and geometries. On a global scale, this leads to individual segments of the lithosphere, or **plates** (**Fig. 3.3**), that are of different sizes, moving at different rates.

Zones of upward-converging convection tend to be sites of **mafic** (magnesium- and iron-rich) volcanism, which reflect the sources of magma derived from the upper mantle. Such volcanism generates new sima-style

crust and commonly occurs in oceanic settings. Up-arching of the crust and accumulation of lava form symmetric mid-ocean ridges and fracture systems, such as the mid-Atlantic rift. Lateral flow away from central rifts, termed **sea-floor spreading**, has been measured to be as rapid as 16 cm/yr along the East Pacific Rise, equivalent to a pencil-length each year. In one million years (a geologic "blink of the eye"), its separation is some 160 km.

Downward-converging convection cells drag lithospheric plates toward one another into collision. Any one of several styles of plate collision can occur, depending on the composition of the crustal segments that are involved and the angle of the collision. Downward dragging of slabs of crust, or **subduction**, generates earthquakes and possible remelting of the crust, leading to volcanism. Because this melt, or **magma**, is formed at least partly from continental crustal materials or from oceanic sediments, the volcanism in subduction zones

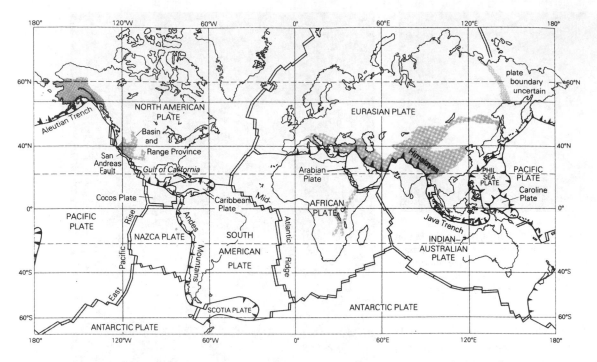

Figure 3.3. The lithosphere of Earth is divided into some seven major plates and more than a dozen smaller segments, most of which are in constant motion. Collisions and differential movement of plates lead to compression and tension and the formation of structural features, such as faults and folds in crustal rocks (courtesy of the US Geological Survey).

tends to be more **silicic** (silica-rich) in comparison with zones of up-welling convection.

In some places, upward convection is concentrated in one place, forming a plume, or **hotspot**. The resulting volcanism can be mafic if the plume rises through oceanic crust, or silicic if the plume rises through continental crust. As will be seen in **Section 3.3**, the composition of the erupted magma has a significant influence on volcanic eruptions and the resulting landforms.

As one can imagine, individual lithospheric plates do not move at the same rate, which causes them to deform by processes such as tension and compression. Coupled with plate collisions, this tectonic activity results in a variety of structures, including faults, joints, and folds. Breaking of rocks (**Fig. 3.4**) can occur as (a) **normal faults**, which result from tensional stresses; (b) **reverse faults**, which result from compressional forces (**thrust faults** are reverse faults involving very-low-angle fault planes); and (c) **strike–slip faults** in which displacement is principally in horizontal directions. The San Andreas fault in California is an example of a strike–slip fault. Both normal and reverse faults involve primarily vertical displacements of rocks along the fault plane. **Joints** also involve fracturing of rocks, but, unlike faulting in which rocks shift along a fault plane, jointing involves just separation of rocks away from the fracture.

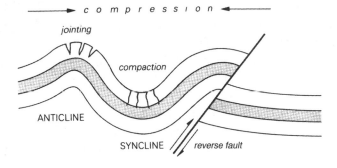

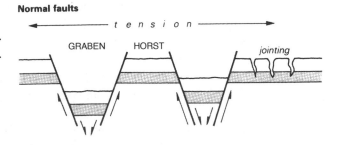

Figure 3.4. Diagrams of features reflecting structural deformation of rocks.

Some rocks subjected to compression yield by bending into folds, rather than breaking along faults. Folds of several types can occur (**Fig. 3.4**), some of which produce topography that directly reflects the form of the fold. Thus, in

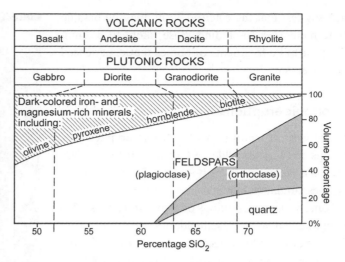

Figure 3.5. A view of the Zagros Mountains, Iran, showing elongate anticlines that resulted from folding of Earth's crust by compression. Most of the mountains mirror the upwarped folded rocks (*Apollo 7* image AS7–1615).

Figure 3.6. Rocks formed from the cooling of magma are either *volcanic* (resulting from eruption of magma to the surface and forming glass or fine-grained minerals) or *plutonic* (resulting from magma cooling from below the surface and forming large-grained materials). The classification of both volcanic and plutonic rocks is based on the mineral content and their percentages. Not all igneous rocks are indicated in this simplified diagram (courtesy of the US Geological Survey).

Fig. 3.5 the mountains coincide with **anticlines** (upwarped rocks). More commonly, however, there is an inversion of topography; during the process of folding, rocks along the axes of the anticlines are subjected to tension, which may cause them to be jointed, or in effect to "open up." Rocks along the axes of **synclines** (downwarped rocks) tend to be compressed, causing pore space and other voids to close. Consequently, weathering and erosion are enhanced along anticlines and are retarded along synclines. With time, net erosion is more rapid along anticlines, leading to the development of valleys, while synclines often form ridges.

Global-scale tectonic processes are generally linked to the thermal evolution and interior characteristics of planets. The smaller terrestrial planets (Mercury, the Moon, and Mars) have thick lithospheres that show little evidence of tectonic deformation over the last 80% of their history. Sometimes called *one-plate* planets, they have crusts that formed early in Solar System history and preserved the terminal period of accretion. Tectonism on these planets is expressed primarily by vertical movements, forming features such as grabens (**Fig. 2.9**), as the crust cooled.

3.3 Volcanic processes

Prior to the middle of the twentieth century, books on volcanism were primarily descriptive. Beginning in the 1970s, a more quantitative approach was taken toward understanding volcanic processes, spurred partly by the discovery that volcanism is important on other planets.

Volcanism involves the generation of magma (molten rock) and its eruption onto the surface, providing clues regarding the thermal evolution and interior characteristics of planets. On Earth, magma forms in the lower crust and upper mantle as a result of heat generated from differentiation of the mantle and core, friction from tectonic processes, and radioactive decay.

Most active volcanoes on Earth form on or near tectonic plate boundaries. As discussed above, crustal spreading zones are typically marked by mafic volcanism, involving iron-rich silicate magmas that commonly produce basaltic rocks (**Fig. 3.6**). Subduction zones typically involve silica-rich magmas, which produce andesitic, dacitic, and rhyolitic rocks. "Hotspots" are also important locations for volcanism. As crustal plates slide across these zones, magma erupts through the plates to the surface, producing chains of volcanoes much like an assembly line, typified by the Hawaiian volcanoes in the Pacific ocean.

Basaltic volcanism dominates the terrestrial planets. It is the principal rock on Earth's sea floors, constitutes the dark areas of Earth's Moon, forms the huge volcanoes of Mars and Venus, and is the rock constituting many of the smooth areas of Mercury. In addition, several lines of evidence suggest that the asteroid Vesta may be basaltic and some meteorites (including those likely to be from Vesta) appear to be fragments of bodies that experienced basaltic volcanism.

3.3.1 Volcanic eruptions

Among the earliest recordings of volcanic eruptions are the writings of Pliny the Elder, a Greek philosopher who witnessed the eruption of Vesuvius in A.D. 79. Through the centuries, an understanding of volcanic processes was gained from similar observations of active volcanoes. This understanding led to a classification scheme based on specific volcanoes, which became type localities for the various classes of eruptions. Today, it is recognized that many of the categories of eruptions are gradational with one another or are variations of very similar processes. Nonetheless, modern literature still carries some of the original classification categories, such as *Strombolian* for the classic volcano in the Mediterranean. In the simplest classification, eruptions are referred to as **explosive** (driven by the release of gasses in the magma as it reaches the surface) or **effusive** (magma released to the surface as liquid lava flows). Explosive eruptions produce **pyroclastic** ("fire-broken") materials, such as ash and cinders. Most modern classifications for the styles of volcanism are based on the characteristics of the products erupted and on the degree of eruption explosivity, as given in **Table 3.1**.

3.3.2 Volcanic morphology

The forms of volcanoes and related terrains result from complex, often interrelated parameters. Planetary volcanologist Whitford-Stark (1982) noted that these factors fall into three groups (**Table 3.2**): planetary variables, magma properties controlling rheology, and intrinsic properties of eruptions.

Planetary variables include those factors that are characteristic for the particular body. For example, the height of an explosive eruption is governed by such considerations as the gravitational acceleration and the presence or absence of an atmosphere. In turn, these factors influence the shape of the resulting volcano; in an airless, low-gravity environment such as the Moon, pyroclastic deposits would be widespread, in contrast to on Earth, where the ejection distance would be retarded by the atmosphere and higher gravity, leading to the formation of cinder cones (**Fig. 3.7**).

Figure 3.7. Cinder cones form by mild explosions (Strombolian eruptions) that eject solid material (pyroclastic fragments) on ballistic trajectories; the pyroclastic materials accumulate at the angle of repose (~34°) to form cones, as seen here in Lassen National Park, California.

Rheology refers to the flow properties of the lavas, of which the viscosity is of primary importance. Viscosity is a function of many variables, including temperature, composition, degree of crystallization, and gas content. High temperatures lead to low-viscosity (**Fig. 3.8**), or runny, liquids (e.g., hot syrup), while low temperatures lead to high-viscosity liquids (e.g., cold syrup). Composition refers to the silica content, in which silica-rich minerals (such as quartz) form complex molecular structures that increase the viscosity and retard flow, while mafic minerals (such as olivine) tend to have simple molecular structures and lower the viscosity. As lava cools, crystallization takes place and the entrained solids also increase the viscosity. Gas content includes not only the amount of volatiles present, but also their state. Generally, volatiles enhance flow, leading to lower viscosities, but, as the gasses come out of solution, the entrained bubbles retard flow and increase the viscosity. Rapid release of gasses is responsible for "driving" explosive eruptions, much like in a can of soda that is shaken (releasing the gasses), which then spews when it is opened. The most violent eruptions on Earth involve explosions of silica-rich magmas of the sort that occurred in southern California 760,000 years ago. This eruption formed the Bishop

Table 3.1. **Classification of volcanic eruptions**

Eruption type	Physical nature of the magma	Character of explosive activity	Nature of effusive activity	Nature of dominant ejecta	Structures built around vent
EFFUSIVE (flow)					
Basaltic flood	Fluid	Very weak ejection of very fluid blebs; little lava fountaining	Voluminous wide-spreading flows of very fluid lava from fissures	Cow-dung bombs and spatter; very little ash	Small spatter cones and ramparts of limited extent; broad lava plain
"Plains" or Icelandic	Fluid	Weak ejection of very fluid blebs; little lava fountaining	Moderately wide-spreading flows of thin sheets, often through lava tubes and channels; aligned vents	Cow-dung bombs and spatter; very little ash	Very broad flat lava cones (shields), which frequently coalesce, some spatter cones
Hawaiian	Fluid	Weak ejection of very fluid blebs; lava fountains	Thin, often extensive flows of fluid lava, frequently involving lava tubes and/or channels	Cow-dung bombs and spatter; very little ash	Shields; spatter cones and ramparts
EXPLOSIVE					
Strombolian	Moderately fluid	Weak to violent ejection of pasty fluid blebs	Thick, short flows of moderately fluid lava; flows may be absent	Spherical to fusiform bombs; cinder; small to large amounts of glassy ash	Cinder cones
Subplinian	Viscous	Moderate to violent ejection of solid or very viscous hot fragments of lava	Flows commonly absent; thick and stubby when present; ash flows rare	Essential, glassy to lithic, blocks and ash; pumice	Ash cones, block cones, block-and-ash cones
Pelean	Viscous	Like subplinian, commonly with glowing avalanches	Domes and/or short very thick flows; may be absent	Like subplinian	Ash and pumice cones; domes
Plinian	Viscous	Paroxysmal ejection of large volumes of ash, often with caldera collapse	Ash flows, small to very voluminous; may be absent	Glassy ash and pumice	Widespread pumice, lapilli, and ash beds; generally no cones
Ultraplinian	Viscous	Relatively small amounts of ash projected upward into the atmosphere	Voluminous wide-spreading ash flows; single flows may have volume of tens of cubic kilometers	Glassy ash and pumice	Flat plain, or broad flat shield, often with caldera
PHREATO-MAGMATIC					
Surtseyan	Fluid	Water plus magma; moderately explosive	Little or none	Juvenile, hydroclastic, accretionary lapilli, base surges	Cones, tuft cones

Table 3.1.   (cont.)

Eruption type	Physical nature of the magma	Character of explosive activity	Nature of effusive activity	Nature of dominant ejecta	Structures built around vent
Phreotoplinian	Viscous	Water plus magma; strongly explosive	Little or none	Accretionary lapilli	Widespread ash deposits
GAS ONLY					
Gas eruption	No magma	Continuous or rhythmic gas release at vent	None	None; or very minor amounts of ash	None
Fumarolic	No magma	Essentially nonexplosive; weak to moderately strong long-continued gas discharge	None	None; or rarely very minor amounts of ash	Generally none; rarely very small ash cones

Table 3.2.   **Factors governing the morphology of volcanic landforms (from Whitford-Stark, 1982)**

Planetary variables	Magma properties controlling rheology	Properties of eruption
Gravity	Viscosity	Eruption rate
Lithostatic pressure	Temperature	Eruption volume
Atmospheric properties	Density	Eruption duration
Surface/subsurface liquids	Composition	Vent characteristics
Planetary radius	Volatiles	Topography
Planetary composition	Amount of solids	Ejection velocity
Temperature	Yield strength, shear strength	

Tuff, a volcanic ash deposit covering some $2,200\,km^2$. Much of the ash was so hot when it was emplaced that it fused together, forming a volcanic rock called **ignimbrite**.

High-viscosity lavas flow short distances from the vent (**Fig. 3.9**) in comparison to low-viscosity lavas and tend to accumulate as steep-sided masses, such as **domes** (**Fig. 3.10**). Low-viscosity lavas are very fluid and can flow long distances, forming broad edifices such as **shield volcanoes** (**Fig. 3.11**), which are often "fed" by **lava channels** (**Fig. 3.12**) and **lava tubes** (**Fig. 3.13**).

Eruption characteristics constitute the third group of factors influencing volcano morphology. For example, low rates of effusion (small volumes of lava erupted per second) produce relatively short flows that tend to accumulate close to the **vent** (where magma reaches the surface), forming lava cones, whereas high rates of effusion form long flows that often produce lava plains.

3.3.3 Volcanic craters

Volcanic craters of diameter larger than about 2 km are termed **calderas**, which can form by collapse, explosion, erosion, or a combination of these processes. Most calderas involve multiple eruptions, leading to nested or overlapping multiple vents (**Fig. 3.14**). Smaller (<1 km) collapse features, termed **pit craters**, commonly form in basaltic lavas and can form without associated eruptions. Still smaller craters include **collapse depressions** (**Fig. 3.15**) that form on basalt flows and the **explosion pits** seen on some silicic flows (**Fig. 3.16**).

Phreatomagmatic explosions involve rising magma that encounters water (either surface or subsurface water or ice), forming **maar craters** (**Fig. 3.17**). Lava flows that cross water-soaked ground, swamps, or lakes can lead to the formation of **pseudocraters**, which are small cones of lava and cinders with summit craters (**Fig. 3.18**). These volcanic

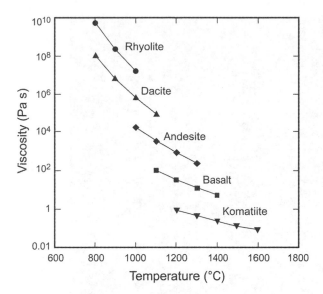

Figure 3.8. Viscosities of common magma compositions at their typical eruption temperatures, showing increasing viscosities at lower temperatures and for increasing silica content ranging from komatiite (rocks with very high mafic content) to rhyolite (this figure was published in *Encyclopedia of Volcanoes*, ed. H. Sigurdsson, B. F. Houghton, S. R. McNutt, H. Rymer and J. Stix, in the article Physical properties of magma, by F. J. Spera, pp. 171–190, copyright Academic Press, 2000).

Figure 3.9. Glass Mountain is in the Medicine Lake highlands of northern California. It is composed of rhyolitic obsidian lava flows that drape amoeba-like over the topography.

Figure 3.10. Lava domes, as seen in this digital elevation image of Mt. Elden in northern Arizona, form from the extrusion of thick, pasty lavas that flow only short distances from their vents or from intrusion of magma. Although typical of silica-rich lavas, they can form in lavas of any composition, including (rarely) basalts. The highways (white lines) visible in this image indicate the general scale (courtesy of the US Geological Survey).

Figure 3.11. Mauna Loa shield volcano (left side on horizon) formed from the eruption of abundant basaltic lava flows; the volcano on the right is Mauna Kea, the summit of which is marked by numerous cinder cones, the reflecting mildly explosive eruptions which followed its primary shield-building phase (US Navy photograph 0066, November 1954).

Figure 3.12. Channels often form in basaltic and some andesitic flows, and transport lavas to the advancing flow front. Constructional levees, seen along these lava channels in Hawaii, are common.

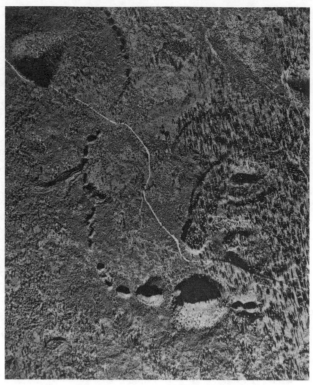

Figure 3.13. Lava tubes are conduits that develop in some basalt flows, either within the body of the flow, or by channels that become roofed; in either case, lava tubes act as extensions of the vent, reducing cooling of the magma, and feeding the flow front. This aerial photograph shows a nearly completely collapsed lava tube indicated by the series of collapse holes on the left side of the image. The lava tube originated from the large (~500 m in diameter) circular pit crater and flowed toward the top of the image.

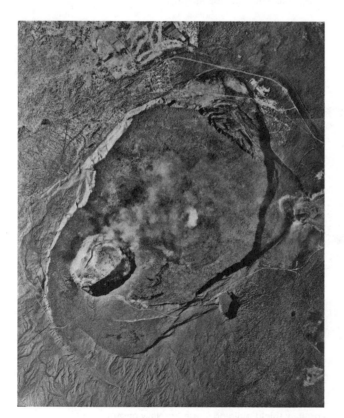

Figure 3.14. Calderas are volcanic features that result from explosions, collapse over magma chambers, erosion, or some combination of these processes. This view is of the caldera at the summit of Kilauea volcano in Hawaii; the area shown is about 3 km by 5 km.

features result from mild explosions generated locally by the interaction of the hot lava and water. The term "pseudo" (meaning false) simply refers to the fact that the features are not directly linked to a vent over a magma chamber.

Figure 3.19 classifies common volcanoes and their relation to some of the more important factors involved in volcanic eruptions. Unfortunately, because data are so limited, the details of volcanic processes on other planets are poorly known. Nonetheless, to some extent, the styles of eruption can be inferred from the morphology of volcanic features.

3.3.4 Intrusive structures

Not all magma reaches the surface to produce volcanoes. Some magma intrudes crustal rocks, cools, crystallizes and may be exposed later through weathering and erosion. Frequently, these intrusive structures are more resistant to erosion than the host rock, and they stand out as topographic features. **Figure 3.20** shows Green Mountain in

Figure 3.15. Collapse depressions are craters typically less than 15 m in diameter that can form on basalt flows as a result of lava draining from beneath a solid or partly solidified crust. This oblique aerial view shows collapse depressions formed on the Amboy lava flow in southern California. Similar-appearing depressions can form by inflation of active lavas surrounding a zone not encroached upon by the lava.

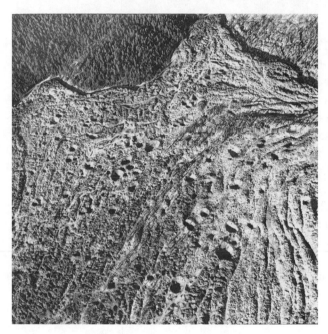

Figure 3.16. An aerial photograph of part of Glass Mountain (see **Fig. 3.9**), showing small (5–10 m) explosion craters formed by the rapid release of volatiles from the viscous obsidian flows.

Wyoming, a dome of sedimentary rocks pushed upward by an intrusion of magma. **Figure 3.21** shows several intersecting ridges composed of cooled magma that intruded as vertical sheets termed **dikes**.

3.4 Impact cratering

Impact cratering involves the nearly instantaneous transfer of energy from an impacting object, called the **bolide**, to a target surface. Bolides can include meteoroids, asteroids, and comets, which impact Earth at velocities of 5–

40 km/s. Using the simple expression for kinetic energy of $KE = 0.5\,mv^2$, in which m is the mass of the bolide and v is its velocity, an average nickel–iron meteoroid 30 m in diameter traveling at 15 km/s could transfer about 1.7×10^{16} J of energy onto a planetary surface, the equivalent of exploding about four million tons of TNT! Such an impact is considered to have been responsible for the formation of Meteor Crater in northern Arizona (**Fig. 2.4**), in which more than 175 million tons of rock were excavated to leave a crater more than 1 km across and 200 m deep. Unlike most geologic processes, impact events are of very short duration. For example, Meteor Crater probably formed in about one minute.

3.4.1 Impact cratering mechanics

Much of our understanding of how impact cratering operates has come from laboratory experiments in which the process could be studied under controlled conditions. From analyses of high-speed motion pictures and detailed analyses of cratered targets, Don Gault and his colleagues at the NASA-Ames Research Center derived the general sequence of the impact process (**Fig. 3.22**). First is the compression stage, in which the projectile contacts the target and penetrates the surface (**Fig. 3.23**), resulting in high-speed jetting of material outward from the zone of contact. At the same time, intense shock waves (**Fig. 3.24**) pass through both the target and the projectile. In this stage of the impact, shock pressures of several megabars are common, exceeding by three to four orders of magnitude the effective strength of common rocks. It is this high shock pressure that sets impact cratering apart from other geologic processes. Impacts result in intensely crushed and broken target material, some of which is so

Figure 3.17. Maar craters result from steam explosions as magma rises through water-saturated host rock, exemplified by the 1.3 km in diameter Crater Elegante, in the Pinnacate lava field of northern Mexico.

Figure 3.18. Pseudocraters, such as these seen at Mývatn in northern Iceland, develop in lava flows that pass over swampy terrain, resulting in local mild explosions that form small cinder and spatter cones with summit depressions.

severely shock-metamorphosed that rock can be melted and even vaporized.

The next stage involves excavation of the crater as shock waves and attendant rarefaction (or decompression) waves set target material into motion. The material excavated from the crater, termed **ejecta** (**Fig. 3.25**), is distributed radially as a blanket of fragmented debris (**continuous ejecta**), which, in turn, grades into zones of **discontinuous ejecta** and **secondary craters**, formed by the impact of ejecta boulders. A particularly important aspect of impact craters is the inversion of stratigraphy in the crater rim. As shown in **Fig. 3.22**, rock strata are overturned as part of the ejection processes. Moreover, the

material from the uppermost layers is thrown farthest from the crater. Thus, if one were to sample ejecta by making a traverse from its outermost extent to the crater rim, the samples would reflect progressively deeper (and hence older) rock layers. In this respect, impact craters represent a drill-hole with the "drill core" laid out on the surface as ejecta.

The final stage includes various post-cratering modifications not directly attributable to shock waves (**Fig. 3.23**). These include slumping of the crater walls, isostatic adjustments of the floor and rim, and erosion and infill of the crater. This stage may continue over long periods of time until the crater is eventually obliterated by gradation and other processes.

3.4.2 Impact craters on Earth

The morning of 30 June 1908 witnessed an unusual event in Siberia near the Tunguska River. A series of blasts knocked local reindeer herders off their feet, broke windows, and hurled one man from his porch. Seismometers recorded the events, while explosions and a fiery cloud were seen some 400 km from the event. No one knew what had happened, and it was many months before an exploration party reached the area. When they arrived at the location identified by seismic records as ground zero, the party found that all of the trees had been flattened into a pattern radial to the blasts, except for a small clump of trees right at ground zero, which were still standing upright but had all the branches stripped away.

The Tunguska event posed a problem that remains today. No crater was found, leading some wags to suggest

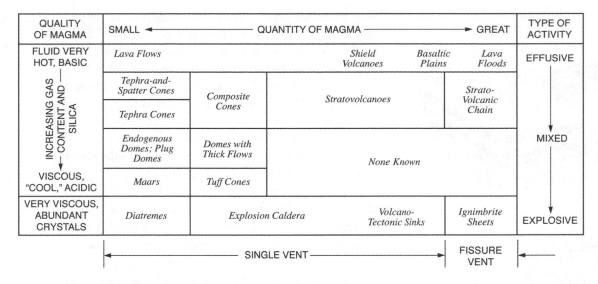

QUALITY OF MAGMA	SMALL ←	QUANTITY OF MAGMA			→ GREAT	TYPE OF ACTIVITY
FLUID VERY HOT, BASIC	*Lava Flows*		*Shield Volcanoes*	*Basaltic Plains*	*Lava Floods*	EFFUSIVE
↑ INCREASING GAS CONTENT AND SILICA	*Tephra-and-Spatter Cones*	*Composite Cones*	*Stratovolcanoes*		*Strato-Volcanic Chain*	MIXED
	Tephra Cones					
	Endogenous Domes; Plug Domes	*Domes with Thick Flows*	*None Known*			
VISCOUS, "COOL," ACIDIC	*Maars*	*Tuff Cones*				
VERY VISCOUS, ABUNDANT CRYSTALS	*Diatremes*	*Explosion Caldera*	*Volcano-Tectonic Sinks*		*Ignimbrite Sheets*	EXPLOSIVE
	← SINGLE VENT			→	FISSURE VENT	←

Figure 3.19. Classification of common volcanic landforms (italicized names), showing the relationships to the quantity of magma, magma composition, type of volcanic activity, and the nature of the vent.

Figure 3.20. Green Mountain, Wyoming, is a dome formed by intrusion of magma into sedimentary rocks (US Department of Agriculture photograph BBU-29–78).

Figure 3.21. These intersecting dikes in the Spanish Peaks area, Colorado, form ridges that stand higher than the now partly eroded rock (US Geological Survey CL-34–71).

the preposterous idea that Earth was hit by a black hole! A more probable explanation suggests that an asteroid or comet broke apart in the atmosphere, forming a shotgun-like blast that left neither a crater nor large remnants of the object itself. Tiny bits of meteoritic material have been recovered from some of the tree trunks, supporting this hypothesis. Current best estimates suggest that the Tunguska event resulted from an object 30–60 m in diameter, releasing energy equivalent to 10–15 megatons of TNT. The energy was released as a series of air-blasts, reflecting multiple objects generated by the breakup of the asteroid or comet, thus accounting for the multiple blasts observed in the town of Kirensk, some 400 km from ground zero.

Even to advocates of impact cratering, prior to 1930 fewer than ten impact structures were acknowledged on Earth. In the early Space Age of the mid 1960s, the number had risen to only about 33, but, after intensive searches and establishment of criteria for the recognition of impact craters (**Fig. 3.26**; **Table 3.3**), by the early part of the twenty-first century nearly 180 craters and related

Figure 3.22. Diagrams showing an impact into layered target materials, as derived from laboratory cratering experiments. Note that, as the crater expands, ejected material is folded over the rim, resulting in overturned stratigraphy of the layers, a characteristic distinctive of impact craters (courtesy of Don Gault).

structures had been documented as resulting from impact processes, plus more than 100 sites of probable impacts.

Compared with the heavily cratered surface of the Moon, 180 craters on Earth seems rather low. However, impact cratering involves chance collisions of planetary objects and, statistically, the longer a surface is exposed, the greater the likelihood that it will be struck by an extraterrestrial object. Thus, we would expect to find impact craters mostly on older surfaces. Because much of Earth's crust has been recycled by plate tectonics and modified by gradation, most impact craters on Earth have been destroyed or obliterated, in contrast to the airless and "waterless" Moon, where most of the early history is preserved in its ancient crust. Thus, to some degree, the number of craters seen on planetary surfaces provides clues as to the degree of surface evolution that the planet has experienced during its geologic history.

Meteor Crater in northern Arizona (**Fig. 2.4**) is one of the best-preserved and most intensely studied impact craters on Earth. Research by Gene Shoemaker, Sue Kieffer, Dave Roddy, and others has shown that the crater formed about 50,000 years ago in flat-lying sedimentary rocks. Detailed field work and analyses of drill cores show that the rocks were highly deformed, with strata in the walls and rocks beneath the crater floor being severely fragmented. The minerals **coesite** and **stishovite** are high-pressure forms of quartz discovered at Meteor Crater. They form only by impact processes and are used as diagnostic criteria in the recognition of impact craters.

Unlike many lunar craters of the same size, Meteor Crater is distinctly polygonal in plan view, demonstrating the importance that rock structure can exert on crater form. The strata surrounding Meteor Crater are jointed, which, at the time of impact, controlled the passage of shock waves and the excavation of the fragmented rocks, leading to the "square" outline of the crater.

Gene Shoemaker predicted that an impact crater the size of Meteor Crater would occur every 50,000 to 100,000 years on Earth. Although the uncertainties in this estimate are numerous, it gives some appreciation of the frequency of impact crater formation spread over geologic time. Analysis of the crater record on the Moon, which is much better documented than the record for Earth, shows that, while smaller impacts are much more frequent, there has been a general decay in the rate of cratering through time (**Fig. 2.10**).

Eroded impact craters (**Fig. 2.5**) yield important clues about rock deformation from impact. For example, the Sierra Madera structure in Texas (**Fig. 3.27**) is estimated to represent a crater ~13 km in diameter that has been deeply eroded. Detailed mapping and reconstructions of the original position of the rocks show that the central zone was uplifted more than 1 km. Such an uplift probably corresponds to the **central uplift**, or **central peak**, of many lunar craters (**Fig. 3.25**). Central uplifts are

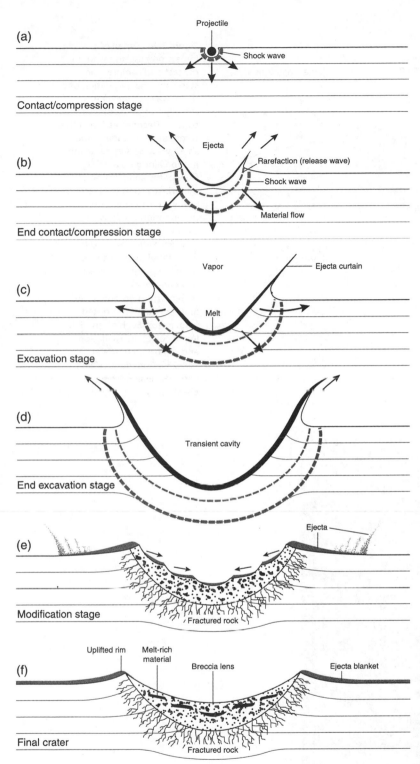

(a)

Projectile

Shock wave

Contact/compression stage

(b)

Ejecta

Rarefaction (release wave)

Shock wave

Material flow

End contact/compression stage

(c)

Vapor

Ejecta curtain

Melt

Excavation stage

(d)

Transient cavity

End excavation stage

(e)

Ejecta

Modification stage

Fractured rock

(f)

Uplifted rim

Melt-rich material

Breccia lens

Ejecta blanket

Final crater

Fractured rock

Figure 3.23. The general sequence of events in an impact involves the initial contact of the bolide with the target, resulting in ejection of target material, vaporization of parts of both the target and the bolide, formation of the transient cavity, and the post-impact modifications, such as slumping of the crater walls and rebound of the crater floor (from French, 1998, after Gault *et al.*, 1968, reprinted with permission of the Lunar and Planetary Institute).

observed only on craters larger than a few kilometers in diameter on Earth (there is no central uplift at Meteor Crater), and it appears that such features require a certain minimum impact energy to form. Although the causes of central uplift are not well defined, they are generally considered to be the result of elastic rebound of the rocks immediately following the excavation of the crater bowl.

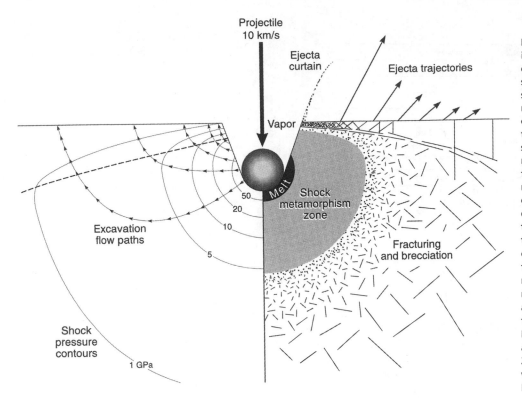

Figure 3.24. Detail of the impact process, showing the contact and compression stage and the generation of a shock wave that expands radially from the point of contact. Decompression of the target after passage of the shock wave sets material into motion along excavation flow-paths and ejection of material to form the crater cavity. The most intense deformation occurs closest to the point of contact, leading to vaporization and melting of material; with distance from the point of contact, rocks are shock metamorphosed, fractured, and brecciated (reprinted from *Icarus*, **59**, Melosh, H. J., Impact ejection, spallation, and the origin of meteorites, 234–260, Copyright (1984), with permission from Elsevier).

Figure 3.25. Lunar crater Euler (27 km in diameter), showing typical impact depression, central peak, terraced walls, and ejecta deposits, including continuous and discontinuous ejecta, and secondary craters formed by blocks ejected from the primary crater (NASA AS 17–2923).

Table 3.3. **Criteria for the recognition of impact craters (modified from Dence 1972)**

Criterion	Characteristics	Reliability
Remote sensing		
Plan view	Distinctly circular; may be modified by slumping, tectonic patterns, or erosion	Fair, but can be attributed to other processes
Rim structure	Inverted stratigraphy	Definitive
Central zone	Floor lower than surrounding plain; may contain central uplift	Fair, but can be attributed to other processes
Geophysical observations		
Gravity anomaly	Generally negative	Supportive, but not conclusive
Magnetic field	Variable; may be distinct anomaly over melt rock	Supportive, but not conclusive
Seismic velocities	Generally lower in brecciated zones	Supportive, but not conclusive
Ground observations		
Presence of meteorites	Rare except in very young craters	Definitive
Shock metamorphism	Features such as high-pressure minerals, impact melt, planar shock features, and shatter cones	Definitive
Brecciation	Observed in ejecta, rim, and floor of craters	May be attributed to other processes

```
0   1   2   3   4   5   6   7   8   9   10
```
Centimeters

Figure 3.26. Shatter cones formed in limestone at the Haughton impact structure, Canada, showing typical striations on the flanks of the cones, the apexes of which are oriented toward the point of impact (courtesy of R. A. F. Grieve).

The 24 km Ries Kessel impact structure in southern Germany provides insight into the mechanics of ejecta emplacement. Extensive mapping and drilling to obtain subsurface core samples have revealed the extent of the ejecta deposit and its properties. The ejecta is composed mostly of fragmented rock, or **breccia**, some of which was mixed with melted rock and volatiles as it was thrown from the transient cavity. This material rained down on the surrounding terrain, where it churned up and mixed with additional local rock. Then the entire mixture continued to slide outward a short distance and settled into its final resting place.

Many other structures on Earth have yielded important data on the morphology of impact craters and the cratering process. For example, the Clearwater Lakes in Canada appear to be a double impact, while the Henbury Craters in Australia consist of at least 13 craters, probably reflecting multiple impacts from the breakup of an incoming bolide.

The largest recognized impact structures on Earth include Sudbury in Ontario, Canada, and Vredefort in South Africa; both are about 140 km in diameter, and they formed at 1.85 Ga and 1.97 Ga, respectively. The Popigai structure in Siberia is 100 km in diameter and involved an impact into limestones that show tiny

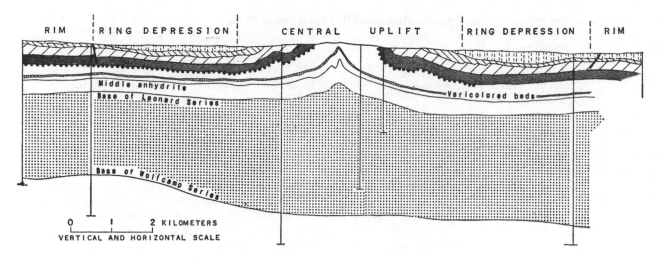

Figure 3.27. Cross-section through the eroded Sierra Madera impact structure, Texas, showing reconstruction of the structure based on drill-holes (indicated by the vertical lines) and evidence for a prominent central uplift, or peak (reprinted from Wilshire, H. G. and Howard, K. A., 1968, Structural patterns in central uplifts of crypto explosive structures, as typified by Sierra Madera, *Science*, **162**, 258–261).

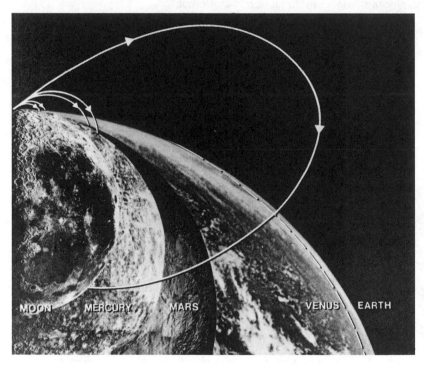

Figure 3.28. A diagram showing the trajectories of ejecta as a function of gravity on Earth, the Moon, Mercury, Venus, and Mars for an impact of equal magnitude, all other factors being equal (courtesy of Peter Schultz).

diamonds, apparently resulting from the high shock pressures. The largest impact found in the United States is the Chesapeake Bay structure at 85 km in diameter and containing a central uplift some 38 km across. Knowledge of this structure comes primarily from drilling and recovered core samples, along with geophysical data. This impact occurred at about 35.5 Ma in a shallow marine environment (Ma = mega-*annum* or 10^6 years).

3.4.3 Impact craters and planetary environments

The primary factors governing the size and shape of impact craters are the impact energy, gravity, and properties of the target, as described by planetary scientist Jay Melosh (1989) of Purdue University. In general, the greater the energy (i.e., larger bolide mass, higher impact velocity), the larger the crater. Gravity affects the cratering process by influencing the dimensions of the excavation bowl, the extent of the ejecta, and various post-impact crater modifications. For equal-size impacts, blocks of ejecta can be lifted and excavated more easily on low-gravity planets, leading to larger craters in comparison with high-gravity environments. Furthermore, in low-gravity environments, ejecta is thrown a greater distance, as shown in **Fig. 3.28**. Thus, we would expect to see a wider (but thinner) zone of ejecta surrounding impact craters on the Moon than on higher-gravity planets such as Mars and Mercury. In the modification stages of impact cratering, gravity also plays a role by influencing the degree of slumping, perhaps governing the size of potential central uplifts.

Target properties influence crater morphology in all stages of the impact process. As noted for Meteor Crater, the structure of the rocks (such as joints) can control the planimetric form, or the outline of the crater. Impacts into soft sediments tend to be larger because less energy is needed to break up the rocks and more energy is available for excavation. Target rocks containing water tend to be fluidized in the impact process, leading to slurry-like ejecta deposits, as seen at the Ries Kessel in Germany and as proposed for many martian craters (**Fig. 3.29**).

Impact craters show a distinctive progression in morphology with increasing size (**Fig. 3.30**). Small craters, such as Meteor Crater, are simple bowl-shaped depressions. Larger craters display central peaks and terraces on their inner walls, and at still larger sizes, clusters of central peaks. The largest impacts form **multi-ringed basins**. The size ranges for these morphologies are different among the planets, being controlled primarily as a function of gravity (**Fig. 3.31**).

The shape of impact craters in plan view is partly controlled by the angle of the incoming projectile. Because impacts involve essentially point-source transfers of energy, both the crater and the distribution of ejecta for most impacts are concentrically symmetric about the point of impact. Although intuition might suggest that oblique angles of impact would cause elongate craters, experiments have shown that only for very low angles (<15°) above the surface do impact craters and ejecta become noticeably non-circular (**Fig. 3.32**).

3.5 Gradation

Gradation is a complex process that begins with weathering and erosion, continues with transport of the weathered debris, and ends with deposition of the material. Think of a road "grader" on a dirt track that cuts off the tops of bumps and fills in the ruts with the debris. Thus, gradation is a "leveling off" process in which topographically high areas are worn down by erosion and low areas are filled by deposition.

Figure 3.29. Many craters on Mars show distinctive flow-lobes that are generally considered to reflect impacts into water- or ice-saturated targets that formed slurry-like ejecta masses (NASA THEMIS mosaic).

3.5.1 Weathering

Weathering is the first step in gradation; it is the process that "softens up" rocks, making them amenable to erosion, either through physical events or by chemical reactions. For example, rocks can be fragmented by impacts, broken to smaller pieces in a stream, or ground into dust by glaciers. On a much smaller scale, grains can be reduced in size by a variety of processes, including salt weathering, in which salt-laden water seeps into tiny cracks, and, as the salt crystallizes and expands, the grains are split into smaller fragments.

Chemical weathering on Earth typically occurs through reactions with water and the atmosphere and includes **oxidation** (the combination of materials with oxygen, forming, for example, rust), **hydration** (the combination of minerals with water molecules), **solution** (in which materials are dissolved, commonly in water), and **carbonation** (a more complicated reaction involving carbon dioxide from the atmosphere and the formation of a weak acid that reacts with minerals). Carbonation is the

(a)

(b)

(c)

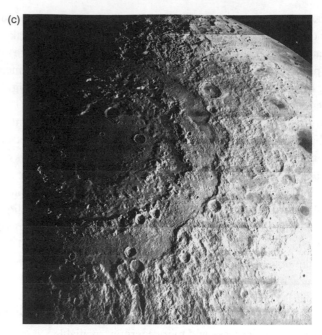

Figure 3.30. Images showing morphology of lunar craters as a function of size: (a) the simple bowl-shaped crater Isidorus D (15 km in diameter), photographed by the *Apollo 16* panoramic camera (NASA AS16–4502); bright streaks on the wall indicate downslope movement of fragmental material; (b) an *Apollo 11* photograph (NASA AS11–44–6611) of the 93 km in diameter crater Daedalus on the lunar far side, showing typical terraced walls, a central mountain peak, and floor deposits; and (c) an oblique view from *Lunar Orbiter IV* (NASA LO IV M-180) of the Orientale basin; this classic multi-ring impact structure is some 930 km across.

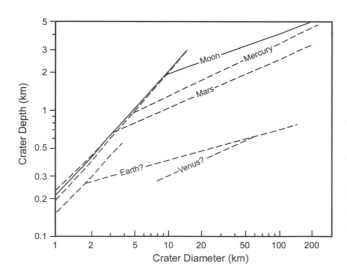

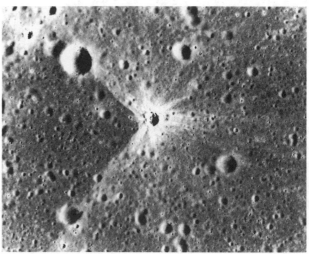

Figure 3.31. Depth-to-diameter ratios for impact craters on terrestrial planets (from Sharpton, 1994).

Figure 3.32. The ejecta from this small (~450 m), fresh impact crater is asymmetric, with the absence of ejecta on the west (left) side representing the incoming direction of a low-angle impact (NASA AS15–9337).

principal process involved in the formation of limestone caves.

Space weathering occurs from the bombardment of planetary surfaces by energy from the Sun and deep space, such as cosmic rays. Being only "skin-deep," this process is more important on airless bodies that are poorly shielded from bombardment. For example, the *Apollo 11* astronauts conducted an experiment to measure the **solar wind** (charged particles streaming from the Sun) impacting the Moon by exposing a sheet of gold foil on the surface, which was returned to Earth at the end of the mission and analyzed. Results enabled the flux of the charged particles from the Sun to be quantified.

Although space weathering is a relatively shallow surface process, it can have a big influence on the exposed rocks and minerals, altering their chemical and physical properties. For example, the molecular structure of ices (as on the outer Solar System satellites) and glasses can be altered, which can influence their signatures in remote sensing data.

Cosmic rays, in addition to "zapping" minerals, can also be used for age-dating rocks, much like counting craters. "Zap" pits and traces of cosmic rays can be counted in mineral grains; from knowledge of the flux of cosmic rays, it is then possible to calculate the length of time that the specimen has been exposed on the surface. Of course, one must assume that the specimen has not been overturned in its history and that the flux has been constant through time. Despite these uncertainties, age determinations based on cosmic-ray abundances have been used to date some lunar samples, as well as some rocks on Earth, as on the rim of Meteor Crater, to help determine the age of the impact.

Once material has been weathered, it is subject to erosion through various agents, such as wind and water. The driving force of these agents is primarily gravity. Through gravity, material is moved by **mass wasting** (such as landslides), flowing liquid water, ice (glaciers), or wind.

3.5.2 Mass wasting

Mass wasting is the downslope movement of rock and debris and is a universal geologic process. Even very small bodies, such as the asteroid Gaspra (**Fig. 1.5**), where gravity is only a tiny fraction that of the Earth, display downslope movement of surface material.

Mass wasting is categorized on the basis of the rate of movement ("slow," or imperceptible to an observer, and "fast"), the types of material that are involved (rock, soil, etc.), and the water content. Water acts in several ways to enhance mass wasting: (a) films of water act as

Figure 3.33. A rock glacier on McCarthy creek, Copper River region, Alaska, showing flow-lobes into the valley. Rock glaciers consist of poorly sorted rocks and fine debris held together by ice (US Geological Survey photograph by F. H. Moffit).

lubricants and can destroy the cohesion between particles; (b) in many materials, particularly the clay minerals, water enters the crystal structure, causing swelling and disrupting of the strength of the material; (c) water adds weight to the potential landslide and thus helps to "push" the mass of rock and soil down hill; and (d) fluid pore pressure can reduce the amount of energy necessary to initiate movement. **Figure 3.33** shows a rock glacier, representing one form of mass wasting on Earth.

3.5.3 Processes associated with the hydrologic cycle

The **hydrologic cycle** defines the movement of water among the atmosphere, surface reservoirs, such as oceans, glaciers, streams, and **groundwater** systems (the term used for water beneath the surface). On Earth, water is a dominant geologic agent, and the hydrologic cycle figures prominently in geomorphic processes. Particularly important are river and stream patterns because they provide clues as to the structure of the underlying rocks and characteristics of the topography (**Fig. 3.34**).

Groundwater accumulates in fractures and the pore space of rocks and, rarely, as bodies of water in larger cavities. Groundwater can move laterally and emerge as springs where it intersects the surface. Erosion produced by groundwater dissolving certain rocks (typically limestones, rock salt, or gypsum) leads to caverns and a terrain termed **karst topography**. Depending upon the stage of evolution, karst topography may display only a few **sinkholes** (collapse pits), numerous sinkholes plus **solution valleys** (a collapsed drainage network), or highly eroded

karst in which only pinnacles and spires remain as erosional remnants.

Landforms with the imprint of former lakes, swamps, and oceans are highly diverse. Typically, these are sites of sedimentary deposition and, with the removal of water, leave flat, broad plains, typified by **playas** (dried lake beds). Shoreline processes may lead to features such as terraces (both erosional and depositional, which may reflect former shorelines), sea cliffs, and beaches. Except for some craters and canyons on Mars, which may have contained ponded water in the past, and the methane lakes on Titan, only Earth appears to display landforms associated with large bodies of water.

3.5.4 Aeolian processes

Aeolian (wind) processes involve the interaction of the atmosphere with the surface. Most deserts, coastal areas, glacial plains, and many semi-arid regions on Earth experience aeolian processes. An atmosphere in motion (wind) possesses energy, and, as the wind moves over a surface, some of that energy is transferred to the surface. If we were to measure the wind velocity at different heights above the surface, we would see that velocity decreases toward the surface, as a reflection of the surface friction. As shown in **Fig. 3.35**, the changing velocity profile defines the **boundary layer**, within which the air flow is turbulent. When plotted on a logarithmic scale, the boundary layer is a straight line, the slope of which is related to a parameter called the **friction velocity**. Although this term is commonly used to describe aeolian processes, it

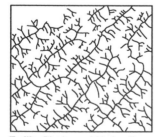

Dendritic. Horizontal sediments or uniformly resistant crystalline rocks. Gentle regional slope at present, or at time of drainage inception.

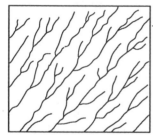

Parallel. Moderate to steep slopes; also in areas of parallel, elongate landforms.

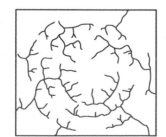

Rectangular. Joints and/or faults at right angles. Streams and divides lack regional continuity.

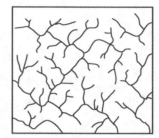

Radial. Volcanoes, domes, and residual erosion features.

Trellis. Dipping or folded sedimentary, volcanic, or low-grade metasedimentary rocks; areas of parallel fractures.

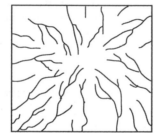

Annular. Structural domes and basins, diatremes, and possibly stocks.

Figure 3.34. Diagrams showing basic stream patterns and relations to the eroded rocks (from Howard, 1967, reprinted by permission of the AAPG, whose permission is required for further use).

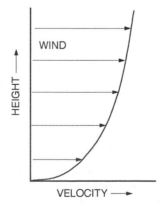

Figure 3.35. A diagram plotting the wind velocity as a function of height above a planetary surface; the increase in velocity with height defines the boundary layer, within which the air movement is turbulent; the velocity decreases toward the surface as energy is transferred by friction along the surface.

is not a true velocity that one would measure but is a function of the boundary layer profile.

As outlined in the classic work by the British army engineer Brigadier Bagnold (1941), wind "threshold" curves (**Fig. 3.36**) define the minimum wind speeds required to initiate movement of different particles and show that the particle size moved by the lowest speed wind is about 100 micrometers (µm) in diameter, or fine sand. The ability of wind to attain threshold is a function primarily of atmospheric density. Thus, the very-low-density atmosphere on Mars (**Table 1.1**) requires wind speeds that are about an order of magnitude stronger than on Earth for particle motion, while relatively gentle winds can move grains in the dense atmosphere of Venus. Once

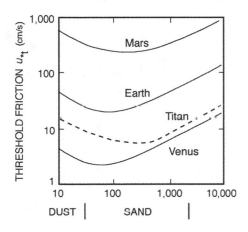

Figure 3.36. Wind friction velocities needed to set particles into motion (termed **threshold wind speed**) as a function of grain size. Friction velocity is not a true wind speed, but rather is a characterization of the wind velocity profile in the turbulent boundary layer near the surface. Note that the grain size moved by the lowest wind is about 100 micrometers in diameter (about the size of fine sand), regardless of planet. Values for Titan are offset because the grains are likely to be ice, rather than higher-density silicate minerals, such as quartz (from Greeley and Iversen, 1985).

they have been set into motion (**Fig. 3.37**), wind transports sediments by **suspension** (mostly silt and clay particles, or "dust," smaller than about 60 µm in diameter), **saltation** (mostly sand-size particles, 60–2,000 µm in diameter), and **surface creep** (particles larger than about 2,000 µm in diameter).

Winds can redistribute enormous quantities of sediment over planetary surfaces, resulting in the formation of landforms large enough to be seen from orbit and deposition of windblown sediments that can be hundreds of meters thick. One of the most useful types of features for interpreting wind processes is the **dune**, a depositional landform (**Fig. 3.38**). Dunes form by sediment transport in saltation and signal the presence of sand-size particles. Both the planimetric shape and the cross-sectional profile of dunes can reflect the prevailing winds in a given area (**Fig. 3.39**). Thus, if certain dune shapes or slopes can be determined, local wind patterns can be inferred for the time of their formation.

On Earth, great quantities of silt and clay are transported in dust storms and eventually deposited as **loess**. However, even where they are relatively young and well exposed on the surface, loess deposits are nearly impossible to identify in remote sensing data. Yet, identification of such deposits could be very important in understanding planetary surfaces, especially on Mars.

Wind erosional features include pits and hollows (called **blowouts**) that form by deflation (the removal of loose particles) and wind-sculpted hills called **yardangs** (**Fig. 3.40**). Yardangs have been likened to inverted boat hulls because of their streamlined shape, the orientation of which indicates the prevailing wind direction at the time of their formation.

Observations of active dust storms, dust devils, and other aeolian features provide direct information on the atmosphere. For example, **variable features** are surface

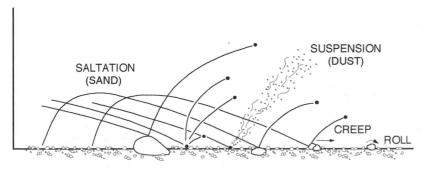

Figure 3.37. Common modes of transport of grains by the wind; saltation applies primarily to sand-size particles, which move in a series of short "hops;" suspension involves finer material such as dust, some of which can be ejected into the atmosphere by the impact of saltating grains; larger materials, such as granules and small gravels, move by surface creep, which can be enhanced by saltation impact. Windblown materials can abrade rocks, forming **ventifacts** (sand-blasted rocks), and can be broken into finer grains upon impact.

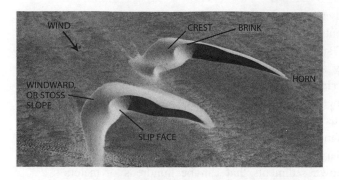

Figure 3.38. Sand dunes form from the accumulation of grains moved primarily in saltation. **Stoss** refers to the upwind side of dunes, while **slipface** refers to the downwind side of the dune formed by sands sliding into their angle of repose from the dune brink; the dune crest refers to the highest part of the dune. This view of two barchan dunes was obtained by E. C. Morris in the Sechura Desert of Peru (courtesy of US Geological Survey).

patterns that form as a result of wind activity and are visible as contrasts in albedo, or surface reflectivity. Repetitive imaging shows that many of them disappear, reappear, or change their size, shape, or position with time. Mapping the orientations of variable features has been used to derive near-surface wind patterns on Mars and Venus.

3.5.5 Periglacial processes

Periglacial refers to processes, conditions, areas, climate, and topographic features in cold regions or in any environment where frost action is important. A review of the various environments in the Solar System shows that all planets and satellites except Venus experience

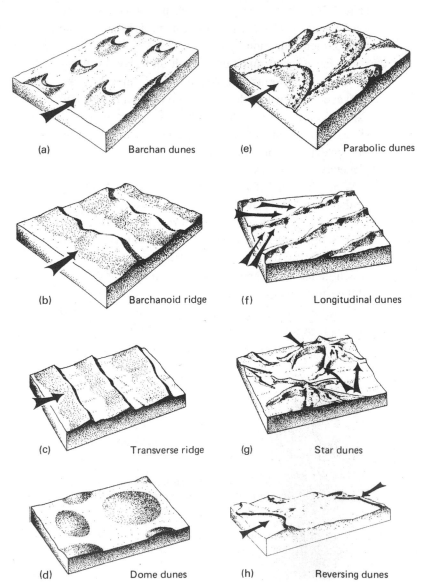

(a) Barchan dunes

(b) Barchanoid ridge

(c) Transverse ridge

(d) Dome dunes

(e) Parabolic dunes

(f) Longitudinal dunes

(g) Star dunes

(h) Reversing dunes

Figure 3.39. Sand dunes are classified primarily on their shape, shown in these diagrams for common dunes on Earth; arrows indicate the prevailing wind directions responsible for their formation (from McKee, 1979).

temperatures below freezing. Water-ice exists on Mars, is a major constituent of most of the outer planet satellites, and is thought to be present in permanently shaded parts of the Moon and Mercury.

Surface features on an icy or ice-rich body result largely from processes of flow and fracture. Although ice is often modeled as a Newtonian viscous fluid, experiments indicate that some ice can be considered a pseudo-plastic fluid that deforms by creep. In a Newtonian fluid, the rate of strain is linearly proportional to the applied stress, and the viscosity is the ratio of the strain to the stress raised to some power. Thus, as the stress level is increased, the material deforms more and more rapidly. The result is that ice appears to become less viscous at higher rates of strain. However, under very rapid strain rates, such as during an impact event, ice behaves more like brittle elastic material than a fluid.

On Earth, precipitation of snow in the **area of accumulation** (the headward part of a glacier) forms a deposit that is about 20% ice and 80% air. Melting and refreezing, plus compaction, converts the snow to spherical ice particles called **firn**. As the firn accumulates, further compaction causes recrystallization to form the main ice mass, typically having less than 10% air. Glaciers are classified as **valley glaciers** or as **piedmont glaciers** (coalesced valley glaciers at the base of a mountain range), or as **ice sheets** (also called continental glaciers, or ice caps) if they are too large to be contained by valleys. All glaciers move downslope or outward, leaving distinctive terrains. A "retreating" glacier simply means that melting and ablation of ice exceed the rate of forward movement by the glacier.

Most glaciers also incorporate rocky materials within the ice mass. This material can include dust and other airborne particles and chunks of rock gouged by the ice

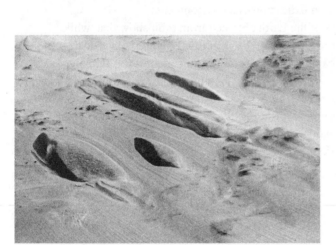

Figure 3.40. Yardangs are wind-sculpted hills that resembled overturned boat hulls, reflecting their streamlined shape, exemplified by these features in Peru (US Geological Survey photograph by J. F. McCauley).

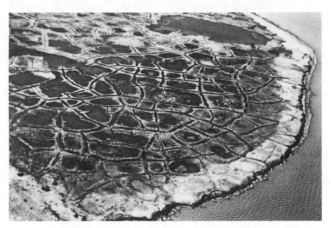

Figure 3.42. These ice-wedge polygons on the Alaska sea coast near Barrow are 7–15 m across and form by periglacial processes working on unconsolidated sediments (US Geological Survey photograph by R. I. Lewellen).

Figure 3.41. "U"-shaped valleys, such as Deadman Canyon shown here in Tulare County, California, are typical of erosion by glaciers (US Geological Survey photograph by F. E. Mathes).

as it moves across a surface or from the accumulation of debris derived from valley walls. Surface material often coalesces into various **moraines**, which are linear deposits. Some material carried by the ice eventually reaches the front of the glacier, where it can be deposited *in situ* or carried away by melt water. The finer material is often transported by the strong winds which are generated along ice margins. Coarser glacial deposits, termed **drift**, may assume a variety of geometries, which provide clues as to the form and position of glaciers after the ice mass has retreated. On Earth, glaciers have effected extensive changes in the landscape. U-shaped valleys (**Fig. 3.41**), grooves, and striations parallel to the flow of ice, and amphitheater-shaped erosional **cirques** in the headward parts of valleys are indicative of glacial erosion.

In periglacial zones (cold regions), **solifluction** processes arising from the slow downslope flow of water-saturated, unconsolidated materials can occur. Periglacial processes involve the erosion of rock or soil by snow and ice, frost action, and chemical weathering, leading to such features as permanently frozen ground (**permafrost**), **patterned ground** (**Fig. 3.42**), **pingos** (dome-shaped ice-cored mounds that can be 70 m high and 600 m across on Earth), and **thermokarst** (melt-eroded) topography. The occurrence of a periglacial region is not genetically related to the proximity of glaciers or continental ice sheets, contrary to what is implied by its etymology. However, the presence of water is essential for most periglacial processes to occur. This broader definition is useful in that it allows us to consider the possible operation of periglacial-type processes on the surfaces of other objects in the Solar System.

3.6 Summary

Planetary surfaces are shaped and modified by four principal processes: tectonism, volcanic activity, impacts, and gradation. Each of these processes produces distinctive landforms on Earth, where most of these processes have been studied in detail. One of the goals of planetary geology is to determine how these landforms might be different in extraterrestrial environments.

Views of the Earth obtained from orbit show that, while some processes can be identified by remote sensing, others cannot, thus introducing uncertainties in the interpretations of images from other planets. Furthermore, with increased knowledge of the outer planet satellites, planetary geologists must assess the validity of applying Earth analogs to bodies composed mostly of ice and having markedly different environments.

Assignments

1. Scan news media sources for the most recent accounts of *active* natural disasters and document one example for each of the four principal geologic processes that shape planetary surfaces. Write a few sentences about the event, identify the process involved, and provide the news media source.

2. Explain why aeolian (wind) processes require stronger winds to move sand grains on Mars than on Venus.

3. Go to NASA websites that contain planetary images and find examples of landforms beyond Earth for the four geologic processes that shape planetary surfaces. Print at least one image for each process (a total of four images); label the mission that obtained the image, identify the planet and the process, and give a short description of the feature.

4. Both impacts and volcanic eruptions can lead to the formation of craters. Discuss the fundamental differences in how these two processes differ in the formation of their respective craters.

5. *Planetary variables* influence the morphology of volcanoes; identify one such variable and explain how a volcano on the Moon might be different from a volcano on Venus, all other factors (such as magma composition and rate of eruption) being equal.

6. Explain why so few impact craters are seen on Earth in comparison with the Moon.

CHAPTER 4

Earth's Moon

4.1 Introduction

Throughout the history of humankind, other than the Sun, no other planetary object has held our attention as much as the Moon. The Moon figures prominently in mythology and literature, with notions of vampires and werewolves that were driven by the phases of the Moon. The very term "lunatic" derives from the idea that mentally unstable individuals are influenced by the Moon. Aside from these aspects, scientifically, the Moon holds much for study, especially in terms of planetary geomorphology. Even with the naked eye, we can see that its surface is not uniform. Some areas are dark and circular (the "eyes" of the Man in the Moon) and other areas are very bright. These characteristics led to the terms **maria** (Latin for seas) for the dark areas for their fanciful resemblance to water areas and **terrae** (Latin for land), or highlands, for the notion that there were continents surrounding the seas.

At 3,476 km, the diameter of the Moon is nearly the width of the United States; its surface area of $3.79 \times 10^7 \, km^2$ is about the same as the land area of Africa and Australia combined. In many ways, the Earth–Moon system is unique in the Solar System, and, because the Moon is comparatively so large, some planetary scientists view Earth–Moon as a "binary" planet. As is true with many natural satellites, our Moon is locked in **synchronous rotation** in its orbit around Earth, meaning that it always shows the same "face," termed the **near side**, toward Earth and hides the **far side** from direct viewing (**Fig. 4.1**). Librations, or "wobbles," in the Moon's movement enable slightly more than a hemisphere to be seen in both polar areas and on the eastern and western sides, or **limbs**, of the Moon.

After more than four decades of analyses of lunar data returned from dozens of successful spacecraft (**Table 4.1**) and the return of nearly a half ton of samples from the Moon, today we recognize that the history of our closest planetary neighbor is inexorably linked to that of our home planet, Earth.

For planetary geology, it is fortunate that the Moon was the first extraterrestrial object to be studied. Because the Moon lacks an atmosphere and appears never to have had liquid water flowing on its surface, the Moon is a relatively simple geologic object and serves as a training ground for studying the formation and evolution of planetary surfaces.

4.2 Lunar exploration

4.2.1 Pre-*Apollo* studies

In the early 1600s, Galileo Galilei turned the newly invented telescope toward the heavens. Although his telescope was a simple tube with a couple of lenses, as we shall see later, he made remarkable discoveries that revolutionized ideas regarding the Solar System. It is, however, surprising that his sketches of the Moon are rather crude (**Fig. 4.2**). Although Galileo showed the distinctive dark and light terrains as well as craters, his drawings do not seem to match well to specific surface features.

Within a few decades of Galileo's observations, improvements in telescopes resulted in maps of the Moon that are rather accurate (**Fig. 4.3**), portraying terrains and individual features in correct positions. Users of telescopic photographs of the Moon should be aware that the images are often inverted in relation to views with the naked eye and in reference to the geographic convention used today, in which north is in an "up" position.

The craters observed on the Moon through telescopes stimulated a great deal of interest and today many of the ideas of lunar crater formation seem rather bizarre. For example, the famous English scientist Sir Robert Hooke

(a)

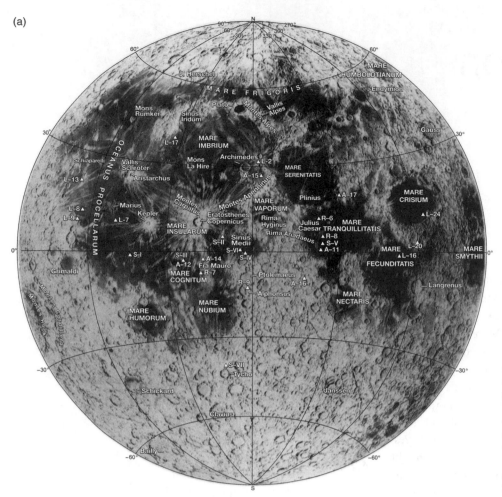

Figure 4.1. Full-disk charts of (a) the near side and (b) the far side of the Moon. Dark maria are smooth, sparsely cratered surfaces indicative of their relatively younger age in comparison with the heavily cratered, bright highlands, or terrae. In addition to selected named features, landing sites are shown (A, *Apollo*; S, *Surveyor*; L, *Luna*; R, *Ranger*).

suggested that the craters were formed by gasses bubbling up from the interior and bursting on the surface, like a pot of boiling mush. Strange as this and some other ideas might seem, the study of lunar craters served as the initial steps in trying to understand the geology of extraterrestrial surfaces. Eventually, two competing ideas emerged for lunar craters, origins by impact and origins by volcanic processes. These and other hypotheses would not be put to rest until the Space Age.

When President Kennedy set the goal of sending astronauts to the Moon and returning them safely to Earth through the *Apollo* program, the newly formed National Aeronautics and Space Administration formulated an extensive set of exploration projects to learn everything that would be needed in order to achieve this national goal. Early on, it was recognized that we knew very little about the characteristics of the Moon that were critical for a successful landing. Debates raged about the engineering properties of the surface, with some scientists advocating that landers would sink out

of sight in a deep layer of fluffy dust. To address these and other concerns, a series of robotic spacecraft was designed to collect much-needed engineering data. This was also the time of the "space race" with the Soviet Union. The Soviets also hoped to send men to the Moon but, unlike the *Apollo* program, they did not formally announce their goal, and it was only after the collapse of the Soviet Union that their program was officially revealed. Nonetheless, the Soviets had a highly successful series of unmanned missions to the Moon, returning data leading to many "firsts," including views of the far side of the Moon from the *Luna 3* spacecraft. These and later images showed that maria are mostly absent on the far side (**Figs. 4.1(b)** and **4.4**).

NASA's unmanned missions involved three projects, *Ranger*, *Surveyor*, and *Lunar Orbiter*. *Ranger* spacecraft were designed to crash on the surface but, on the way down, to return a series of progressively higher-resolution images to give close-up views of the surface. Although not all of the missions were successful, *Ranger* images

(b)

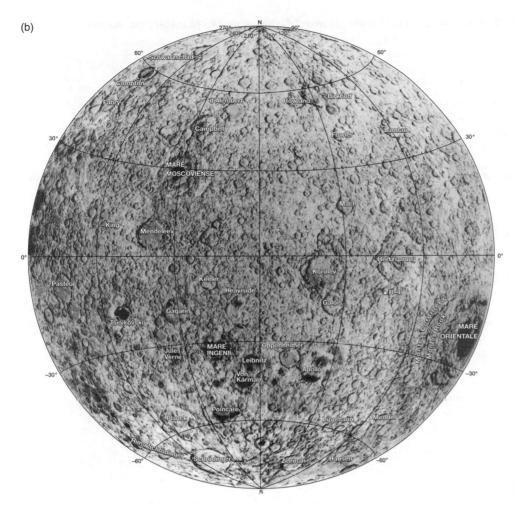

Figure 4.1. (*cont.*)

showed a continuum of crater sizes right down to the moment of impact. They also included views of features that were correctly attributed to volcanism **(Fig. 4.5)**, adding fuel to the debate regarding volcanism versus impact cratering as agents of lunar surface evolution.

The *Surveyor* project involved "soft" landers **(Fig. 4.6)**, five of which were successful. These missions demonstrated that the surface was sufficiently solid to support landers and provided engineering data for the lunar soils. They also returned compositional information showing a predominance of iron-rich materials in mare regions, reflecting their basaltic lava composition, and revealed craters only tens of centimeters across. Surveyor landers also showed that the surface of the Moon is covered with debris of a wide range of sizes from boulders to fine dust. The term **regolith** was borrowed from terrestrial soil sciences and applied to this material, most of which is generated by impact processes. At about the same time, experiments using the impact facility at NASA's Ames Research Center **(Fig. 2.2)** were conducted to gain some

idea of the amount of fragmental material that can result from impact. As shown in **Fig. 4.7**, even a pea-size object can yield abundant rock fragments when impacting at high speed.

The *Lunar Orbiter* (*LO*) series consisted of five spacecraft designed to return images for choosing *Apollo* landing sites. The primary payload consisted of a medium-resolution camera and a high-resolution camera that took frames nested within the medium-resolution frames at ten times higher resolution. The cameras used film systems that enabled on-board chemical processing, after which the film was scanned, and the data returned electronically to Earth for reconstruction. *LO* frames can be identified by the "strips" of images, reflecting the way in which the data were reconstructed **(Fig. 2.20)**.

The first three *LO* spacecraft were placed in equatorial orbits to photograph the surface at resolutions as high as 2 m on the lunar near side. These orbits were chosen because the *Apollo* landings would have to occur in the near-side equatorial band to allow communication with

Table 4.1. **Selected successful missions to the Moon that have returned data relevant to planetary geoscience (NASA, unless noted otherwise)**

Spacecraft	Encounter date	Mission	Encounter characteristics
Luna 3[a]	Oct. 1959	Flyby	First (indistinct) photos of far side of Moon
Ranger 7	July 1964	Hard lander	4,300 High-resolution images with about 2,000 times better definition than Earth-based photography; impacted in Mare Cognitum
Luna 10[a]	Apr. 1966	Orbiter	First object to orbit Moon; measured lunar magnetism and radiation
Surveyor 1	June 1966	Lander	Soft landing in Oceanus Procellarum; transmitted 11,240 TV images
Lunar Orbiter 1	Aug. 1966	Orbiter	Obtained 216 images, including 11 of the lunar far side. Medium-resolution pictures good, high-resolution smeared.
Apollo 11	July 1969	Lander	First manned lunar landing; soft landing in Mare Tranquillitatis; 1,359 photographs, 22 kg of samples returned
Luna 17[a]	Nov. 1970	Lander	Soft landing in western Mare Imbrium; *Lunokhod* I roving surface vehicle traversed 20 km
Luna 20[a]	Feb. 1972	Lander	Soft landing in Apollonius Highlands; 30 g of samples returned to Earth
Apollo 17	Dec. 1972	Lander	Soft landing in Taurus–Littrow Valley; 30 km (lunar rover) traverse; deployed scientific experiments; 5,807 photographs plus 4,710 mapping photographs (from orbit); 111 kg of samples returned
Luna 24[a]	Aug. 1976	Lander	Soft landing in Mare Crisium; 160 cm core sample returned
Clementine	Feb. 1994	Orbiter	Global mapping, altimetry, radar
Lunar Prospector	Jan. 1998	Orbiter	Global spectroscopy, magnetometry
SMART 1[b]	Nov. 2003	Orbiter	Imaging, spectroscopy
Kaguya[c]	Oct. 2007	Orbiter	Global mapping (with relay satellites)
Chang'e 1[d]	Nov. 2007	Orbiter	Global mapping
Chandrayaan[e]	Nov. 2008	Orbiter	Global mapping
Lunar Reconnaissance Orbiter	June 2009	Orbiter	Global mapping
LCROSS	June 2009	Impactor	Water detection

[a] Soviet missions.
[b] European Space Agency.
[c] Japan Aerospace Exploration Agency.
[d] China National Space Administration.
[e] Indian Space Research Organization.

Earth. *LO I, II*, and *III* were so successful that the two remaining orbiters (*LO IV* and *V*) were placed in polar orbits to collect data contributing to the general knowledge of the Moon and to photograph sites of geologic interest, such as the Copernicus crater (**Fig. 4.8**) and the Marius Hills (**Fig. 4.9**).

Collectively, the NASA and Soviet unmanned missions returned an incredible wealth of engineering and scientific data, much of which remains a valuable resource today. This period of lunar exploration was particularly important for planetary geology because the techniques that are commonly used to study the Solar System were developed at this time. Analyses and interpretations of surface features, studies of geologic processes in an environment different from that of the Earth, and planetary geologic

mapping all had their beginning in the 1960s and were tested through the *Apollo* Moon landings.

4.2.2 The *Apollo* era

Prior to the successful landing on the Moon by the *Apollo 11* astronauts, a series of pre-landing *Apollo* missions was conducted to test critical engineering components of the spacecraft system. This included the Christmas Eve 1968 trip around the Moon by the *Apollo 8* crew without landing, which resulted in the famous image of "Earth-rise" (**Fig. 4.10**).

Each of the six successful *Apollo* landings involved crews of three astronauts, two who went to the surface, and a third who remained in orbit in the command module.

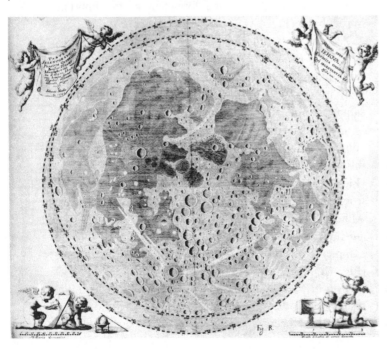

Figure 4.4. An image of the lunar far side obtained by the Soviet spacecraft *Zond 8*, showing the general lack of mare regions. Crater Aitken, near the middle of the image, is about 135 km in diameter and its floor is partly flooded with mare lavas.

Figure 4.2. Sketches of the Moon prepared by Galileo Galilei and published in 1610.

Figure 4.3. A map of the Moon prepared by Johannes Hevelius in 1647; improvements in the telescope enabled fairly accurate portrayal of surface features. Overlapping disks show that more than half of the Moon can be seen from Earth due to the Moon's libration.

(a)

(b)

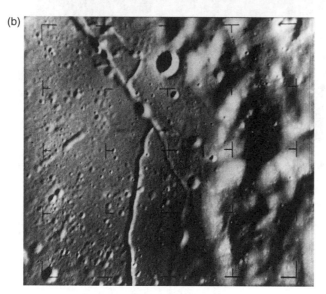

Figure 4.6. *Apollo 12* astronaut Pete Conrad examined the *Surveyor 3* television camera, as shown in this photograph taken by astronaut Al Bean. The Surveyor camera and the scoop were returned to Earth for analysis (NASA AS 12–48–7133).

Figure 4.5. (a) An *Apollo 16* photograph of Alphonsus crater (~125 km in diameter) showing the same features as seen earlier from *Ranger IX* (shown in part (b)), as well as other, similar features. The dark haloes are considered to be pyroclastic deposits (NASA AS 2478). (b) A *Ranger IX* image (Frame B-75) of part of the floor of Alphonsus crater moments before impact of the spacecraft, showing elongate, dark-halo volcanic craters associated with fractures.

A total of 382 kg of lunar samples was returned to Earth, more than half of which remains untouched, awaiting new analytical techniques. This means that substantial samples are still available for study by scientists around the world. In addition to the return of samples, diverse other *Apollo* experiments were conducted, including establishing a seismometer network, measurements of the remnant magnetic field, and remote sensing of the surface from orbit.

Apollos 11 and *12*, while returning new critical data, were geared primarily toward engineering tests and involved landings in relatively smooth mare sites. As the *Apollo* series progressed, the time spent on the Moon and the array of scientific investigations increased. The next *Apollo* mission was slated for a more complex highland site but, as well documented by the movie of the same name, *Apollo 13* experienced an explosion on the way to the Moon, and it was only through the heroic work of the astronauts and the ground crew that disaster was averted. The scientific goals of the *Apollo 13* mission, which were to study and sample deposits associated with the formation of the Imbrium impact basin (**Fig. 4.11**), carried over to *Apollo 14*.

The payloads and goals for the last three *Apollo* missions (*15, 16,* and *17*) were substantially enhanced with the addition of roving vehicles, giving the astronauts mobility to carry out longer traverses (**Fig. 4.12**) and to sample more diverse terrains.

In parallel with the *Apollo* program, the Soviet Union continued a series of highly successful unmanned lunar missions, including the first use of robotic roving vehicles (**Fig. 4.13**) and the autonomous return of samples to Earth from three sites on the Moon, including core samples. Although the total mass of their returned samples was small, they provided critical data on lunar regions to complement those visited by the *Apollo* astronauts (**Fig. 4.1**).

With the return of the *Apollo 11* samples and the initial results which they revealed, NASA organized the first of what would become the premier scientific meeting for

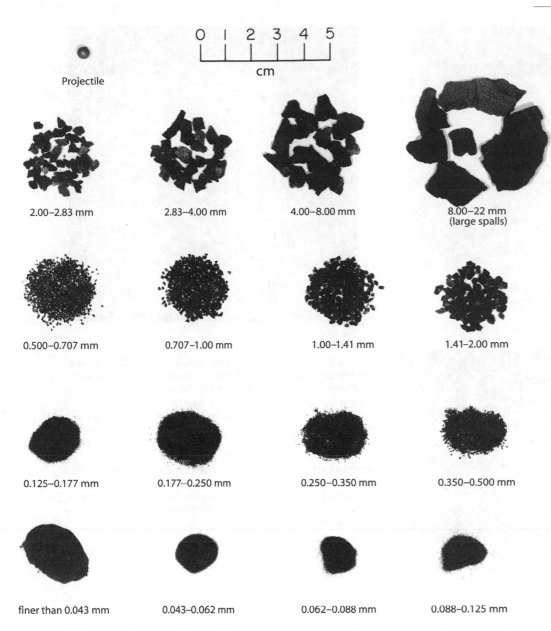

Projectile

0 1 2 3 4 5
cm

2.00–2.83 mm

2.83–4.00 mm

4.00–8.00 mm

8.00–22 mm
(large spalls)

0.500–0.707 mm

0.707–1.00 mm

1.00–1.41 mm

1.41–2.00 mm

0.125–0.177 mm

0.177–0.250 mm

0.250–0.350 mm

0.350–0.500 mm

finer than 0.043 mm

0.043–0.062 mm

0.062–0.088 mm

0.088–0.125 mm

Figure 4.7. Photographs showing rock fragments, sorted by size, that resulted from a single impact by a small ball-bearing projectile into a block of solid basalt (courtesy of Don Gault, NASA-Ames Research Center).

planetary geoscience, the Lunar (and Planetary) Science Conference (LPSC), held in Houston, Texas. The purpose of the first meeting, held in 1970, was for the science teams working on the *Apollo* samples to share their findings with each other and the public. In subsequent years, the conference was expanded to include all solid-surface planetary objects. Among the many surprises revealed in the *Apollo 11* samples were the great ages of the mare basalts. Although they had been considered to be geologically young because of the paucity of superposed impact craters, the lavas were found to exceed 3 Ga in age. However, it should also be noted that, well before the return of samples, planetary scientist Bill Hartman had predicted an age of 3.6

Ga for lunar mare regions on the basis of his crater counts and estimates of the rate of impacts.

4.2.3 Post-*Apollo* exploration

Following the successes of *Apollo* and the Soviet programs (and the brief flyby of *Mariner 10* in 1973 on the way to Mercury), analysis of lunar samples and other data continued, and the understanding of the Moon began to mature, including the formulation of the now widely accepted hypothesis of its origin. However, it would be more than 14 years until the next spacecraft would encounter the Moon. Although not designed or intended to collect data

(a)

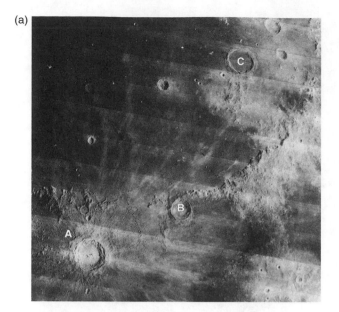

(b)

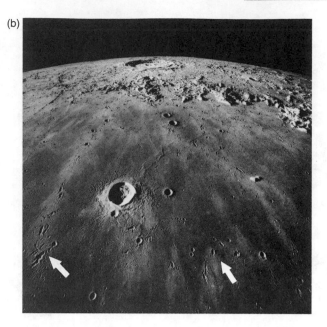

(c)

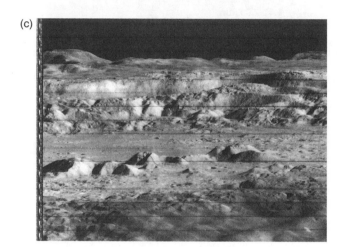

Figure 4.8. (a) A *Lunar Orbiter* photograph (LO IV M-126) showing Copernicus crater (A), Eratosthenes crater (B), and Archimedes crater (C). Copernicus is 107 km in diameter and is one of the youngest large impacts on the Moon, serving as the type locality for the stratigraphic Copernican System. (b) An oblique *Apollo 17* view (NASA AS 17–2444) of Copernicus Crater (on the horizon); the crater in the middle foreground is the 20 km in diameter Pytheas crater. Small clusters and aligned craters (indicated by arrows) are secondaries, forming V-shaped (or herringbone) patterns that point back to their source, Copernicus. (c) An oblique *Lunar Orbiter II* photograph (NASA LO II H-162) of Copernicus Crater's central peak (mountains rising > 300 m in the middle foreground) and the terraced northern wall of the crater.

for the Moon, the *Galileo* spacecraft bound for Jupiter was sent on a trajectory that carried it past the Moon in 1990 and 1992, providing the first new data from modern (at the time!) instruments, including a near-infrared (NIR) multispectral imager. The first flyby provided views of the lunar far side and the Orientale basin (**Fig. 4.14**), while the second pass flew over the North Polar region. *Galileo* CCD images and NIR data revealed the presence of ancient, thinly mantled lava flows, which planetary geologists Jim Head and Lionel Wilson termed **cryptomaria**, showing that volcanism was more extensive than previously thought. This concept built on the recognition of **dark-halo craters** seen on earlier images and interpreted as representing small impacts that penetrated through mantles to excavate underlying, darker basaltic lavas (**Fig. 4.15**).

In 1990, Japan launched the *Hiten* spacecraft (also called *Muses-A*), which was primarily to test various engineering concepts, including the release of a small orbiter around the Moon, which failed. After a series of tests, including of the concept of aerobraking using Earth's atmosphere, *Hiten* was placed in lunar orbit in 1991. The only science experiment on board was the Munich Dust Counter, provided by Germany, to measure cosmic dust between the Earth and the Moon. Results showed that some dust particles traveled at speeds of nearly 100 km/s, suggesting that they came from beyond the Solar System.

In January 1994, the *Clementine* spacecraft was launched and placed in orbit around the Moon. This mission was a joint Department of Defense and NASA project that carried a scientific payload that included an

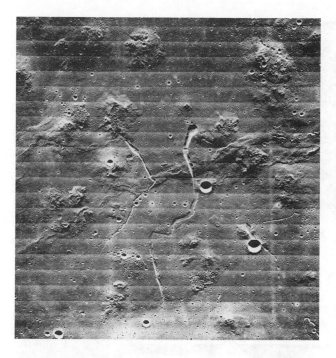

Figure 4.9. The Marius Hills in Oceanus Procellarum, seen in this *Lunar Orbiter V* photograph, is a prominent volcanic region that includes sinuous rilles and volcanic domes; the area shown is 75 km by 85 km (NASA LO VM-214).

Figure 4.10. A view of "Earthrise" over the Moon obtained by the *Apollo 8* crew in their historic journey around the Moon in December 1968; this mission was a precursor to the *Apollo* landings as a test of the flight system (NASA AS 8–14–2383).

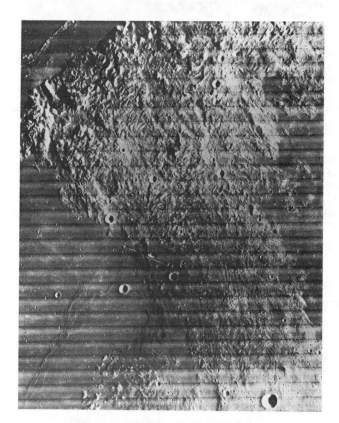

Figure 4.11. The Fra Mauro Formation consists of ejecta deposits (rugged terrain, upper half of image) from the Imbrium basin, the rim of which is visible in the upper left. This formation marks the base of the Imbrium System and is one of the principal stratigraphic horizons on the lunar surface. The area shown is about 325 km by 250 km (NASA LO IVH-109).

Figure 4.12. *Apollos 15, 16*, and *17* each were equipped with lunar rovers in which astronauts could travel many kilometers from the lander to carry out scientific investigations and collect diverse samples. This view is from *Apollo 15*, showing astronaut Jim Irwin with the mountain Hadley Delta in the background (NASA AS 15–82–11121).

Figure 4.13. The Soviet *Lunokhod 1* was the first robotic rover used in planetary exploration. It was carried to the Moon by *Luna 17* and then traveled more than 10 km across the western Mare Imbrium.

Figured 4.14. A *Galileo* image of the lunar far side, showing the Orientale basin (center), Oceanus Procellarum (upper right), and part of the South Pole–Aiken basin (lower left), obtained by the flyby of the spacecraft in 1990 (NASA PIA00077).

(a)

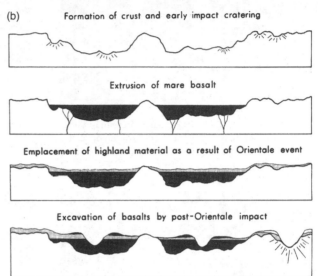

(b)

Figure 4.15. (a) Dark-halo craters, as seen here in the Orientale basin, are impacts that penetrate through thin mantles of high-albedo material into underlying darker material, which is excavated as ejecta; these craters are ~600 m in diameter (NASA LO IV 195-H2). (b) The diagrams illustrate how dark-halo craters can be used to infer the presence of buried mare deposits (courtesy of B. Ray Hawke).

ultraviolet–visible CCD camera, a near-infrared camera, a long-wavelength infrared camera, a laser altimeter, a radar system, and a high-resolution imager. Placed in a slightly inclined polar orbit, the remote sensing data provided the first near-global multispectral data for the

Moon and showed that the surface compositions are much more diverse than had been revealed by previous missions, reflecting the complex evolution of the Moon. The *Clementine* team also wished to test the hypothesis that ice is present in some areas of the Moon. The concept is that parts of some polar craters are permanently shaded, precluding solar heating. On the assumption that cometary impacts would implant ice in these areas, over several billion years of lunar history substantial ice could accumulate and be preserved. The *Clementine* radar system was focused on some of these areas with the reflected beams picked up by the Earth-based radar observatory in Puerto Rico. Although the results were not definitive, the radar signature gave hints that ice could be present.

The *Lunar Prospector* mission was the first of the scientist Principal Investigator-led projects in NASA's Discovery Program. Carrying a payload that included a gamma-ray spectrometer, a neutron spectrometer, and a magnetotmeter and launched in 1998, this mission had the primary objectives of searching for traces of water, mapping surface compositions, and characterizing the signatures of magnetic fields preserved in the rocks. Initial results from the neutron spectrometer revealed hydrogen in some areas, which was interpreted to be due to water-ice contained within the regolith. At the end of the mission, it was decided to crash the spacecraft into the south polar highlands and observe the impact from Earth in the hope of detecting released water vapor. Although the results were negative, it is generally thought that the impact energy was insufficient to release significant volatiles and that the Earth-based telescopic resolution was too low for their detection.

In the fall of 2003, the European Space Agency launched the *SMART-1* (*Small Missions for Advanced Research in Technology*) spacecraft to the Moon. Although, as the name implies, this was primarily an engineering mission to test solar-electric propulsion and a set of small instruments, part of the payload included an imaging system and spectrometers to collect scientific information for the lunar surface. The mission was an engineering success and returned new data, including excellent images, such as those of the poorly understood Reiner Gamma Formation (**Fig. 4.16**) on the western near side of the Moon. This bright swirling pattern has been a matter of controversy for decades, with ideas including the possibility that transient magnetic fields generated by impacts caused sorting and redistribution of fine lunar dust. The feature had been imaged previously and planetologist Peter Schultz suggested that a low-density comet impacted the area at a low angle, with the debris mixing with local lunar surface materials.

The *SMART-1* mission ended with its planned crash on the near side in September 2006. The light flash generated by the impact was recorded at various Earth-based observatories with the hope of detecting water vapor but, again, the results were inconclusive.

The early part of the twenty-first century saw renewed interest in the Moon, with many nations entering the "deep space club." Following the successes of previous missions,

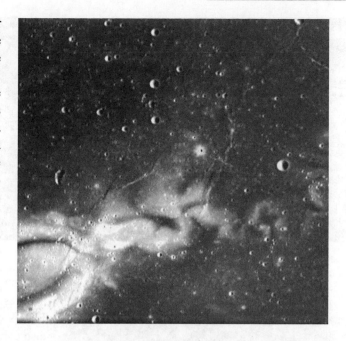

Figure 4.16. The swirling bright pattern seen in the lower half of this *SMART-1* image is part of the Reiner Gamma Formation in Oceanus Procellarum. This pattern could represent the impact of a comet, with the debris altering the lunar surface soils, as suggested by Pete Schultz, Patrick Pinet, and others. The area shown is about 30 km by 60 km (ESA *SMART-1* image SEMWQCVNFGLE).

Japan, China, India, and other nations began plans for robotic exploration of the Moon. Launched in 2007 by the Japan Aerospace Exploration Agency (JAXA), the *Kaguya* (also referred to as *SELENE* in earlier years) spacecraft carried 14 instruments into orbit, including an imaging system (**Fig. 4.17**), and began returning data in October of that year. The mission also included two sub-satellites to enable communication from the lunar far side; careful tracking of the three craft enabled the generation of precise gravity maps. A radar sounder system provided subsurface profiling to detect the boundary between mare lava flows and the underlying regolith. The *Kaguya* mission came to an end in June 2009 with the intended crash on the Moon. The year 2007 also saw the launch of the Chinese *Chang'e 1* orbiter, which began returning images from its 200 km orbit in late 2007 (**Fig. 4.18**). This mission was primarily for engineering purposes and for collecting images for subsequent landed missions to the Moon. It, too, ended with a planned crash on the Moon in March 2009.

India's entry to lunar exploration was marked with the launch of *Chandayaan-1* in October 2008. This mission had both engineering and scientific objectives, and carried

Figure 4.17. An oblique view across the lunar north polar cratered terrain obtained by the Japanese *Kaguya* spacecraft. The smoothly rounded appearance of the terrain results from repeated bombardment by impacts of a wide range of sizes, reflecting the effectiveness of impact as an agent of surface modification by the erosion of high-standing areas, such as crater rims, and filling-in of low areas, such as crater floors (© Japan Aerospace Exploration Agency [JAXA]).

Figure 4.18. One of the first image mosaics produced from *Chang'e 1*, China's lunar orbiter, which began operation in late 2007. The area shown covers mostly highlands of the south polar region and includes part of Mare Australe (upper right). This mosaic consists of 19 images of resolution about 120 m per pixel.

11 instruments that included contributions from NASA, such as the Moon Mineralogy Mapper for which Carlé Pieters of Brown University was the PI. The spacecraft also carried the Moon Impact Probe, which was released to crash into Shackleton Crater in the south polar region to provide signs of water. The mission was ended in August 2009 when various components failed to operate, probably due to overheating when the spacecraft was in full sunlight.

NASA's return to the Moon in the twenty-first century began with the launch of the *Lunar Reconnaissance Orbiter* (*LRO*) in June 2009. Although its primary goal is to return detailed information to support ambitious human landings, its payload is yielding a wealth of data of direct scientific interest, including altimetry, remote sensing data, and high-resolution images under the direction of Mark Robinson at Arizona State University. In addition to mapping the Moon, NASA's lunar missions included *LCROSS* (*Lunar Crater Observation and Sensing Satellite*), which used cameras and spectrometers to watch the impact of its upper stage in the south polar region. The resulting crater was 28 m across, and the impact formed an ejecta plume in which clear evidence of water was revealed.

In the past decade, there has been a growing body of evidence that substantial amounts of water are present on the Moon. Many of the recent lunar missions involved crashing objects onto the surface to observe the resulting impact ejecta to detect signatures of water, while instruments on orbiters have been used to map the presence of hydrogen and hydroxyls as surrogates for water. Global mapping from *Chandraayan-1* shows that such signatures increase toward the poles, as expected for water implanted by comets into permanently shadowed craters, but, in addition, occurrences elsewhere suggest chemical reactions of the solar wind with lunar minerals to form water-bearing materials.

In addition to the search for water on the Moon, attention is also being given to characterizing the lunar interior. In 2011, two small spacecraft that completed objectives to study space weathering and solar processes were repositioned to collect geophysical data for the Moon. The mission, termed *ARTEMIS* (*Acceleration, Reconnection, Turbulence and Electrodynamics of the Moon's Interaction with the Sun*), places the spacecraft in orbit around the Moon, coming as close as 100 km above the surface, and will operate for seven to ten years. In parallel, the *Gravity Recovery and Interior Laboratory* (*GRAIL*) mission was launched in 2011. It involves two spacecraft in near-circular polar orbits that are mapping the detailed gravity of the Moon to determine the structure of the crust and lithosphere, as well as to gain insight into the deep interior. *GRAIL* is a Discovery project led by MIT geophysicist Maria Zuber.

4.3 Interior characteristics

Information on the interior of the Moon is derived primarily from the *Apollo* network of seismometers, gravity measurements, considerations of the Moon's weak magnetic field, and other geophysical data. Each of the six *Apollo* landings included placement of seismometers on the surface. The *Apollo 11* station used solar-powered batteries and was of limited duration, whereas *Apollos 12* and *14* through *17* used

RTG (radioisotope thermoelectric generator) power sources that enabled longer lifetimes. In addition, *Apollos 14, 16*, and *17* involved active experiments in which small explosive charges were detonated to generate seismic events in order to assess shallow subsurface structure.

The natural seismic events on the Moon are rather different from those seen on Earth. Typically, they register less than 3 on the Richter scale, and, although more than 3,000 events were seen each year, the total annual energy was less than 2×10^{13} ergs, compared with the annual energy released on Earth by earthquakes of 10^{24}–10^{25} ergs. Thus, the total annual energy generated from "moonquakes" is equivalent to only about 500 grams of TNT, or slightly more than the detonation of about one pound of explosives.

Moonquakes generate signals that have been described as "ringing," in which the magnitude is low but of long duration. Except for seismic events from impacts, natural moonquakes tend to occur in the same areas on the Moon and are considered to result from stresses generated by tidal interactions with Earth. High-frequency teleseisms are shallow moonquakes that manifest releases of energy in the lunar crust. Recent applications of array-processing techniques by Renee Weber *et al.* (2011) to the Apollo data suggest that the Moon has a solid inner and a fluid outer core overlain by a zone of partial melt and a thick mantle (**Fig. 4.19**). Although there is very little information on the Moon's core, the overall density of the Moon (**Table 1.2**) suggests that it is depleted in iron in comparison with the

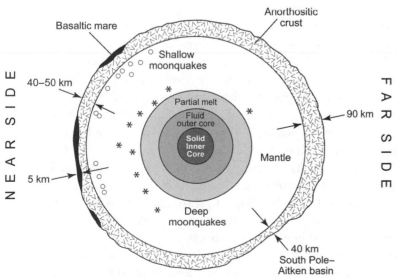

Figure 4.19. The interior configuration of the Moon (not to scale), based on geophysical data and models; anorthositic rocks compose a crust ranging in thickness from 90 km on the far side to 40–50 km on the near side, with relatively thin mare deposits primarily on the near side; the greatest mass is in the mantle, with shallow moonquakes (high-frequency teleseisms) occurring near the crust–mantle boundary and deep moonquakes near the lower mantle boundary; the zone of partial melt has a radius of ~480 km, while the outer fluid core has a radius of ~330 km; the inner solid core has a radius of ~240 km.

other terrestrial planets. It must be remembered that the *Apollo* network was located in the low-latitude, near-side band dictated by the constraints of the landing sites and was not ideal. Refinements to the model of the lunar interior must await new measurements obtained from better-sited stations.

One of the unexpected benefits of the *Lunar Orbiter* missions came from careful tracking of the spacecraft in their orbits around the Moon. It was found that in some places the spacecraft dipped closer to the surface and speeded up, while in other places the spacecraft rose in altitude and slowed down. Although such changes in orbit were small (changes in velocity as small as 1 mm/s could be detected), by applying the laws of physics to these motions, it was inferred that some locations on the Moon consist of higher-density materials, causing the spacecraft to be gravitationally accelerated, or "pulled," toward them. These areas were dubbed **mascons** (short for mass concentrations) for positive gravity anomalies. Negative gravity anomalies were identified over areas where the spacecraft rose in altitude.

When later compared with terrains and geologic maps, most mascons were found to correlate with circular patches of maria in lunar impact basins, such as Imbrium, Serenitatis, and Crisium, as well as mare deposits in some smaller craters. They also occur in association with some large mountain ranges around basins, as well as in the volcanic Marius Hills (**Fig. 4.9**). Initially, it was thought that mascons reflected buried impact projectiles, but it was realized that bolides are generally fragmented upon impact and mostly disbursed with the ejecta. High-resolution gravity data from the *Kaguya* mission support the earlier idea that mascons result from a combination of mare flooding (perhaps accompanied by high-density iron and titanium oxide minerals that settled to the bottom of flood-lava lakes) and uplift of high-density mantle material following the impact. As described by Namiki *et al.* (2009), gravity data indicate that the basins on the lunar far side are supported by a rigid lithosphere, whereas the near-side basins deformed during the eruption and emplacement of the mare lavas.

One of the early discoveries of the pre-*Apollo* landings was the existence of a very weak magnetic field. Subsequent measurements both from orbit and from the surface showed that the field is not uniform on the Moon. The source of the magnetic signature remains debatable, but some suggest that it was somehow imposed externally by the Earth or the Sun or that it represents an ancient intrinsic field from a time when the Moon's interior was sufficiently molten to enable the operation of a dynamo similar to that responsible for Earth's magnetic field. Others have noted that many of the circular basins show stronger magnetic signatures and have suggested that massive impacts can generate local magnetic fields. This idea is at least partly supported by laboratory impact experiments by Pete Schultz that have recorded slight transient magnetic fields.

On Earth, measurements of the amount of heat reaching the surface from the interior provide important insight into the interior of our planet. Such measurements were attempted on *Apollos 15, 16,* and *17*. Unfortunately, when the heat-flow experiment was deployed on *Apollo 16*, an astronaut tripped over the wire and disconnected it from the main station. It probably did not matter, however, as the results from *Apollos 15* and *17* are now generally regarded as inconclusive. Heat-flow experiments require deployment into the subsurface (the deeper, the better) and sufficient time for the surrounding rock and soil to stabilize thermally, after the disturbances caused by their emplacement. The Apollo experiments were placed < 3 m below the surface and did not record for a sufficient length of time for thermal stabilization.

4.4 Surface composition

Information on the composition of the lunar surface comes from *in situ* measurements made from unmanned landers, remote sensing from orbiters and Earth-based observatories, samples returned to Earth from the *Apollo* and Soviet missions, and from dozens of meteorites considered to have been blasted from the Moon by impacts and sent on trajectories to Earth.

NASA's *Surveyor* landers revealed iron-rich (mafic) compositions for the mare deposits, which, when combined with images of features suggestive of flows, were correctly identified as basaltic lavas. The spectral signatures of the titanium-rich lavas of *Apollo 11* were later mapped in remote sensing data and extrapolated to other parts of the Moon, but they were found to be relatively restricted in distribution in comparison with most of the mare basalts. In fact, Paul Spudis of the Lunar and Planetary Institute has noted that the basalts in the returned samples (**Fig. 4.20(a)**) represent only about one-third of the total number of varieties of mare basalts suggested in the *Clementine* data.

The so-called "genesis" highland rock from the *Apollo 15* site was found to be nearly pure **anorthosite (Fig. 4.20(b))**,

Figure 4.20. (a) Vesicular mare basalt collected by the *Apollo 15* astronauts at the Hadley Rille site (sample 15016, about 15 cm wide; NASA S71–46632). (b) A view of the "genesis rock," a sample of anorthosite collected at the *Apollo 15* site (sample 15415; NASA 571–42951). (c) A scanning electron micrograph of a composite volcanic orange glass droplet from *Apollo 17* (from McKay *et al.*, 1991). (d) Lunar breccia from the *Apollo 16* landing site (sample 67015; NASA 72–37216).

an igneous rock composed of plagioclase feldspar. Similar occurrences were found at highland sites, such as *Apollo 14*. Extension of these "ground-truth" sites to the global *Clementine* maps shows that most of the highlands on the near side and far side are composed of anorthosite. Dated at 4.2 Ga, anorthosite represents much of the ancient lunar crust and is considered to have formed from the cooling of a massive "magma ocean" that covered the Moon to a depth of more than 100 km early in its history. In this model, the lower-density feldspar crystals rose toward the surface of the ocean to form a crust, while the denser olivine and pyroxene crystals sank to a lower level and served as the source for later-stage eruptions of basalt.

One of the discoveries in the lunar samples is the widespread occurrence of a chemical group of incompatible elements, termed **KREEP** (*K* for potassium, *REE* for rare-earth elements, and *P* for phosphorus). The age of KREEP is uniformly found to be 4.35 Ga and is considered to reflect the crystallization of the lunar magma ocean.

In addition to mare basalts and anorthositic crustal materials, **dark mantle deposits** (**Figs. 2.8** and **4.21**) are seen in many areas of the Moon, including the margins of the Serenitatis basin and the *Apollo 17* landing site. *Apollo 17* samples collected by geologist-astronaut Harrison "Jack" Schmitt revealed that this material consists of orange glass beads that are coated with volatile elements such as sulfur and chlorine, which are indicative of volcanic origins (**Fig. 4.20(c)**). Thus, dark mantle deposits were probably produced from pyroclastic eruptions similar to the fire-fountain eruptions seen in Hawaii. In the low-gravity, airless environment of the Moon, such eruptions would have spread the pyroclastics over wide areas.

In addition to basaltic maria, anorthositic highlands, and pyroclastic deposits, other surface materials include breccias and impact melts. Glassy beads in lunar soils and glasses incorporated in other samples reflect melted rocks

generated during impacts, some of which "splashed" onto the surrounding terrain (**Fig. 4.22**). Unlike the orange pyroclastic glass beads, impact melt materials lack the coatings of volatile elements.

Breccias are composed of angular rock fragments, which, on the Moon, result primarily from impact processes that both break up and consolidate rocks (**Fig. 4.20(d)**). Many of the lunar samples show breccias within breccias, indicative of repeated breakup and consolidation by multiple impact events.

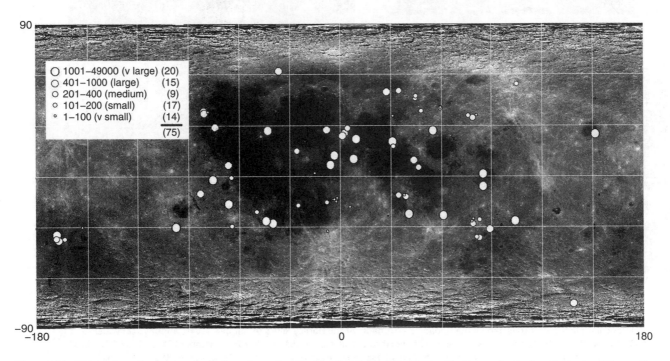

○ 1001–49000 (v large)	(20)
○ 401–1000 (large)	(15)
○ 201–400 (medium)	(9)
○ 101–200 (small)	(17)
° 1–100 (v small)	(14)
	(75)

Figure 4.21. Distribution of lunar pyroclastic deposits analyzed by Lisa Gaddis *et al*. (2003), classified by areal extent on the basis of *Clementine* UV – VIS data (reprinted from *Icarus*, **161**, Gaddis, L. R. *et al*., Compositional analyses of lunar pyroclastic deposits, 262–280, 2003, with permission from Elsevier).

(a)

©JAXA/SELENE

Figure 4.22. (a) A view of the central peak of Tycho from the *Kaguya* orbiter; Tycho is a very young, bright-rayed crater in the near-side southern hemisphere and has a diameter of 85 km; as seen in this image, the crater floor surrounding the central peak is hummocky and fractured, considered to be fall-back melt deposits from the impact. (b) A high-resolution image from the *Lunar Reconnaissance Orbiter* camera showing melt deposits on Tycho's floor and the direction of motion as the melt slid into place (NASA/GSFC/Arizona State University).

(b)

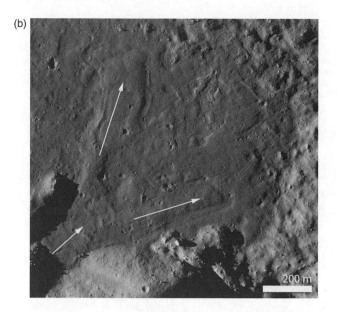

200 m

Figure 4.22. (cont.)

4.5 Geomorphology

The original two-fold classification of the lunar surface into smooth, dark maria and rugged, light highlands or terrae is still valid, but sub-units are now recognized, including terrains associated with impact basins, highlands plains, and various features on the maria. In addition, features associated with tectonic and gradation processes are seen throughout the lunar surface.

4.5.1 Impact craters and basins

Impact craters are the dominant landform on the Moon; interest in them ultimately contributed to the foundation of planetary geology as a discipline. Lunar craters range in size from features only microns across, seen on lunar samples, to the enormous "basins" that are hundreds of kilometers across. Lunar craters serve as the basis for describing the morphology of impact craters in general, although the geometries (such as the depth-to-diameter ratios, **Fig. 3.31**) vary with planetary environment as described in **Section 3.4.3**.

Planetary scientists Bill Hartmann and Gerard Kuiper of Arizona coined the term **basins** for large impact structures on the Moon. In addition to the large size (> 220 km), lunar basins are typified by concentric rings of mountain ranges.

Maps of the Moon (**Fig. 4.23**) show that basins are randomly distributed over the entire surface and are not preferentially located in any one hemisphere, although the north polar region appears to have a paucity of impact features > 300 km in diameter, as described by NASA scientist Herb Frey (2011). Samples returned by the *Apollo* astronauts, coupled with photogeologic mapping, show that most of the basins formed prior to 3.8 Ga ago. Basin-related geology dominates the lunar surface in several ways, including (a) mountain ranges that are segments of basin rings, (b) lunar crustal fractures that controlled the eruptions of some lunar lava flows, and (c) ejecta of deposits that blanket much of the older surface.

Given the critical role of basins in the evolution of the lunar surface, many of the *Apollo* and *Luna* landing sites were selected to provide insight into these structures, including one of the most prominent features, the Imbrium basin. The American geologist G. K. Gilbert was keenly interested in the geology of the Moon and studied its surface telescopically, especially the area around Imbrium. As noted in **Section 2.2**, he also conducted experiments to simulate impact processes. His results led Gilbert to suggest that the Imbrium basin was a huge impact scar, some 1,140 km across. This feature dominates the near side of the Moon, and Gilbert recognized the distinctive radial grooves and furrows, which he termed the Imbrium sculpture (**Fig. 4.24**). Geologic mapping shows that the Imbrium basin includes three rings, with the main ring defined by the Apennine Mountains. An intermediate ring 850 km across is marked by the lunar Alps, part of the Sinus Iridium rim, and smaller isolated mountains such as La Hire. An inner ring about 570 km across is defined only by arc-like ridges seen on the mare lavas, which have been interpreted as representing flooding by lava flows over now-buried basin structure.

Ejecta deposits from the Imbrium impact are spread over much of the lunar near side. Its distinctive appearance led to the definition by early lunar geologic mappers of the Fra Mauro Formation (**Fig. 4.11**), which is as thick as 1 km some 600 km from the basin. This unit serves as one of the primary index markers, or datum planes, for lunar stratigraphy, in which other units are dated relatively by superposition and cross-cutting relations. Because of its critical importance, the Fra Mauro Formation. was targeted as an *Apollo* site early in the series of landings. Samples returned from *Apollo 14* showed that the formation consists of highly brecciated rocks and provided a date for the Imbrium impact of 3.85 Ga.

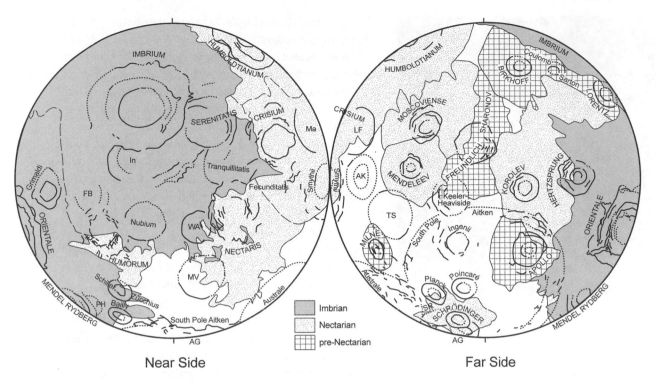

Near Side

Far Side

Figure 4.23. A map of lunar basins; rings and ring-arcs are shown solid where exposed, dotted where buried or inferred. Names of basins having mapped deposits are in capital letters. Names of other definite basins are written out in lower case. Initials refer to indefinite basins: AG, Amundsen–Ganswindt; AK, Al Khwarizmi–King; FB, Flamsteed–Billy; In, Insularum; LF, Lomonosov–Fleming; Ma, Marginis; MV, Mutus–Vlacq; PH, Pingré–Hausen; SR, Sikorsky–Rittenhouse; TS, Tsiolkovsky–Stark; and WA, Werner–Airy (from Wilhelms, 1980).

Figure 4.24. The *Apollo 16* astronauts took this photograph of the "Imbrium sculpture" northwest of the crater Ptolemaeus; the texture resulted from the gouging by ejecta from the Imbrium basin, which is visible on the horizon (NASA AS 16–1412).

Orientale (**Fig. 4.25**) is the youngest impact basin on the Moon. So-named from the older system of lunar geography in which telescopic images invert positions, the "eastern" (i.e., "orient") basin is actually on the western limb of the Moon. Although it has not been dated directly, superposition of the ejecta deposits from Orientale, the Hevelius Formation, shows that it is younger than the Imbrium basin. Unlike most of the impact basins on the near side, the Orientale basin is only partly flooded by younger mare lavas, enabling its structure to be studied. Mapping by Jack McCauley and others identified the key formations and basin rings (**Fig. 4.26**), including the Inner Rook, Outer Rook, and Cordillera Mountains. The Maunder Formation lies between the Inner and Outer Rook Mountains. Its fissured, high-albedo appearance suggests that it is melted rock generated from the Orientale impact. The Montes Rook Formation (**Fig. 4.27**) is found between the Outer Rook Mountains and the Cordilleras and is also thought to be impact melt, but includes non-melted materials, probably representing material formed late in the ejection process. Topographic mapping of Orientale, as well

Figure 4.25. The 930 km in diameter Orientale basin, imaged by the *Lunar Reconnaissance Orbiter*, is marked by the Inner Rook Mountains (A), the Outer Rook Mountains (B), and the Cordillera Mountains (C), which form concentric rings. The outer two rings rise some 3 km above the surrounding terrain and are among the highest features on the Moon (NASA PIA 13225).

as the rest of the Moon, from the Lunar (Reconnaissance) Orbiter Laser Altimeter on the *LRO* spacecraft and from the *Kaguya* mission enables accurate assessments of the elevations of the mountains and the intervening low-lying areas.

There has been considerable controversy regarding the size of the original transient cavity of the Orientale impact, as well as the mechanisms of basin-ring formation in general. Many researchers favor the Outer Rook Mountains to define the Orientale transient cavity, on the basis of observations that most of the ejecta lies beyond the scarp formed by this range. The Cordillera Mountains are thought to be a "mega-terrace," formed by the inward slump of material into the transient cavity. However, an alternative model suggests that the Cordilleras represent the transient cavity, because of the knobby texture of the Montes Rook Formation, which lies beyond the Cordilleras and is suggested to be the primary ejecta. In either case, the inner rings and scarps probably represent adjustments and rebound of the lunar crust following the impact.

4.5.2 Highland plains

Many areas in the highlands display relatively smooth, bright plains. Early in photogeologic mapping of the Moon, these plains were interpreted as being either a form of highland volcanism (such as silicic ash flows) or ejecta deposits from large impacts. Named the Cayley Formation, this unit was given a high priority for Apollo exploration to resolve the issue. For example, if the plains represented silicic volcanism, then models for magma evolution on the Moon would need substantial revision from the favored basalt-dominated models. With this goal in mind, *Apollo 16* was sited to land on the Cayley Formation (**Fig. 4.28**) near the rim of the Nectaris basin. This unit is now widely regarded as ejecta deposits associated with one or more basin-forming impact. Stimulated by the initial controversy of the origin of the Cayley Formation, Verne Oberbeck of NASA-Ames Research Center conducted impact experiments and modeled the ejecta process. As shown in **Fig. 4.29**, he demonstrated that ejecta deposits consist of progressively higher percentages of locally derived materials with increasing distance from the primary impact. In considering impact cratering mechanics, this makes sense; the blocks of ejecta thrown the greatest distance impact at the highest speeds; thus, they would transfer the highest energies to the surface to form secondary craters and generate local ejecta.

4.5.3 Mare terrains

Lunar maria and basins are often erroneously considered to be synonymous, even by some planetary scientists. As noted above, basins are impact structures, while maria are lava flows, most of which are younger than the formation of the basins. Mare lavas tended to accumulate in topographically low areas, which typically are in the basins. Thus, the outlines of many maria are circular, conforming to the shape of the impact basins that contain them.

Maria cover about 17% of the surface of the Moon and are found mostly on the near side, whereas basins are randomly distributed over the Moon (**Fig. 4.23**). If cryptomaria are taken into account, the areal extent of mare lavas increases to about 20% of surface and near-surface materials (Antonenko *et al.*, 1995). Although difficult to determine precisely, since estimates are based on the degree of flooding and partial flooding of

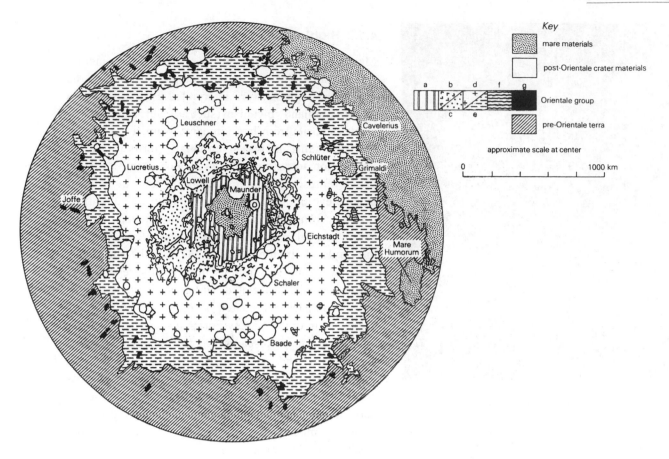

Figure 4.26. A geologic map of the Orientale basin, showing the main ejecta deposits and mare units. The units of the Orientale Group are (a) Maunder Formation, (b) Montes Rook Formation, knobby facies, (c) Montes Rook Formation, massif facies, (d) Hevelius Formation, inner facies, (e) Hevelius Formation, outer facies, (f) Hevelius Formation, transverse facies, and (g) Hevelius Formation, secondary crater facies (reprinted from *Phys. Earth Planet. Inter.*, **15**, McCauley, J. F., Orientale and Caloris, 1977, 220–250, with permission from Elsevier).

impact craters by mare lavas, the thicknesses of the lava flows are estimated to be generally less than a few kilometers. Data from recent missions to the Moon demonstrate a complex stratigraphic history of volcanism, as reviewed by Hiesinger *et al.* (2011), including estimates of dates based on sophisticated crater-counts. Results show that volcanism apparently ceased at about 1.2 Ga.

Lunar lavas are very similar to terrestrial basalts but have slightly higher abundances of iron, magnesium, and titanium in some areas. Analysis of lunar samples suggests that the basalts were derived from the mantle at depths of 150–450 km below the surface.

Even before the return of samples, photogeologic evidence suggested that the lunar lavas were very fluid at the time of their eruption, as exemplified by the flows in Mare Imbrium (**Fig. 4.30**). These flows extend more than 1,200 km but are less than 10–65 m thick. When the lava

compositions were determined from the Apollo samples, synthetic batches of lunar lavas were found to be extremely fluid (about 10 poise), equivalent to motor oil at room temperature. This is substantially more fluid than typical basalt flows on Earth and can account for the great lengths of the flows on the Moon.

For the most part, the source vents for the lunar lava flows are not known, but are mostly inferred to be associated with impact-basin fractures. Similar to flood eruptions on Earth, most of the mare lava flows probably buried any vestiges of their vent structures. There are, however, some well-known volcanic vents, such as those associated with lunar sinuous rilles.

4.5.4 Sinuous rilles

Sinuous rilles are channel-like features found on some mare surfaces and (rarely) in highland terrains. The

Figure 4.27. A *Lunar Orbiter* photograph of part of the interior of the Orientale basin, showing the dark smooth mare lavas (top) of the basin center, fissured and hummocky deposits of the Maunder Formation (middle of the image), and the knobby Montes Rook Formation; the area shown is about 280 km by 312 km (NASA LO IV H-195).

Figure 4.28. The *Apollo 16* landing site (X) was chosen to sample the Cayley Plains, the extensive, smooth unit found in much of the highlands (the area shown is about 95 km by 110 km; NASA AS 16–0439).

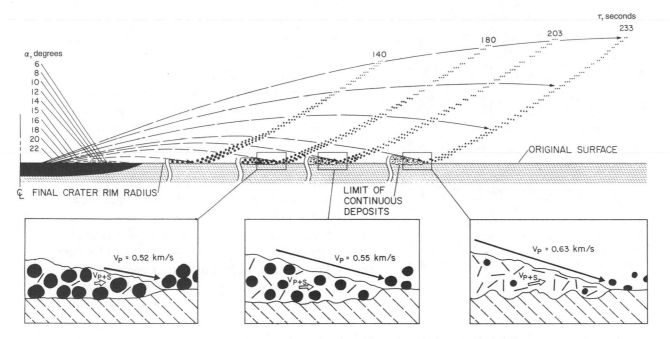

Figure 4.29. A diagram showing the relative proportions of material ejected from an impact crater (solid black) and locally derived material (dashed lines) as a function of distance from the crater. The initial ejecta consists of the finest material which travels the greatest distance and impacts at the highest velocities, resulting in generation of locally derived secondary ejecta deposits; late-stage ejecta consists of large fragments deposited closest to the crater and proportionately is more abundant than locally derived material (from Oberbeck, 1975).

Figure 4.30. A view across Mare Imbrium showing lobate lava flows, some of which are inferred to have flowed 1200 km from their source; the flows seen here are estimated to be only 10–65 m thick, suggesting extremely low viscosity at the time of their emplacement (NASA AS 15–1555).

Figure 4.31. The volcanic Aristarchus Plateau (the 40 km in diameter impact crater Aristarchus is on the left) is the source for Schröter's Valley, the large sinuous rille extending toward the right side of the image and emptying into Oceanus Procellarum (NASA AS 15–2611).

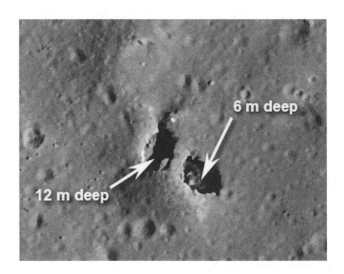

Figure 4.32. A high-resolution image of two collapse features considered to be "skylights" into a lava tube; the image was taken in early morning light (sun shining from right to left) by the *Lunar Reconnaissance Orbiter* camera and covers an area ~200 m across (NASA/GSFC/Arizona State University).

largest, Schröter's Valley (**Fig. 4.31**), was discovered telescopically in the 1700s near Aristarchus crater. *Lunar Orbiter* images (**Fig. 4.9**) stimulated a great deal of interest in sinuous rilles and their origin. Ideas included suggestions that they are ancient channels carved by water or cut by silicic ash flows, or that they are remnants of lava channels or lava tubes.

When the *Apollo* samples showed a lack of evidence for flowing water and extensive silicic materials on the Moon, the ideas involving lava channels and lava tubes were quickly accepted. As noted in **Section 3.2**, these volcanic flow features are common in basalt flows and are effective conduits for transporting lavas to the advancing flow front. Tubes can either form directly in flows or develop from channels that become roofed, either completely or in segments. Once the molten lava has drained, lava tube roofs often collapse (**Fig. 4.32**), leaving a series of discontinuous depressions. Thus, the observations that many sinuous rilles include discontinuous segments is consistent with their being lava channels/tubes. For example, parts of the Hadley Rille, site of the *Apollo 15* landing (**Fig. 4.33**), appear to include roofed segments. Views of the wall of the rille taken by the astronauts show that the basalts consist of multiple lava flows (**Fig. 4.34**) characteristic of the types of eruptions in which lava channels and tubes commonly form.

The margins of some mare regions show distinctive benches where the lava is in contact with hills or

Figure 4.33. A view of the Hadley Rille, site of the *Apollo 15* landing site (A). The Montes Apennines (H) form the rugged terrain on the right and are part of the ring structure for the Imbrium basin (the area shown is about 60 km by 100 km; morning illumination is from the right; NASA AS 15–0414).

Figure 4.34. A view of the Hadley Rille wall photographed by the *Apollo 15* astronauts, showing a cross-section of at least three basaltic lava flows, overlain by fragmental regolith and talus deposits on the lower slopes; these are the only outcrops visited thus far in the exploration of the Moon; the Montes Apennines are visible on the horizon (NASA H-12115).

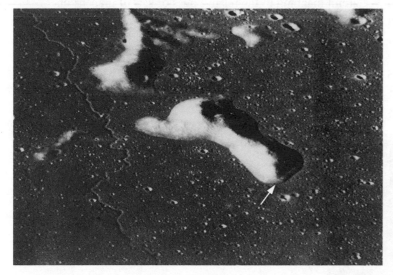

Figure 4.35. The terrace, or bench (arrow), at the base of this hill is thought to represent a high-standing mark of the mare lavas, the surface of which was subsequently lowered by drainage through lava channel(s), degassing, or a combination of these processes. In this oblique view of the Hergonius region, the near-field view is about 30 km across (NASA 16–19140).

highland terrains, and many are found in mare deposits that contain sinuous rilles (**Fig. 4.35**). Some benches are 20 m high and can be traced for several kilometers. The benches probably represent high-stands of the lava flows that subsequently drained downslope, possibly aided by flow through lava tubes and channels. However, degassing of the basalts during cooling could also contribute to the lowering of the mare surface. Some of the *Apollo* basalt samples (**Fig. 4.20(a)**) are quite vesicular, attesting to the high gas content of the lavas in some areas.

4.5.5 Volcanic constructs

Although the Moon lacks large classic shield volcanoes and composite cones, it does show evidence of several styles of "central" volcanism in the forms of domes, small shield volcanoes, and cones. For example, the Marius Hills (**Fig. 4.9**) represent more than 200 domes that have been identified on mare surfaces. Such **mare domes** range in diameter from 2 to 25 km and can be as high as 300 m. Their association with sinuous rilles would suggest basaltic compositions.

In contrast to the Marius Hills, the steeper slopes of the Gruithuisen domes (**Fig. 4.36**) appear to reflect the eruption of lavas with rheological properties more akin to those of silicic magmas than to those of basalts, as postulated by Lionel Wilson and Jim Head. For example, the Gruithuisen domes are as large as 20 km across and stand a kilometer high, suggesting formation from extrusion of viscous lavas.

Small shield volcanoes, called **low shields** because of their small height-to-diameter aspect ratio (~0.008 compared with 0.03 for Hawaiian shield volcanoes), have been identified in the mare deposits of the Orientale basin (**Fig. 4.37**). These features are only about 10 km across and are similar to features seen on Earth, Mars, and Venus.

In a few areas of the Moon (**Fig. 4.38**), rows of small cones aligned on inferred fissures are thought to be spatter cones, formed by the ejection of clots of pasty lava. Other aligned features include the much larger rimless craters of the Hyginus Rille (**Fig. 4.39**); the occurrence of rimless craters within a linear rille (a probable graben) precludes

their origin as secondary impact craters and suggests a volcanic origin probably similar to that of collapse-pit craters.

While not strictly volcanic constructs, several areas of the Moon are centers of extensive volcanism. For example, the Aristarchus Plateau (**Fig. 4.31**), the source of Schröter's Valley, displays spectral signatures indicative of a variety of basaltic lavas and pyroclastic deposits and is probably one of the largest volcanic centers on the lunar near side. Similarly, the Harbinger Mountains are the source region for numerous large sinuous rilles (**Fig. 4.40**). Recent analysis of *LRO* altimetry by Paul Spudis *et al.* (2011) of the Lunar and Planetary Institute suggests that some of the volcanic centers, such as the Marius Hills and the Mayer-Hortensius dome west of Copernicus crater, could be very large shield-like volcanoes that lack central calderas.

4.5.6 Tectonic features

Most tectonic features on the Moon appear to be associated with crustal adjustments in response to large impacts or adjustments of mare deposits within basins. For example, the Alpine Valley (**Fig. 4.41**) is 10 km wide and 150 km long, and is oriented radially to the Imbrium basin. **Linear rilles**, such as Rima Ariadaeus (**Fig. 2.9**), are grabens that cut across highlands, maria, and existing impact craters, reflecting deep-seated tectonic deformation of the crust. Careful mapping of the relative ages of sets of grabens on the Moon suggests that they post-date the emplacement of most maria and appear to represent reactivation of structures associated with impact basins.

Floor-fractured craters are large impacts with floors that have been extensively modified (**Fig. 4.42(a)**). Studies

Figure 4.36. An oblique view of the Gruithuisen domes, obtained by the *Apollo 15* astronauts. These non-mare domes are thought to represent non-basaltic volcanism (NASA AS I5–93–12711).

Figure 4.37. These low-shield volcanoes are about 10 km across and occur in mare deposits of the Orientale basin; they represent a style of basaltic volcanism typified by eruptions of low-volume fluid lavas (NASA LO IVH-181).

Figure 4.38. This row of small cones in Oceanus Procellarum intersects the crater Hortensius D and is thought to consist of volcanic spatter; each cone is less than a few hundred meters across (NASA LO IV H-133).

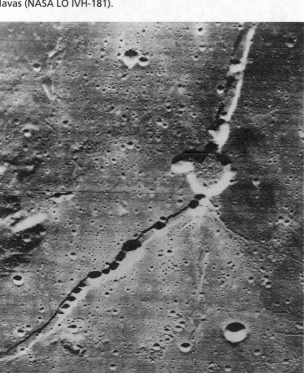

Figure 4.39. The Hyginus Rille is more than 200 km long and contains a series of rimless craters, some as large as 12 km across. The rille is likely to be a graben, while the craters are probably of endogenic origin, possibly analogous to volcanic pit craters formed by collapse (NASA LO V M-97).

Figure 4.40. A view of the Harbinger Mountains (left+side), viewed south across the mare plains of Oceanus Procellarum and east of the crater, Aristarchus (on the right). The numerous sinuous rilles indicate extensive volcanism in this part of the Moon. Note the 46 km in diameter crater Prinz that has been partly flooded by mare basalts (NASA AS 15–2607).

by Pete Schultz suggest that they represent intrusion of magma along fracture systems generated by the impact which lifts the floor units, including central peak structures, to higher-than-normal elevations (**Fig. 4.42(b)**).

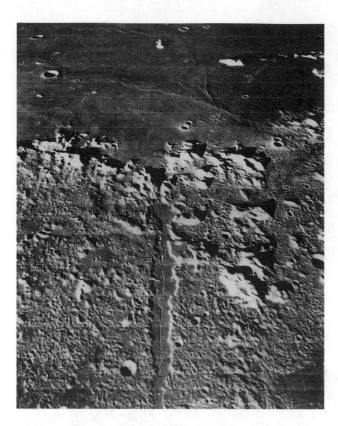

Figure 4.41. The Alpine Valley is a structural graben 150 km long by 10 km wide radial to the Imbrium basin, seen toward the top of the image. The floor of the graben has been flooded with mare deposits emplaced by a rille in the center of the valley (NASA LO IV M-102).

(a)

Figure 4.42. (a) Floor-fractured craters bordering western Oceanus Procellarum; largest crater is ~75 km across (NASA LO IV H-189). (b) Diagrams showing the possible formation and evolution of floor-fractured craters through the intrusion of magma and uplift of the crater floor zone (with permission from Springer Science+Business Media: *The Moon*, Floor-fractured lunar craters, **15**, 1976, 241–273, Schultz, P. H., Fig. 10).

Mare ridges constitute some of the most common tectonic features on the Moon. Also called **"wrinkle ridges"** (an apt name given their appearance; **Fig. 4.43**), these features extend tens of kilometers in length across mare surfaces. Mare ridges typically consist of a broad, gentle arch surmounted by a steeper-sided, narrow crenulated ridge crest. While most of the ridges are found on mare surfaces, many extend into highland terrains (**Fig. 4.44**). As shown in **Fig. 4.45**, most mare ridges fractured the basalts well after the emplacement of the lavas (at least after a crust of sufficient thickness had formed to support the preservation of the crater shown in the figure). However, in some places, mare ridges also exhibit flow lobes suggesting volcanic extrusions; thus, some ridges probably formed on mare surfaces before complete solidification of the lavas. The consensus is that most mare ridges are predominantly structural features that reflect deformation of basaltic rocks.

4.5.7 Gradational features

In the absence of wind and flowing water, gradation on the Moon occurs in the form of "space weathering" (see **Section 3.5**) and downslope mass wasting of debris under the influence of gravity. Mass wasting of several forms is seen in most areas of the Moon. These include landslides (**Fig. 4.44**) and individual rocks that have rolled down slopes, leaving tracks (**Fig. 4.46**).

The physical breakup and fusing of surface materials occurs by impacts at all scales. The formation of soil **agglutinates** is particularly important. This material involves gasses formed by micrometeoroid bombardment, gasses implanted by the solar wind, and subsequent fusing of soil grains. With time, the amount of agglutinates in the regolith increases, leading to soil maturation. Understanding the effects of soil maturity is critical in the interpretation of

(b)

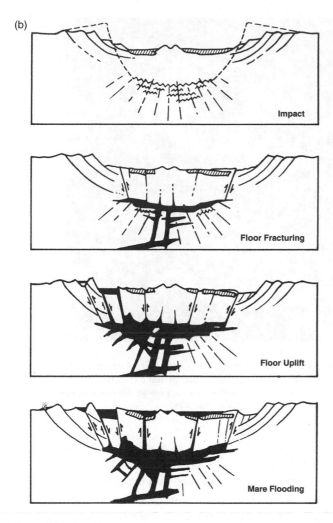

Figure 4.42. (cont.)

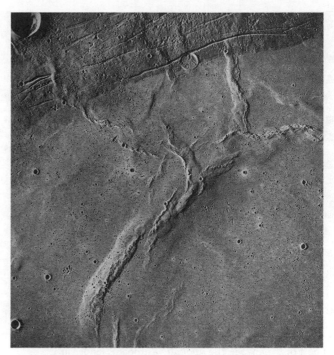

Figure 4.43. The mare ridge (also called "wrinkle ridge") in Mare Serenitatis, showing the typical broad basal component and upper, sharper ridge component; note that the ridge system cuts across both higher-and lower-albedo mare units. Mare ridges are considered to be primarily structural features resulting from tectonic processes, although they might also be accompanied by igneous activity in some cases; the area shown is 150 km by 170 km (NASA AS 17–0451).

remote sensing data for the Moon because it influences the spectral signatures.

Impact-generated debris, or regolith, is pervasive on the lunar surface and increases with time in response to the frequency and size of impacts. Verne Oberbeck and Bill Quaide of NASA studied the morphology of small impact craters that formed on unconsolidated debris overlying more coherent bedrock. As shown in **Fig. 4.47**, they found that normal bowl-shaped craters formed in thick debris layers (through which the transient crater did not penetrate), flat-floored craters formed in intermediate thicknesses of debris, and concentric craters formed in thin debris layers. Analysis enabled derivation of equations relating the geometries of the crater diameters to the thickness of the debris layer. They applied this technique by analyzing the geometry of craters at the *Apollo* landing sites and found that the youngest mare surfaces on the

Moon had regolith thicknesses of about 2 m, while the older surfaces were as thick as 16 m.

It is important to note that the term **megaregolith** is also used. This refers to the part of the lunar crust that was deeply fractured during heavy bombardment in the final stages of Solar System formation and after the lunar crust had solidified. The megaregolith is thought to be many kilometers thick.

4.6 Geologic history of the Moon

After more than four decades of intensive study, a general hypothesis for the origin of the Moon has emerged. Sometimes referred to as the "Big Whack" model, the concept involves the collision of a Mars-size object with the proto-Earth in the final stages of Solar System formation. Both objects had already differentiated to form early mantles and cores. Computer models developed by planetologist Jay Melosh suggest that the Mars-size object collided with the proto-Earth at an oblique angle, ejecting

Figure 4.44. A view of the Taurus–Littrow valley on the eastern margin of Mare Serenitatis, showing a bright avalanche deposit that slid from South Massif across the valley toward North Massif (in the upper right part of the image). The *Apollo 17* astronauts landed in the cluster of craters east (left) of the landslide; using the lunar rover, they traversed to the craters at the base of North Massif, across the valley floor, and across the landslide deposit to the base of South Massif. Note the "wrinkle ridge" that cuts across the mare-filled valley and into the highlands, which is indicative of a tectonic origin; the area shown is about 22 km by 22 km (NASA M-1220).

Figure 4.45. This 6 km in diameter impact crater in Mare Cognitum has been cut by a small mare ridge, showing that a crust had formed on the mare lavas of sufficient strength to preserve the crater at the time of ridge formation (NASA AS 16–5429).

Figure 4.46. The *Apollo 17* astronauts photographed these tracks left by boulders that had rolled down the slope of North Massif in the Taurus–Littrow Valley; the boulder on the right is about 5 m across (NASA AS 17–144–21991).

a mass of superheated silicate-rich gasses derived from the mantles of both objects, but with most of the mass of the Mars-size object being incorporated into the proto-Earth. Most of the cloud of silicates and ejected debris was gravitationally bound to Earth and condensed to form the Moon.

The rapid accretion of material to form the Moon produced sufficient heat to generate a magma ocean, which subsequently cooled to form the anorthositic crust seen today (**Fig. 4.48**). As portrayed by Don Davis and Don Wilhelms of the US Geological Survey, the Moon continued to be bombarded by large objects, the records of which are preserved as impact basins (**Fig. 4.49(a)**). Early stages of volcanism, represented by some KREEP materials, occurred in this early phase of lunar history. However, most volcanism occurred during and following the final stages of basin-forming impacts through partial melting of the lithosphere to generate the mare basalts (**Fig. 4.49(b)**). With time, the sources for mare volcanism deepened and eventually magmas were unable to reach the surface, thus ending volcanic activity by about 2 Ga. Subsequently, most of the geologic activity on the Moon has been reduced to minor impact cratering, modest moonquakes in response to tidal flexing, and surface modifications by gradation (**Fig. 4.49c**).

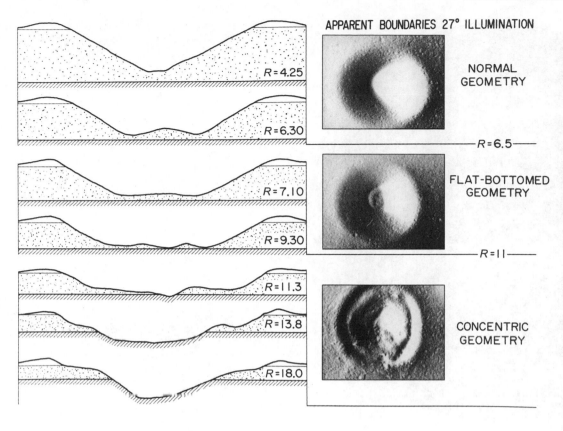

APPARENT BOUNDARIES 27° ILLUMINATION

R = 4.25

R = 6.30

— *R* = 6.5 —

NORMAL GEOMETRY

R = 7.10

R = 9.30

— *R* = 11 —

FLAT-BOTTOMED GEOMETRY

R = 11.3

R = 13.8

R = 18.0

CONCENTRIC GEOMETRY

Figure 4.47. Diagrams showing crater morphology as a function of simulated regolith (stippled pattern) thickness overlying bedrock. Normal craters form when impacts fail to penetrate the thick debris layer; flat-bottomed craters form when impact impinges bedrock but does not penetrate; concentric craters form as impact penetrates through regolith and excavates bedrock. The cross-sections and crater images are from experiments. *R* is the crater diameter/layer thickness. The photographs are representative morphologies with boundary *R* values (from Oberbeck and Quaide, 1967).

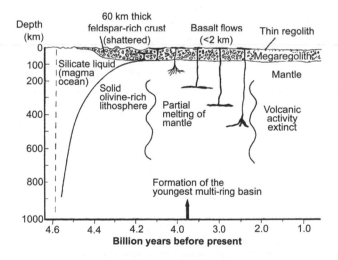

Figure 4.48. Evolution of the lunar crust through time, beginning with the formation of the magma ocean, through the development of the anorthositic crust and the eruption of basaltic lava flows (with permission from Springer Science +Business Media: Stratigraphy and isotope ages of lunar geologic units: chronological standard for the inner Solar System, *Space Science Reviews*, **96**, 2001, 9–54, Stoffler, D. and Ryder, G.).

A generalized geologic time scale for the Moon has been developed. It is based on mapping that began with Earth-based telescopic observations (**Fig. 2.7**), integration of spacecraft data, and incorporation of information from samples, including ages for key geologic rock units (**Fig. 2.10**). As codified by Don Wilhelms in his classic reference *The Geologic History of the Moon*, lunar history is divided into five periods (**Fig. 4.50**). The pre-Nectarian Period is represented by the ancient heavily cratered highlands. No specific formal geologic formations have been mapped for this interval of time. As Wilhelms noted, operational constraints precluded *Apollo* and *Luna* landings in pre-Nectarian sites, and untangling this episode of lunar history remains a challenge, but new insight is currently being gained from the recent lunar missions.

The Nectarian Period is marked by the formation of the Nectaris impact basin and the emplacement of its ejecta blanket, the Janssen Formation. The Nectaris basin is about 860 km in diameter and probably resembled the Orientale basin (**Fig. 4.25**) shortly after its formation. Samples of Nectarian-age materials are thought to be

(a)

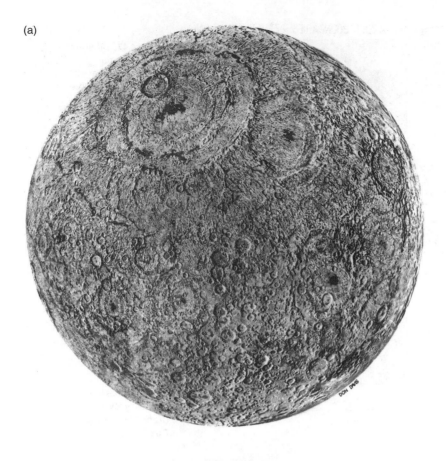

(b)

Figure 4.49. A series of artist's renditions of the lunar near side showing its paleogeology. From (a) the middle of the Imbrian Period after the formation of most of the large impact basins but before extensive eruptions of mare lavas. (b) The lunar near side after emplacement of most mare lavas but before the impacts to form large craters such as Copernicus, Aristarchus, and Tycho. (c) The lunar near side as it appears today. Reprinted from *Icarus*, **15**, Wilhelms, D. E. and Davis, D. E., Two former faces of the Moon, 368–372, Copyright 1971, with permission from Elsevier.

(c)

Figure 4.49. (*cont.*)

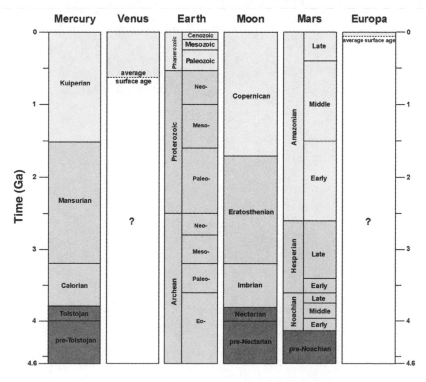

Figure 4.50. Geologic time scales have been derived for the terrestrial planets on the basis of radiogenic dates from samples for Earth and Moon and extrapolations to other planets based on superposed impact crater frequencies. Venus and Jupiter's satellite Europa have very young surfaces, indicative of extensive processes of resurfacing in the geologically recent past.

represented in samples from the *Apollo 17* mission, the *Luna 20* site, and possibly from the *Apollo 16* site. Other large basins formed during the Nectarian Period include Crisium and Humorum. High-aluminum mare lavas and lavas rich in KREEP components were erupted at this time as well.

The Imbrian Period began with the formation of the Imbrium basin at ~3.85 Ga and the emplacement of its ejecta, the Fra Mauro Formation. The Orientale basin, formed during this time, marked the cessation of large impacts on the Moon (and probably the inner Solar System in general). This period of time saw extensive eruptions of mare basalts, flooding the floors of near side basins and embaying heavily cratered highland terrains. Mapping the relative ages of mare materials in accord with this system shows that lunar volcanism began at ~3.9 to 4 Ga to ~1.2 Ga, with most eruptions occurring in the late Imbrian Period.

The Eratosthenian and Copernican Periods were originally based on mapping and relative ages of crater deposits from the impact craters Eratosthenes and Copernicus. Eratosthenian-age craters are "fresh," with sharp crater rims and clearly identified secondary craters, although lacking bright rays, while Copernican-age craters are sufficiently young to still show bright rays (**Fig. 4.8**).

Assignments

1. Compare and contrast the *Apollo* lunar landing sites with those of the former Soviet Union; be specific with regard to the type(s) of terrain (mare, highland, etc.) and the key scientific objectives.

2. Refer to Chapter 3 and the relations of the morphology of volcanic features; consider lunar maria, lunar volcanic domes, and lunar sinuous rilles and infer the key parameters (composition, rate of effusion, etc.) involved in their formation, taking into account the lunar environment.

3. Discuss what might be expected in ejecta deposits found in a lunar mare setting 500 km from the source impact crater in the lunar highlands.

4. Explain the concept behind the discovery of lunar *mascons*.

5. Compare the figures showing the distribution of lunar maria with the diagram for the interior of the Moon and offer an explanation for why maria are concentrated on the near side.

6. Discuss how water-ice can be present on and near the surface of the Moon.

CHAPTER 5

Mercury

5.1 Introduction

As a planetary "wanderer," Mercury probably has been recognized as long as the heavens have been viewed. Although many cultural mythologies refer to this planet, the first recorded mention of Mercury was in 265 B.C. by the Greek Timocharis. With the advent of telescopes and their use in science, careful observations by Giovanni Zupus in 1639 revealed that Mercury goes through phases similar to the Moon. In the 1800s, several well-known planetary observers noted various aspects of Mercury, including supposed surface markings. For example, Giovanni Schiaparelli and Percival Lowell, both of Mars fame, made simple maps of Mercury and named features that they thought could be seen. While most of these features turned out not to exist, some of the names are still used (**Fig. 5.1**).

5.2 Mercury exploration

In some ways, Mercury has been the forgotten planet. Until recently, only NASA's *Mariner 10* spacecraft, flown in the early 1970s, had returned data from this, the closest planet to the Sun. Because of its orbit within the inner Solar System, Mercury is difficult to observe from Earth telescopically, never being more than 28° from the Sun. In fact, many observers are reluctant to train their telescopes in the direction of Mercury for fear that stray light from the Sun would damage the instruments. Nonetheless, some cautious observations were made, which provided key data on the physical properties and astronomical characteristics of the planet. In addition, Earth-based radar observations provided insight into Mercury, including hints of some very large surface features.

By the dawn of the Space Age and before the flight of *Mariner 10*, the major characteristics of Mercury were

known (**Table 1.1**), including its mass, as estimated from Mercury's interactions with Venus and other bodies. In the mid 1960s, radar experts Gordon Pettengill of MIT and Rolf Dyce tracked Mercury using the Arecibo radar facility in Puerto Rico and determined that the planet's spin rate equaled a rotation period of 59 days, a value later refined to 58.6 days. This is exactly two-thirds of Mercury's orbital period, which means that its spin and orbit are locked into a 3:2 resonance. Thus, the planet spins on its axis three times for every two orbits around the Sun. If one were to stand on the surface, sunrise would be repeated every two orbits or 176 days, and the temperature would vary by 600 °C.

Most of our knowledge of the geologic aspects of Mercury comes from the *Mariner 10* and the *MESSENGER* (MErcury Surface Space Environment, GEochemistry, and Ranging) missions. *Mariner 10* was launched in late 1973 and, after a flyby of Venus and a journey of nearly 5 months, flew past Mercury in March 1974, returning the first close-up views of this Sun-baked planet. Although this was not in the original plan, talented engineers at the Jet Propulsion Laboratory, the NASA facility responsible for the mission, analyzed the trajectory of the spacecraft and found that, after looping around the Sun, *Mariner 10* could make a second and then a third flyby of Mercury (**Fig. 5.2**). This resulted in a tremendous increase in data return over the original plan, all nearly for the price of one flyby. *Mariner 10* returned more than 2,700 useful images, covering about 45% the surface of Mercury, as well as other data. It should be noted, however, that the resolution of these images ranges from about 100 m to 4 km per pixel, comparable to that for Earth-based images of our Moon. Consequently, the analysis of Mercury's geomorphology was similar to that conducted for the Moon using Earth-based telescopes prior to the Space Age.

MESSENGER is a Discovery-class mission with Sean Solomon as the Principal Investigator who leads an

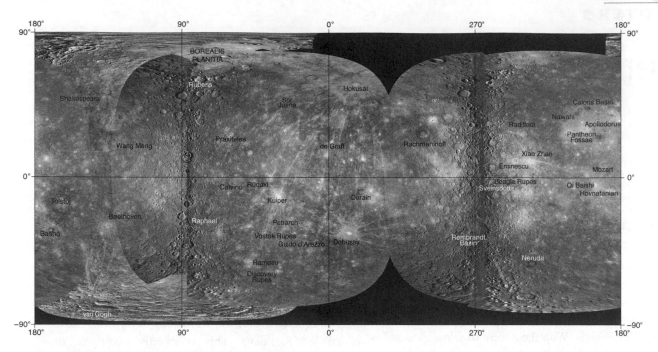

Figure 5.1. A mosaic of *MESSENGER* images for Mercury with some key named features (image from NASA/Johns Hopkins University Applied Physics Laboratory/Carnegie Institution of Washington).

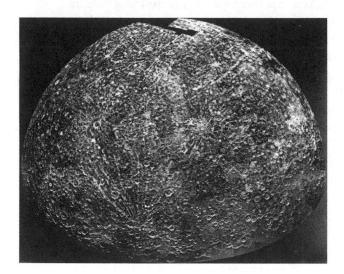

Figure 5.2. A mosaic of *Mariner 10* images of the south polar region of Mercury, obtained during the third flyby.

outstanding team of planetary scientists. The spacecraft was launched in August 2004 (some 33 years since *Mariner 10*) to begin its roughly seven-year journey before it went into orbit around Mercury in March 2011. Along the way, it made three successful flybys of Mercury, returning unparalleled data on the geology and other characteristics of the planet (Solomon *et al.*, 2008). Operated by the Applied Physics Laboratory of Johns Hopkins University, *MESSENGER* carries a sophisticated array of instruments,

including the Mercury Dual Imaging System (MDIS) with a wide-angle camera (WAC) and a narrow-angle camera (NAC), spectrometers, a laser altimeter, and a magnetometer. Thousands of images have been returned (covering ~98% of the surface), including versions in color, and topographic data along with spectra for determining the compositions of the surface and tenuous atmosphere.

5.3 Interior characteristics

The magnetometer experiment on *Mariner 10* detected a weak magnetic field around Mercury, setting off a debate about the properties and configuration of the planet's interior. Because of its high density, Mercury is thought to have a large core that is 65%–70% iron, making it unique in the Solar System (**Fig. 5.3**) and accounting for 60% of the planet's mass. With the discovery of the magnetic field, it was suggested that parts of the interior might be molten today, enabling the same sort of "dynamo" as on Earth to generate the magnetic field. On the other hand, Mercury's slow rotation poses a problem because it might be insufficient to drive a dynamo, and some researchers suggested that *Mariner 10* detected a remnant magnetic field, rather than an active field. If this is the case, it would suggest that Mercury was significantly different in the past (perhaps with a greater spin rate) or

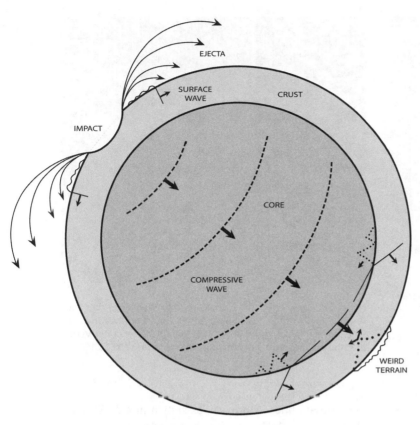

Figure 5.3. Cross-section of Mercury showing its large iron-rich core (shaded) and a diagram of a large impact. Such an impact is thought to have generated seismic energy (both surface waves and body waves through the planet) that was focused in the antipode region, resulting in jostling of the upper crust to form hilly and lineated terrain, or "weird terrain;" although not shown here, Mercury could have a mantle (courtesy of Peter Schultz).

that it reflects some unknown event in its history that might have generated a magnetic field.

In 2007, planetary scientists Jean-Luc Margot, Stan Peale, and their colleagues analyzed Earth-based radar signals bounced off Mercury over a six-year period and found that slight variations in the spin were more than should be expected if Mercury were completely solid. They suggested that the spin variation results from a lithosphere that is decoupled from the interior by a liquid zone, meaning that at least part of the interior is molten today. How could such a small planet as Mercury remain molten since its formation? One suggestion is that Mercury's iron core contains sulfur, which would lower the melting temperature and keep the core molten. Although preliminary *MESSENGER* results are not conclusive, data from the flybys suggest that at least part of the magnetic signature results from an active dynamo, which would be consistent with a molten interior zone.

5.4 Surface composition

Information on the composition of Mercury's surface is derived primarily from *MESSENGER*'s spectrometers and color-imaging data, coupled with *Mariner 10* data and Earth-based telescopic observations in the visible to mid-IR parts of the electromagnetic spectrum (**Fig. 2.14**). In the 1990s, Mark Robinson recalibrated the *Mariner 10* color data and derived new insight into surface compositions; those new insights are being confirmed with *MESSENGER* data. For example, some local heterogeneities in color and albedo can be correlated with the geology, including inferred volcanic materials described below.

Mercury's surface appears to be compositionally heterogeneous, with a wide variety of silicate materials and rocks such as low-iron basalt, as reviewed in the book *Exploring Mercury: The Iron Planet* by Robert Strom and Ann Sprague (2003). *MESSENGER* data at global scales confirm this conclusion. In contrast with the lunar highlands, Mercury's crust does not appear to be dominated by anorthosite but instead seems to have olivine and low-iron pyroxene. Thus, in comparison with the iron-rich lavas of the other terrestrial planets, the FeO content of the mercurian surface is lower by factors of 4–12. However, albedo and soil maturity related to space weathering influence the spectral signatures of the regolith on the Moon, and the same considerations must be taken into account for remote sensing data for Mercury.

Figure 5.4. An oblique image taken by *Mariner 10*, showing heavily cratered terrain and patches of more sparsely impacted intercrater plains (NASA *Mariner 10* FDS 27328).

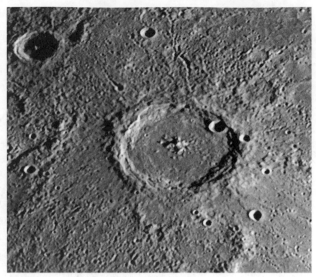

Figure 5.5. Large craters on Mercury tend to have lower rims and shallower floors than do comparable-size craters on the Moon, as seen in this oblique image of the 160 km in diameter crater Verdi in the Shakespeare region of Mercury (NASA *Mariner 10* FDS 166).

5.5 Geomorphology

At first glance, Mercury can be confused with Earth's Moon. On a global scale, it is cratered with patches of plains, and, as a small, airless body, it appears to have experienced similar surface processes to the Moon. Upon closer inspection, however, Mercury is seen to have its own unique surface characteristics at all scales.

5.5.1 General physiography

The principal terrains on Mercury were defined by the *Mariner 10* imaging team as the *heavily cratered highlands, intercrater plains* (Fig. 5.4), and *smooth plains*. While these superficially resemble the lunar highlands and mare regions, the team recognized that there are fundamental differences between the lunar and mercurian terrains.

The heavily cratered terrain on Mercury consists of abundant large, overlapping impact craters representing the final stages of intense bombardment in the inner Solar System as are also seen on the Moon (**Fig. 4.4**). Although initial *Mariner 10* analysis suggested that there are fewer large craters per unit area than on the Moon, subsequent study shows this not to be the case.

Individual ejecta deposits and secondary craters are rarely seen around the older mercurian craters in the heavily cratered terrain. This could be due to gravitational effects, mantling by younger deposits, or a combination of

these possibilities. Moreover, as shown in *MESSENGER* altimeter data, the large craters (**Fig. 5.5**) have lower rims and shallower floors in comparison with craters of similar size on the Moon. This can be explained in part by the higher surface gravity of Mercury, which would inhibit rim uplift during the excavation stage and result in lower topographic relief. In addition, the floors of many of the large craters are filled with smooth plains (described below), which also contribute to the lower depth-to-diameter ratio.

Intercrater plains occur as irregular-shaped patches of level-to-gently rolling terrain within the heavily cratered regions (**Fig. 5.6**). Unlike many of the lunar maria, most mercurian intercrater plains are not found in circular patches associated with obvious large impact basins. What are the intercrater plains and how did they form? These questions have puzzled planetary geologists for more than three decades. Early ideas suggested that the intercrater plains represent a primordial surface preserved since the solidification of the crust or that they are comparable to the highland plains seen on the Moon, which are attributed to ejecta deposits from large impact basins. However, specific large impacts have not been identified as sources on Mercury. Plains are now recognized to span a wide range of ages. Some are relatively ancient, which is evidenced by large superposed impact craters, while embayment by intercrater plains into heavily cratered

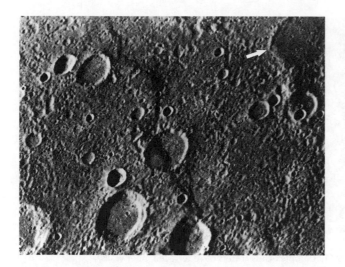

Figure 5.6. A *Mariner 10* image of typical intercrater plains and their embayment of a 90 km impact crater (arrow), showing that some intercrater plains post-date the heavily cratered terrain. The lobate scarp, Santa Maria Rupes, is in the middle of the image and cross-cuts the terrain, indicating that this tectonic deformation post-dated the emplacement of intercrater plains. The area shown is about 400 km across (from Strom, 1979; NASA *Mariner 10* FDS 27448).

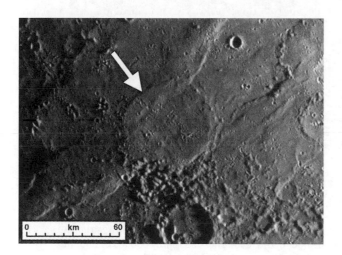

Figure 5.7. A circular ridge, or "ghost crater," similar to those seen on the Moon and Mars that are considered to represent impact craters buried by lava flows. In this model, Head *et al.* (2008) estimate that flows of ~2.7 km thickness would be required in order to form the feature shown here on Mercury (NASA *MESSENGER* image EN108827047M).

terrain shows that some plains are relatively young. "Ghost" craters, similar to those seen on the Moon, suggest that some intercrater plains represent a relatively thin mantle that buries craters (**Fig. 5.7**), perhaps analogous to lunar volcanic flood lavas. Thus, the intercrater plains are likely to be of volcanic origin.

Figure 5.8. An oblique view of smooth plains showing the characteristic ridges (similar to mare ridges on the Moon) and the relative lack of superposed impact craters; the area shown is about 600 km wide (NASA *MESSENGER* NAC 162744209).

Smooth plains constitute about 40% of the surface of Mercury. They are generally flat or gently rolling and are relatively sparsely cratered, making them the youngest of the principal physiographic units. Smooth plains are found as irregular-shaped patches within the heavily cratered terrain and intercrater plains, as well as within impact craters and basins. Most smooth plains are characterized by ridges that resemble lunar mare ridges (**Fig. 5.8**). Embayment by smooth plains into heavily cratered terrains suggests that they are of volcanic origin (**Fig. 5.9**) and are comparable in size to flood lavas on Earth and the Moon. Alternatively, some investigators suggest that the Mercury smooth plains are ejecta deposits similar to the Cayley Plains on the Moon; however, as with the intercrater plains, there are no obvious primary source craters and the smooth plains are also thought to be volcanic.

5.5.2 Impact craters

Impact craters on Mercury range in size from the 1,560 km in diameter Borealis basin down to the limit of recognition on *MESSENGER* images of tens of meters. No doubt even smaller craters exist in the airless environment of Mercury. As on the Moon, mercurian craters increase in complexity with size, from simple

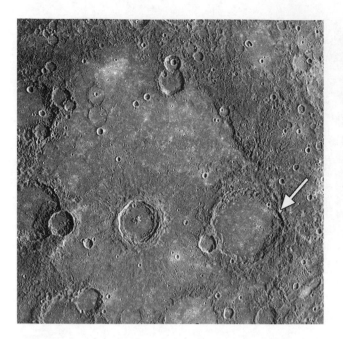

Figure 5.9. Embayment by smooth plains into the cratered terrain to the right and partial flooding of the 120 km in diameter Rudaki crater (indicated by the arrow) suggest a volcanic origin for the plains; crater Calvino (see **Fig. 5.11**) is to the left of Rudaki (NASA PIA 11400).

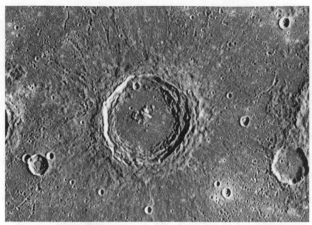

Figure 5.11. Complex-class craters have terraced inner walls, scalloped rim crests, central peaks, and hummocky floors, as seen in the ~85 km in diameter Calvino crater in the Renoir region of Mercury (see **Fig. 5.9**) (NASA PIA 11400).

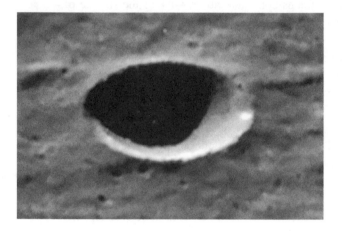

Figure 5.10. An oblique view of a simple-class crater on Mercury (~10 km in diameter), showing the typical bowl-shape and relatively smooth rim crest (NASA *Mariner 10* FDS 27475).

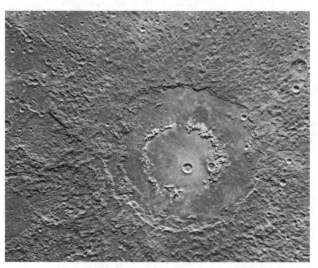

Figure 5.12. An image of Raditlandi, a double-ring basin ~260 km in diameter; the floor of this feature has been partly flooded by smooth plains of possible volcanic origin (NASA PIA 10378).

craters to complex craters, double-ring basins, and multiring basins. However, the transitions from one type to the next tend to occur at smaller sizes than on the Moon.

Simple craters are bowl-shaped with smooth, circular rim crests (**Fig. 5.10**). They are typically smaller than ~10 km in diameter and have depth-to-diameter ratios of less than 1 to 5 (**Fig. 3.31**). Complex craters are characterized by scalloped rims, terraced walls (representing slumps), hummocky floors, and central peaks (**Fig. 5.11**). They range in size from about 10 to ~110 km in diameter and have lower depth-to-diameter ratios than simple craters, becoming flatter with increasing diameter. Double-ring basins range in size from ~110 to ~400 km in diameter. They consist of an outer crater rim and an inner ring, which somewhat resembles the peak-ring craters on the Moon, but the inner ring is much better defined on Mercury (**Fig. 5.12**). The onset of multi-ring (three or more rings) basins (**Fig. 5.13**) occurs at diameters of ~400 km.

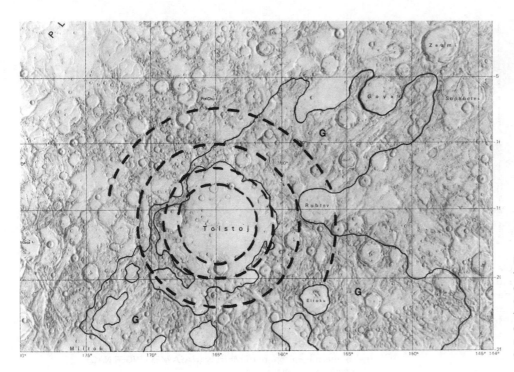

Figure 5.13. A map showing the Tolstoj multi-ring basin, the distribution of its ejecta (the Goya Formation, G), and the suggested ring locations determined by geologic mapping (from Spudis and Guest, 1988).

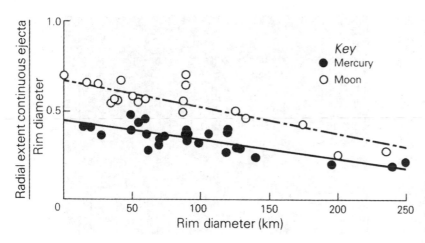

Figure 5.14. Average radial extent of continuous ejecta deposits for fresh impact craters on Mercury and the Moon (after Gault *et al.*, 1975, copyright American Geophysical Union).

Analysis of the ejecta around mercurian impact craters shows that both the continuous and the discontinuous ejecta are deposited closer to the crater rim than is seen on the Moon. The continuous ejecta typically occurs within 0.5 crater diameter, whereas on the Moon the rule of thumb is deposition within 0.7–1 crater diameter (**Fig. 5.14**). Similarly, secondary craters and crater chains occur closer to the primary crater rim (**Figs. 5.15** and **5.16**), with some occurring within the continuous ejecta deposits. These differences are at least partly the result of the higher-gravity environment on Mercury (**Fig. 3.28**). For example, because the surface gravity on Mercury is

370 cm/s^2, or more than twice that of the Moon's 162 cm/s^2, a block of ejecta might travel only half the distance from the primary crater on Mercury.

On the other hand, the bright rays from some craters on Mercury extend farther than those on the Moon (**Fig. 5.17**). For example, one crater ray imaged by *MESSENGER* is >4,500 km long, compared with the longest ray from the lunar crater Tycho, which is ~2,000 km long. This suggests that the mercurian crater rays are not as degraded as Tycho and might be younger, or that "space weathering" effects (which obliterate albedo features) are less efficient on Mercury, or that there is some unknown

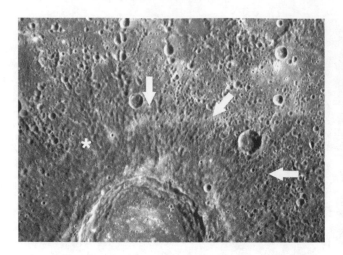

Figure 5.15. This ~95 km in diameter complex crater shows that the continuous ejecta deposits (outer boundary marked with arrows) fall within a distance about half the crater diameter and that some secondary crater chains (asterisk) occur within the ejecta deposits (NASA *MESSENGER* images 108826040 and 108826045).

Figure 5.17. The bright-rayed 80 km in diameter crater Debussy in the southern hemisphere of Mercury (NASA *MESSENGER* NAC 131773947, NASA PIA 11371).

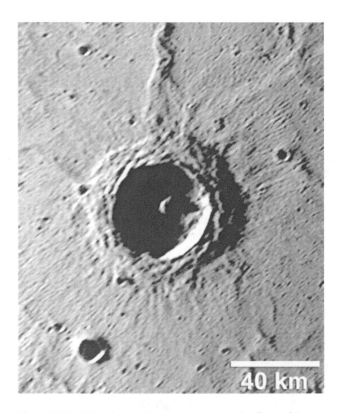

Figure 5.16. This 44 km in diameter crater is on the floor of the Rembrandt basin; the very low Sun-angle illumination emphasizes the continuous ejecta and its proximity to the rim (NASA *MESSENGER* NAC 131766401M).

Figure 5.18. The asymmetric bright-ray pattern around the 15 km in diameter Qi Baishi crater (left side) suggests a low-angle impact by an object traveling from west to east (left to right); the "butterfly" ejecta pattern around the 13 km in diameter Hovnatanian crater (lower right) also suggests an oblique impact, but it is not known whether the object was traveling from north to south, or from south to north (NASA *MESSENGER* image, NASA PIA 12039).

process involved in the formation and/or preservation of bright rays in general.

Statistically, most impacts occur at angles that are not normal (90°) with respect to planetary surfaces. However, as discussed in **Section 3.4.3**, differences in crater morphology and ejecta deposits are seen only when the angle of impact is

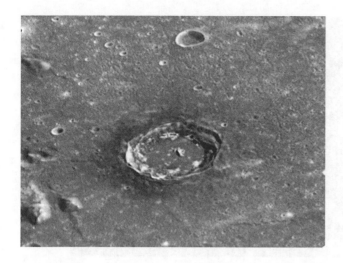

Figure 5.19. An oblique view of Nawahi crater (34 km in diameter) on the floor of the Caloris basin. Its dark ejecta deposit suggests an impact through a thin mantle of relatively bright materials to excavate darker materials from the subsurface (NASA *MESSENGER* NAC 108826682).

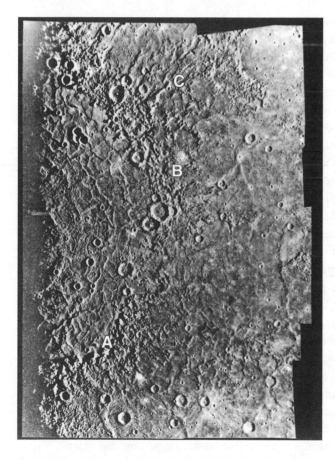

Figure 5.20. A mosaic of *Mariner 10* images showing the eastern part of the Caloris basin with its prominent ring named the Caloris Montes (A), the Odin Formation. (B), the Van Eyck Formation. (C), and the plains that fill the basin floor to the left (NASA Jet Propulsion Laboratory).

lower than about 15°. **Figure 5.18** shows two such impacts on Mercury, craters Qi Baishi and Hovnatanian, both of which have asymmetric ejecta patterns.

In many areas, some impact craters display ejecta deposits that are distinctly darker than the surrounding plains, suggesting penetration through a bright mantle to excavate darker underlying materials (**Fig. 5.19**). Like dark-halo craters on the Moon, these craters on Mercury might reflect mafic volcanic materials thinly mantled with brighter material, comparable to lunar cryptomaria.

5.5.3 Multi-ring basins

Multi-ring basins are common on Mercury and most pre-date the intercrater plains deposits. Two of the best preserved multi-ring basins are the Caloris and Rembrandt impact structures. The Caloris basin was discovered and described from *Mariner 10* images, but with less than half of it being seen (**Fig. 5.20**). Caloris was fully imaged by *MESSENGER* (Murchie *et al.*, 2008; **Fig. 5.21**), and was found to have a main ring zone 1,550 km in diameter by as wide as 250 km. This ring zone includes massifs that stand 1–2 km above the surrounding plains. The massifs are considered to be blocks of crust uplifted by the impact and are defined as the Caloris Montes Formation (**Fig. 5.22**). Plains units are found among the massifs and are thought to be fall-back ejecta, and perhaps melt deposits, from the impact that formed the basin. Radially

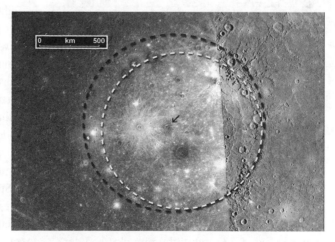

Figure 5.21 A mosaic of the Caloris basin using *Mariner 10* images on the east (right) side that were taken under low Sun-angle illumination and *MESSENGER* images (left) under near full-Sun illumination. The inner dashed circle shows the basin diameter inferred from the *Mariner 10* data, while the larger dashed circle shows the main ring diameter of 1,550 km derived from the more complete data. The arrow marks "the spider," officially named Pantheon Fossae, shown in **Fig. 5.25** (NASA *MESSENGER* images).

Figure 5.22. A *Mariner 10* view of the Caloris Montes, the main ring of the basin; area shown is about 730 km across (NASA *Mariner 10* FDS 229).

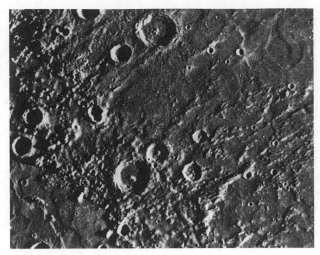

Figure 5.23. A view of the northeast side of the Caloris basin, showing the radial lineated terrain of the Van Eyck Formation, one of the ejecta facies of Caloris; the area shown is about 590 km across (NASA *Mariner 10* FDS 193).

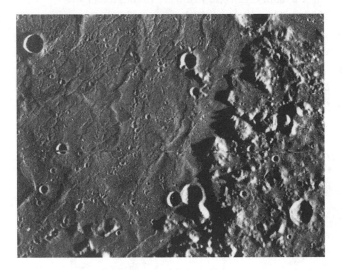

Figure 5.24. Smooth plains fill the floor of the Caloris basin (left half of image) and embay the Caloris Montes (right side). The plains show ridges, troughs, and fractures reflecting deformation of the inferred volcanic deposits; the area shown is 275 km across (NASA *Mariner 10* FDS 110).

textured terrain to the northeast of the basin marks the Van Eyck Formation (**Fig. 5.23**), while hummocky and dark knobby terrain to the east comprises the Odin Formation; both units are considered to be facies of the Caloris ejecta deposits. *MESSENGER* data show that the rim deposits, intermontane materials, and ejecta facies all have the same spectral properties, suggesting that they have the same composition.

The interior of the Caloris basin is floored with plains materials that commonly have ridges, troughs, and fractures (**Fig. 5.24**), similar in morphology to the smooth plains found elsewhere on Mercury. Unlike lunar mare deposits inside basins on the Moon, the Caloris plains are high in albedo and appear to lack iron oxide. The ridges are 50–300 km long and 1–12 km wide, by 100–500 m high, and form roughly polygonal patterns. Many of the troughs have flat floors, suggestive of grabens, and tend to be either concentric with the Caloris rim or radial to the center of the basin. The origin of the basin floor plains is somewhat controversial, although most workers agree that the material is not associated with the Caloris impact because the plains are much younger, as indicated by low superposed impact crater frequencies. It is likely that the plains represent flooding by volcanic materials that deformed upon cooling to form the ridges and fractures.

Figure 5.25 shows Pantheon Fossae, informally termed "the Spider," because of the radial pattern of troughs (i.e., *fossae*). Superposed on Pantheon Fossae is the 40 km in diameter impact crater Apollodorus, indicating that the fossae pre-date the impact. The troughs are considered to be grabens and about 230 have been mapped that converge near the center of the basin, suggesting that extension of the basin-fill took place, perhaps as a consequence of magma intrusion.

Analysis of images from *Mariner 10* led to the discovery of an unusual terrain in the region **antipodal** (on the

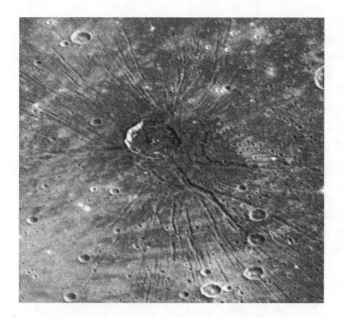

Figure 5.25. An oblique view of Pantheon Fossae, defined by the set of troughs that converge near the center of Caloris basin. The troughs are probably grabens, which would reflect extension of the smooth plains on the floor of the basin. Ejecta from the 41 km in diameter impact crater, Apollodorus, is superposed on the troughs and is dark, suggesting that it has excavated a material different from that of the smooth plains (NASA PIA 10635).

Figure 5.26. Hilly and lineated ("weird") terrain imaged by *Mariner 10* at the antipode to the Caloris impact basin. Crater rims are broken into massifs, and the intervening terrain consists of hills and valleys thought to have resulted from focusing of seismic energy. The smooth plains that fill the 150 km Petrarch crater (left side) have not been disrupted, indicating their emplacement after the Caloris impact (NASA *Mariner 10* FDS 27370).

opposite side) to the Caloris basin. Informally termed "weird terrain" by the science team, and now known as hilly and lineated terrain, this landscape covers some 250,000 km^2 and consists of various hills and depressions, along with crater rims that have been broken into large blocks (**Fig. 5.26**). Pete Schultz and Don Gault (1975) estimated the energy that was generated from the Caloris impact and noted that seismic surface and body waves would travel around and through Mercury to focus at the antipode (**Fig. 5.3**). In this region, the crust would have been highly disrupted, with the surface lifted tens of meters vertically, breaking up the crust and leading to the jostled appearance of the terrain.

The Rembrandt basin was discovered during the second flyby of Mercury by *MESSENGER* (**Fig. 5.27**). As documented by Smithsonian planetary geologist Tom Waters and his *MESSENGER* science team colleagues (Waters *et al.*, 2009), this 715 km in diameter impact structure is about the same age as the Caloris basin but shows distinctive features indicative of a complex tectonic history. As with Caloris, the main ring is identified by an inward-facing scarp and massifs that are more than 1 km high. Hummocky and radially textured ejecta are seen beyond the ring, especially to

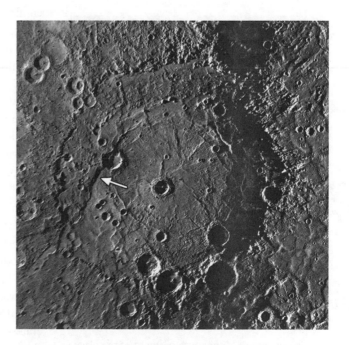

Figure 5.27. A mosaic of *MESSENGER* images for Rembrandt basin, a 715 km impact structure comparable in age to the Caloris basin. Note the lobate scarp (arrow) that cuts across the basin (NASA *MESSENGER* mosaic).

the north and northeast, where illumination favors their detection on the available images.

The interior of the Rembrandt basin is covered with smooth plains of inferred volcanic origin, although they might also be impact melt. However, multispectral data from *MESSENGER* show that the plains do not have the same properties as the rim and ejecta, suggesting a different composition, and more closely matching other plains units on Mercury that are considered to be volcanic. Regardless of origin, Waters and colleagues estimate that the amount of floor-fill must be at least 2 km thick, from the observation that ejecta from a crater 44 km in diameter on the basin floor (**Fig. 5.16**) has the same spectral properties as the plains, which are different from those of the basin rim and basin ejecta deposits; thus, the crater did not excavate through the plains. The estimation of a minimum thickness of 2 km is based on the depth of excavation resulting from a 44 km crater.

The interior plains of Rembrandt display a well-developed pattern of tectonic deformation in the form of concentric ridges and radial troughs (inferred to be grabens) and ridges (**Fig. 5.28**). As with the Caloris basin, these features probably formed as the volcanic plains materials cooled. Superposition of smooth, lightly

cratered plains over some of these features near the center of the basin suggests that plains emplacement and tectonic deformation were interspersed. The basin was later deformed by global-scale tectonics, as is evidenced by the lobate scarp that cuts across the northwest part of Rembrandt (**Fig. 5.27**) and is more than 1,000 km long.

5.5.4 Volcanic features

The low resolution of the *Mariner 10* images precluded the definitive identification of volcanism on Mercury. The smooth plains were inferred to be volcanic, not on the basis of distinctive features such as flows and vents, but more by default because there seemed to be no other mechanism to explain their relative youth. For example, while impact melt and ejecta deposits can form smooth deposits, the smooth plains in the Caloris basin are much younger than the impact that formed the basin, and there are no young, large impact structures that could account for their emplacement as ejecta.

Hints of volcanic origins for some smooth plains came from carefully processed *Mariner 10* color data, which show that some smooth plains embay older terrains and have different spectral properties. These results are now confirmed with *MESSENGER* multispectral data and images that clearly show numerous morphologic features indicative of volcanism, including vents and lobate lava flows. For example, **Fig. 5.29** shows a feature analyzed by Jim Head of Brown University and colleagues (Head *et al.*, 2008) in the Caloris basin. It includes a dome or shield-shaped structure about 80 km by 60 km that includes irregular-shaped depressions surrounded by bright material with diffuse margins. This feature is interpreted as a volcano with multiple vents, some of which erupted bright pyroclastic materials. This is only one of several similar structures imaged in the southern part of the Caloris basin. Similar features are seen in other areas of Mercury, including pits that could represent collapse over magma chambers and explosive vents (**Fig. 5.30**).

Intrusive activity also has probably taken place in Mercury's crust, as possibly reflected in floor-fractured craters similar to those seen on the Moon. Some of the ridge structures, described below, could be the surface expression of dikes, as might be the troughs of Pantheon Fossae (**Fig. 5.25**).

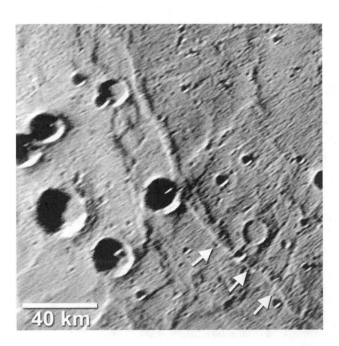

Figure 5.28. An image mosaic of the southwest part of the plains in Rembrandt crater showing the radial troughs (arrows) that cut across concentric ridges; the area shown is 160 km across (NASA *MESSENGER* NAC 131766401M and 131766417M).

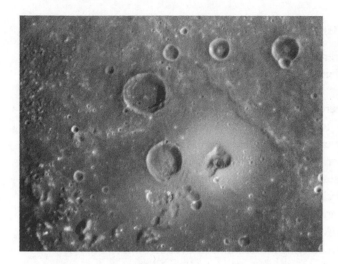

Figure 5.29. A volcano and vent complex in the Caloris basin, as mapped by Head *et al.* (2008), showing a dome or shield structure and various irregular-shaped depressions (vents). The bright halo is thought to be a deposit of pyroclastic materials (NASA *MESSENGER* EN0108826812M and EN108822877M).

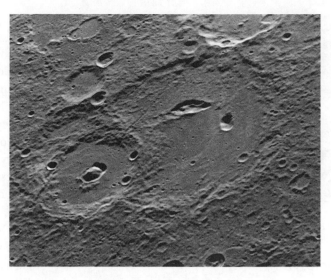

Figure 5.30. An oblique view of irregularly shaped pits on the floors of two craters, suggested to represent collapse over magma chambers; the area shown is about 200 km across (NASA *MESSENGER* NAC 162744290, NASA PIA 12284).

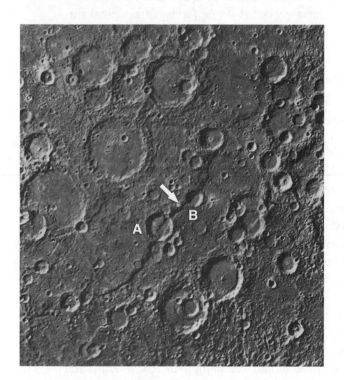

Figure 5.31. Discovery Rupes is more than 500 km long and cuts across the terrain including craters Rameau (A, 55 km in diameter) and crater B; the arrow points to a small dome that could be a volcanic construct (NASA *Mariner 10*, from Davies *et al.*, 1976).

Although conclusions await full analysis of *MESSENGER* data, initial results suggest that the style of mercurian volcanism was similar to that on the Moon in that it involved massive eruptions of flood lavas (but with compositions low in iron and titanium), accompanied by local pyroclastic activity. Volcanism on Mercury is set apart from volcanism on Earth, Venus, and Mars by the apparent absence of large shield volcanoes.

5.5.5 Tectonic features

Mercury's crust has been subjected to tectonic deformation that has led to the formation of a variety of surface features, including scarps and ridges of various shapes, including arcuate, lobate, and linear forms. Mapping these features and assessing their ages in relation to the terrain that they cut can provide clues as to the style and timing of tectonic processes in different regions of Mercury.

Lobate scarps form tongue-like margins, range in length from 100 to >1,000 km, and can be as high as 2 km. Although some lobate scarps superficially resemble lava flows, most investigators consider the scarps to be faults, perhaps controlled by earlier tectonic patterns. Other scarps are arcuate in planform or consist of short linear segments. **Figure 5.31** shows Discovery Rupes, a typical mercurian scarp that cuts across large craters. Scarps, such as Vostock Rupes, can shorten the circumference of the craters they cut across, suggesting thrust-fault movement (**Fig. 5.32**).

In general, the various types of scarps are thought to represent compressional deformation that occurred as Mercury cooled and shrank. Careful mapping and independent analytical modeling resulted in the same

Figure 5.32. Vostock Rupes (arrow) cuts across the 65 km in diameter crater Guido D'Arezzo, foreshortening the rim and suggesting 5–7 km thrust-fault displacement resulting from compression (NASA *Mariner 10* FDS 27380).

conclusion that cooling and shrinkage of Mercury's massive iron core led to a 1–2 km decrease in planetary radius and the subsequent foreshortening of the crust. Alternatively, it has been suggested that tidal "despinning" could have generated compressional stresses to form the scarps. In this case, compressional stresses would have been east–west, with northwest–northeast linear shear and north–south thrusting in the crust.

Ridges occur in many regions and geologic settings on Mercury. They are positive relief features (in contrast with the "stepped" scarps) that are as long as 400 km, with heights that can exceed 700 m and widths of up to 35 km, and can closely resemble mare ridges on the Moon (**Fig. 5.8**). Although some ridges are found in intercrater plains, most occur in smooth plains deposits and within the floor-filling materials of impact basins. Like scarps, ridges also are considered to represent tectonic compression, although perhaps on a more local scale. For example, ridges within Caloris and other basins could reflect settling of lava or impact melt toward the center of the basin, leading to concentric ridge patterns. Ridges not associated with basin interiors could have formed in response to the proposed global shrinkage of Mercury.

Many investigators have mapped the distribution of the types and ages of the various scarps and ridges using *Mariner 10* data in attempts to place constraints on the interior evolution of Mercury. These studies, however, have been hampered by the lack of a global data set and the low resolution of the data. No doubt, these studies will be revisited using *MESSENGER* global data and the ability to derive better age relations among the structures and the associated units.

5.5.6 Gradation features

An extremely thin "cloud" of sodium, oxygen, potassium, and calcium is found above Mercury's surface, constituting an "atmosphere" (perhaps more properly called an exosphere) around the planet. However, it is so thin that in terms of geologic processes it is insignificant. Moreover, there is no evidence to suggest that Mercury has ever had an atmosphere that could support surface processes related to wind or liquid water. Consequently, Mercury's surface is the result primarily of impact, volcanic, and tectonic processes, much like the surface of Earth's Moon. However, some gradation does occur by space weathering as described in **Section 3.5.1**, a process that is ubiquitous on airless bodies. Moreover, images show features resembling landslides in some areas, such as on the walls of some impact craters.

Impact cratering at all scales leads to gradation, especially in the airless environment of Mercury. As on the Moon, impact-generated debris forms a fragmental surface layer of regolith.

5.6 Geologic history

Geologic mapping of Mercury led to the establishment of a formal five-fold time–stratigraphic sequence (**Table 5.1**). From oldest to youngest, the sequence consists of the Pre-Tolstojan System, Tolstojan System, Calorian System, Mansurian System, and Kuiperian System, with names derived from the type localities for each system.

Mercury, along with its sibling terrestrial planets, formed by accretion of smaller bodies, leading to global heating, the likely formation of a magma ocean, and subsequent differentiation into a core, crust, and possible mantle. The accretion of a high percentage of iron led to Mercury's very large core. As on the Moon, the earliest geologic record on Mercury (pre-Tolstojan) consists of the heavy cratered terrain, reflecting sufficient solidification of the crust over the inferred magma ocean to record the heavy bombardment. Some intercrater plains were probably emplaced at this time. Cooling and solidification of Mercury could have established some of the earliest tectonic patterns of the evolving crust. The Tolstoj impact basin marks the base of the Tolstojan System. Throughout this system, the formation of numerous large craters and basins, such as Beethoven, reflects the waning stages of heavy bombardment in the inner Solar System.

Table 5.1. Time–stratigraphic sequence for Mercury (after *Spudis and Guest, 1988*)

System	Major units[a]	Age of base of system[a]	Lunar counterpart[b]
Kuiperian	Crater materials	1.0 Ga	Copernican
Mansurian	Crater materials, smooth plains	3.0–3.5 Ga	Eratosthenian
Calorian	Caloris Group; plains, crater, small-basin materials	3.9 Ga	Imbrian
Tolstojan	Goya Formation; crater, small-basin, plains materials	3.9–4.0 Ga	Nectarian
Pre-Tolstojan	Intercrater plains, multi-ring basin, crater materials	pre-4.0 Ga	Pre-Nectarian

[a] Approximate ages based on the assumption of a lunar-type impact flux history on Mercury.
[b] Included for reference only; no implication of exact time correlation is intended.

Ejecta from the Caloris impact basin (the Van Eyck and Odin formations) serves as the primary stratigraphic horizon for Mercury's geologic history (similarly to the Fra Mauro Formation on the Moon) and the start of the Calorian Period. The hilly and lineated terrain in the region antipodal to the Caloris basin is considered to have resulted from focusing of seismic energy from the impact. The Caloris and Rembrandt basins reflect the termination of impacts by large objects. Continued cooling and contraction of Mercury generated additional compressional features, such as the scarps (*rupes*) and mare-type ridges.

Fresh craters lacking bright rays and volcanic floods of lava that formed many of the smooth plains on Mercury record the Mansurian Period, the start of which is defined by the impact of crater Mansur. The paucity of superposed impact craters on the smooth plains indicates the relative youth of Mansurian-age materials and is roughly analogous to the lunar Eratosthenian System in its characteristics. The youngest geologic units are keyed to young bright-rayed craters, for which Kuiper is the type example, and mark the initiation of the Kuiper Period. This period is modeled on the lunar Copernicus Period.

This sketch of Mercury's geologic history no doubt will be refined with the analysis of *MESSENGER* data in the coming years. It is likely that many of the units and terrains identified on *Mariner 10* data will be seen to be more complex and subject to subdivision. Not only will units be better mapped and characterized using higher-resolution images and compositional data, but also the ability to obtain precise altimetry and crater counts will provide a better understanding of the age relations among the units.

Assignments

1. Compare and contrast the mercurian smooth plains with lunar maria.

2. Discuss the missions that have returned data from Mercury (flyby, orbiter, lander, etc., dates of operation, scientific payload, and principal scientific results).

3. Describe the differences in morphology between *rupes* on Mercury, *ridges* on Mercury, and "*wrinkle ridges*" on the Moon.

4. Explain how a global geologic map could be derived for Mercury with currently available data; include how units could be defined and placed in a stratigraphic sequence.

5. Discuss the evidence for the hypothesis that Mercury "shrank" in size during its evolution and explain how such shrinkage could have occurred.

6. Mercury and the Moon are both "airless" bodies, yet impact craters of the same diameter have different morphologies. Describe these differences and offer an explanation for the differences.

CHAPTER 6

Venus

6.1 Introduction

After the Sun and the Moon, Venus is the brightest object in the sky, a consequence of sunlight being reflected from its dense clouds. The planet's diameter, mass, and gravity are nearly the same as those of Earth **(Table 1.1)**. Along with the presence of an atmosphere, these characteristics led some observers to refer to Venus as Earth's sister planet. Even as late as the 1960s, some serious researchers thought that the surface of Venus was a wet, tropical environment, possibly teaming with life.

With the dawn of the Space Age, Venus was revealed to be substantially different from Earth. The surface temperature is a hellish 480 °C and exceeds the melting point of lead, while the dense carbon dioxide atmosphere is laced with droplets of sulfuric acid that form dense clouds and exerts a surface pressure of 95 bars, comparable to being underwater on the sea floor of Earth at a depth of 900 m. This leads some wags to refer to Venus as Earth's *evil* sister. On the other hand, the geomorphology of Venus displays features indicative of extensive tectonic and volcanic processes **(Fig. 6.1)**, some of which are similar to those on Earth and could be active today, and a surface age of no older than about 750 Ma, much like most of the surface of Earth.

6.2 Venus exploration

In the early 1600s, Galileo trained his primitive telescope on Venus and noted that the planet has lunar-like phases, lending further credence to the Copernican Sun-centered model of the Solar System. In the 1880s, the Russian astronomer Lomonosov noted that Venus exhibits a gray halo when viewed against the Sun, and he correctly inferred that Venus has an atmosphere. Improvements in telescopes and the use of spectroscopy in the middle of the twentieth century demonstrated that the atmosphere is predominantly carbon dioxide with small amounts of water, but that the clouds were unlikely to be composed of water droplets or ice crystals.

Earth-based radar observations enabled estimates of the extremely high surface temperatures that were later confirmed by spacecraft. This led to the formulation of the **runaway greenhouse model** by Carl Sagan and Jim Pollack. In this model, solar energy penetrates the clouds and is reflected from the surface but cannot escape back to space, which is similar to the processes that keep greenhouses on Earth warm. Radar data also enabled refinement of knowledge of the orbital characteristics of Venus, showing that the planet rotates extremely slowly on its axis; one day on Venus is equal to more than 116 Earth days. Moreover, the direction of rotation is the opposite of its orbit around the Sun, meaning that Venus is in retrograde rotation.

Because of its close proximity to Earth, a great many missions have flown either to Venus or past the planet on the way to other destinations **(Table 6.1)**. Because visible imaging of the surface is precluded by clouds, most of these missions focused on atmospheric sciences rather than geoscience. Some of the earliest missions of direct geologic interest were the successful Soviet *Venera* landers. Although the landers operated for less than two hours each, these were remarkable engineering successes, given the extremely hostile surface environment. For example, in the 1970s *Veneras 8* and *9* were soft landers and returned the first pictures directly from the surface **(Fig. 6.2)**, along with data on rock compositions. So little was known about the surface conditions that lights were carried on the landers to illuminate the terrain for taking pictures because of the possibility that little sunlight would reach the surface. As subsequent results showed, the lights were unnecessary, but it was a wise precaution nonetheless.

Table 6.1. Selected missions to Venus (all by the USSR except where indicated otherwise)

Spacecraft	Arrival date	Type of vehicle
Venera 1	May 1961	Flyby
Mariner 2[a]	Dec. 1962	Flyby
Venera 4	Oct. 1967	Descent vehicle and flyby
Venera 5	May 1969	Descent vehicle and flyby
Venera 8	Jul. 1972	Soft lander and flyby
Mariner 10[a]	Feb. 1974	Flyby
Venera 9	Oct. 1975	Soft lander and orbiter
Venera 10	Oct. 1975	Soft lander and orbiter
Pioneer Venus probes[a]	Dec. 1978	One large and three small probes and orbiter
Venera 11	Dec. 1978	Lander and flyby
Venera 12	Dec. 1978	Lander and flyby
Venera 13 and *14*	Mar. 1982	Landers and flybys
Venera 15 and *16*	Oct. 1983	Orbiters
Vega 1 and *2*	Jun. 1985	Landers and flybys
Magellan[a]	Aug. 1990	Orbiter
Venus Express[b]	Nov. 2006	Orbiter

[a] NASA.
[b] ESA.

Figure 6.1. Mosaics of *Magellan* radar images with selected named features and regions (PIA 00157, 158, 159, and 160 as base images).

Figure 6.1. (*cont.*)

From orbit, only the long wavelength of radar (**Fig. 2.14**) can penetrate the thick clouds of Venus to reveal the surface. In 1978, the NASA *Pioneer Venus* spacecraft began orbiting the planet and mapping the topography with a radar altimeter, giving the first near-global perspective of Venus. **Figure 6.3** shows the general physiography and some of the principal named features. Because Venus is in retrograde motion, longitude increases toward the east, as set by conventions of the International Astronomical Union. The prime meridian for Venus, or the planet-fixed reference for longitude, passes through the central peak of crater Ariadne, southwest of Alpha Regio. *Pioneer Venus* also obtained information on gravity distributions and sub-meter surface roughness related to rocks, lava textures, and fractures.

Pioneer Venus data showed important differences in the distribution of elevations in comparison with Earth, suggesting some fundamental differences in interior and surface processes (**Fig. 6.4**). Analysis of the topography enabled major terrains to be discerned, including rift belts similar to those on Earth.

In addition to providing new data from orbit, *Pioneer Venus* ejected four entry probes that plunged through the dense atmosphere, measuring the temperature and pressure as a function of altitude and providing some information on atmospheric compositions and near-surface winds. Of particular note was the measurement of a key isotopic ratio, D/H (deuterium-to-hydrogen), indicating that Venus might once have had extensive water on the surface (Donahue *et al.*, 1982), perhaps including vast oceans, which has subsequently been lost.

The first clues to the geomorphology of individual surface features came from the Soviet *Venera 15* and *16* spacecraft in the early 1980s. These orbiters carried synthetic aperture radar (SAR, **Fig. 6.5**) imaging systems operating at wavelength 8 cm and giving images of resolution 1–2 km for about 25% of the surface. *Venera 15* and *16* provided the first clues as to the variety of volcanic landforms, impact craters, and tectonic features. However, the data suggested the absence of the signatures of plate tectonics. A few years later, the Soviet *Vega 1* and *2* spacecraft flew past

Figure 6.1. (*cont.*)

Venus on the way to rendezvous with Comet Halley. Both *Vega 1* and *Vega 2* ejected balloons into the venusian atmosphere and provided additional information on atmospheric characteristics, including wind patterns through careful tracking of the balloon paths. *Vega 1* and *2* also sent landers to the surface, which provided additional measurements of surface compositions.

NASA's *Magellan* spacecraft began orbiting Venus in 1990 and returned 12.6 cm wavelength SAR images with spatial resolution 120–300 m. A spare antenna from the earlier *Voyager* mission was used both for radar mapping and for data return to Earth. As shown in **Fig. 6.6**, the spacecraft was in a near-polar orbit, with radar mapping taking place when the *Magellan* spacecraft was closest to the planet and data return to Earth occurring for the remainder of each orbit. In addition to high-quality images, data were returned on surface emissivity (which is partly a function of composition), surface roughness at the sub-meter scale, and ground elevations with 80 m

vertical resolution and ~10 km spatial resolution. The obtaining of "right-looking" SAR data on one orbit and "left-looking" SAR data on a subsequent orbit meant that some images can be viewed stereoscopically and used for detailed topographic and stratigraphic studies. At the conclusion of the mission, more than 98% of the planet had been imaged, and the absence of signs of plate tectonics was confirmed. The *Magellan* data remain the primary source for understanding the geomorphology of Venus.

Missions following *Magellan* include the ESA's *Venus Express* and flybys of Venus by the *MESSENGER* spacecraft, both of which returned additional information on the atmosphere but little geologic data for the surface. However, the Visible Infrared Thermal Imaging Spectrometer (VIRTIS) on *Venus Express* returned low-spatial-resolution information for surface compositions over some areas. In May 2010, JAXA launched the *Venus Climate Orbiter*, named *Akatsaki*, but in 2011 the spacecraft failed to go into orbit.

270°E

Guinevere Planitia   Sedna Planitia

Kawelu Planitia   Guinevere Planitia

Beta
Regio

Ulfrun Regio   Devana   Undine
Chasma   Planitia

0°   Hinemoa Planitia   Devana Chasma   0°

Lengdin   Navak
Corona   Phoebe   Planitia
Regio

Maram   Dione
Corona   Parga Chasmata   Regio

Wawalag
Planitia   Sabin

Helen Planitia

270°E

Figure 6.1. (*cont.*)

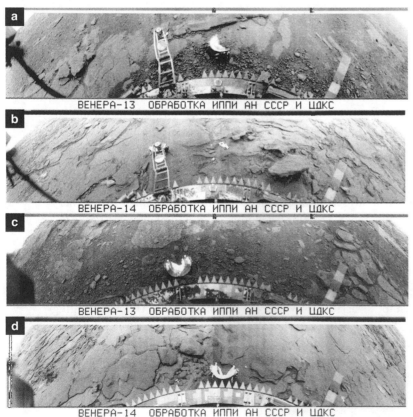

a   ВЕНЕРА-13 ОБРАБОТКА ИППИ АН СССР И ЦДКС

b   ВЕНЕРА-14 ОБРАБОТКА ИППИ АН СССР И ЦДКС

c   ВЕНЕРА-13 ОБРАБОТКА ИППИ АН СССР И ЦДКС

d   ВЕНЕРА-14 ОБРАБОТКА ИППИ АН СССР И ЦДКС

Figure 6.2. Views of the surface of Venus from the Soviet Venera landers: (a) *Venera 9* site on the northeastern flank of Beta Regio, showing moderately rounded rocks set in finer-grained soils; the horizon is visible in the upper right; (b) *Venera 10* on the eastern flank of Beta Regio; (c) *Venera 13*; and (d) *Venera 14*, both in the southern part of Beta Regio. The *Venera 10, 13,* and *14* sites show flat outcrop surfaces, with some loose, platey slabs. Compositional measurements suggest mafic rocks at all four sites. The *Venera 13* and *14* images were obtained in color, in which the terrain and sky have a yellow–greenish tint, as a result of sunlight passing through the sulfuric acid clouds. In these views, the edge of the spacecraft is visible and the spacing between the "teeth" is 5 cm.

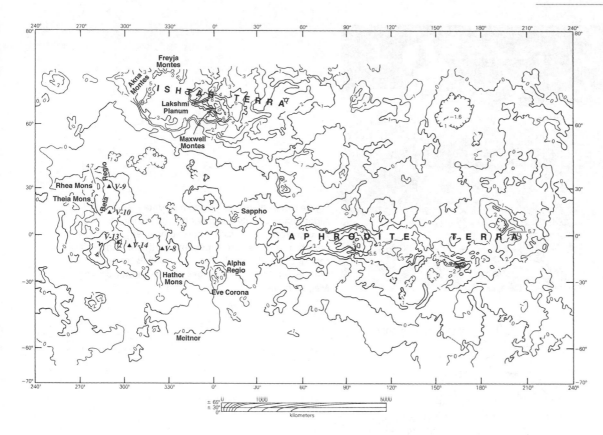

Figure 6.3. A topographic map of Venus derived from *Pioneer Venus* altimetry, showing names of major regions, including highland plateaus such as Aphrodite Terra, and the location of the *Venera* and *Vega* (*V*) landing sites.

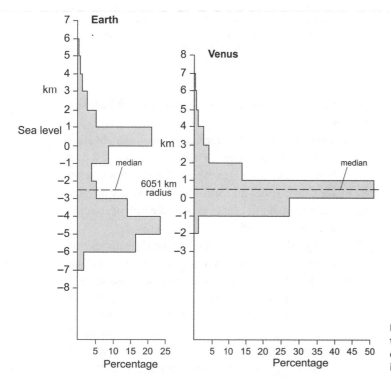

Figure 6.4. Hypsometric diagrams of elevations showing the unimodal distribution on Venus and the bimodal distribution on Earth, suggesting fundamental differences in internal processes.

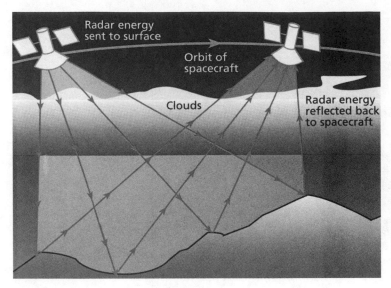

Figure 6.5. A diagram showing how data are collected for radar imaging; pulses of radar energy are sent from the spacecraft to the surface, where some energy is reflected back to the spacecraft, depending on the terrain, surface roughness, and other factors. The signal is then synthesized to produce images, such as those from the *Magellan* mission.

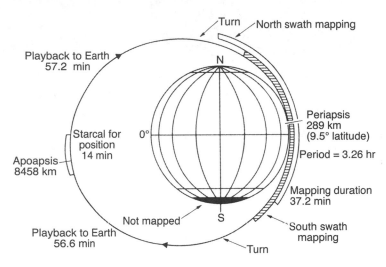

Figure 6.6. *Magellan* used one antenna for radar mapping and return of data to Earth; this diagram shows a typical orbit for the mapping phase and the "downlink" phase to Earth, with each phase separated by a short time when the spacecraft was turned to face either Venus (for mapping) or Earth (for downlink); observations of known stars were used to determine the precise position of the spacecraft, a technique commonly used in planetary missions.

6.3 Interior characteristics

Information is very limited for the interior of Venus. In the absence of seismic data, we must use geophysical models that are based on (1) planetary characteristics (such as orbit and spin periods), (2) gravity distributions, (3) topography, and (4) surface compositions. For example, compositional measurements from the *Venera* and *Vega* landers indicate that basaltic plains and possibly more silica-rich uplands exist, supporting the notion that Venus has experienced differentiation. Thus, because it is a rocky object of nearly the same size as Earth, it is likely that Venus has a core, mantle, and crust and that radioactive materials generate interior heat. Some planetologists suggest that the core is solid, while others consider core formation and solidification to still be taking place now.

Given the overall interior configuration, we might speculate that the planet has a magnetic field, but in fact it does not. Although it is tempting to suggest that the slow rotation rate of Venus precludes an Earth-like dynamo, geophysical models indicate that this cannot explain the lack of an active magnetic field, and its absence remains a puzzle.

The topography of the venusian upland plateaus appears to be isostatically compensated, suggesting crustal thicknesses of 20–40 km. As reviewed by planetary geophysicist Sue Smrekar and colleagues (Smrekar *et al.*, 2007), the very low water content inferred for the interior from the *Pioneer Venus* data is thought to have a large influence on the rheology of the interior materials. For example, the low water content would suggest a much stronger lithosphere than on

Earth and that the "slippery" zone, or asthenosphere, of the upper mantle is probably absent. This would reduce the possibility for plate tectonics and perhaps result in a stagnant "lid" for capping heat loss from the interior, which could lead to the formation of individual plumes, or hot spots.

6.4 Surface compositions

Information on venusian surface compositions comes mostly from the Soviet missions and from the *Venus Express* VIRTIS mapper. Because of operational constraints, the *Venera* and *Vega* landing sites are restricted mostly to the equatorial regions **(Fig. 6.3)**. The Venera sites are on the eastern flanks of Beta Regio, a major upland region, while the *Vega* sites are on Rusalka Planitia near Aphrodite, another major upland. Measurements from most of the lander sites show that the rocks are similar to basalts found on Earth's sea floor **(Fig. 6.7)**. It is important to note that the 300 km radius uncertainty in the exact location of the *Venera* and *Vega* landing sites makes the geologic context poorly constrained.

The *Venera 8* site is particularly important because it has a very high thorium content, which is indicative of significant chemical differentiation. In addition, its high potassium content is comparable to that of felsic rocks on Earth, suggesting to some investigators the presence of granite-like materials and the idea that the upland plateaus might be roughly comparable to Earth-like continents. However, the Soviet instruments used to measure surface compositions were developed decades ago, and the quality of the data is not as good as can be obtained with current technologies. Nonetheless, the generally basaltic compositions for most of the landing sites are consistent with the recent VIRTIS data from the *Venus Express* orbiter. Until new missions are flown to the surface of Venus, the *Venera* and *Vega* landers provide the only data from direct surface measurements.

Radar data reveal upland regions, such as Maxwell Montes, to have surprisingly low emissivities. These regions are thought to represent some unusual surface compositions. One possibility is that the rocks are coated with materials that form only where the temperatures and pressures are lower at the high elevations on Venus. Although intuitively strange, the term metallic "frosts" has been used to suggest that volatile metals such as tellurium and bismuth might be deposited on surface materials, which would then yield low radar emissivities. Others have suggested that concentrations of minerals such as pyrite could explain the radar signatures and that wind-winnowing and removal of lower-density grains could leave a lag surface of higher-density minerals; however, remarkably large amounts of pyrite would be required in order to explain the radar signatures.

6.5. Geomorphology

6.5.1 General physiography

The first hints of the major terrains on Venus came from Earth-based radar data **(Fig. 6.8)** in which certain distinctive

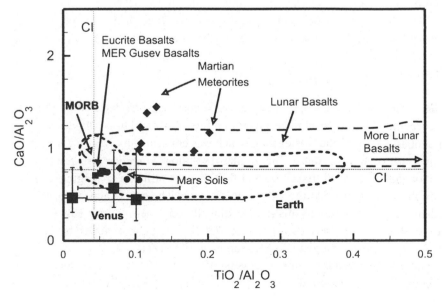

Figure 6.7. Venus compositions from *Venera* and *Vega* measurements compared with Earth mid-ocean-ridge basalts (MORB), basalts in Gusev crater on Mars, martian meteorites, and lunar basalts (after Treiman, 2007).

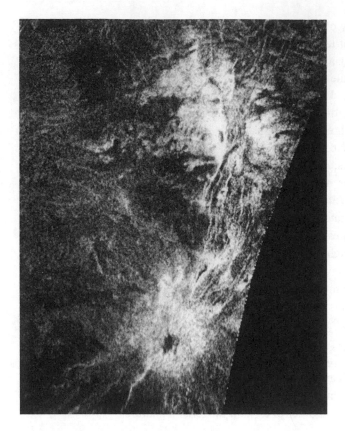

Figure 6.8. An Earth-based radar image showing Beta Regio provided the first hints of tectonic and volcanic features, suggested by the linear radar-bright fractures connecting zones with radial patterns thought to be volcanoes (Arecibo Observatory image courtesy of Don Campbell of Cornell University).

radar-bright terrains were identified and named, such as the "Alpha" and "Beta" regions. Except for these and a few other features, names on Venus are of feminine derivation.

Key to understanding the general geomorphology and surface processes of any planet are topographic data, as provided by the *Pioneer. Venus* and *Magellan* spacecraft. Because Venus lacks an ocean today, the topographic reference datum is the mean planetary radius. **Figure 6.4** graphs the distributions of elevations for Venus compared to Earth. The bimodal distribution on Earth reflects the continents and the sea floors, as well as differences in densities between the felsic and mafic crusts. In contrast, the distribution on Venus is unimodal and shows that about 60% of the topography is within 500 m of the datum, with less than 5% being more than 2 km above datum. This suggests that there are some fundamental aspects of mountain-building and tectonic processes that are different from those of Earth.

The late planetary geologist Hal Masursky and his colleagues defined three principal terrains for Venus on the basis of topographic mapping: *highlands*, which stand above 2 km elevation and make up 8% of the planet, *upland rolling plains*, which are between 0 and 2 km elevation and constitute about 65% of Venus, and *lowland areas*, which are below the 0 km datum and represent about 27% of the surface. Both the highlands and parts of the rolling plains are generally radar-bright, while the lowland plains are typically radar-dark, suggesting low-backscatter surface materials such as smooth lava flows and fine-grained sediments.

The mountains Maxwell Montes at 65.2° N, 3.3° E are nearly 800 km across and rise in elevation to 11.5 km, making this the highest feature on Venus. It is part of Ishtar Terra, a plateau that is a few kilometers above datum. The other large upland region is Aphrodite Terra, which stretches along the equator eastward from Ovda Regio more than 10,000 km to the east. This region rises gradually from the lowland plains and in places nearly reaches 6 km high.

Chasmata, or rift valleys, are found along the equator and in the southern hemisphere and represent crustal extension. The lowest area on Venus is Diana Chasma, a trench 2 km deep centered at 14° S, 156° E. Venus, therefore, has a maximum relief of about 13 km, which can be compared with the relief on Earth of 20 km (measured from the peak of Mount Everest to the bottom of the Mariana Trench on the floor of the Pacific Ocean). Because surface temperature and atmospheric pressure vary with elevation, the venusian summit regions average 374 °C at 41 bars, while the lowlands are at 465 °C and 96 bars. This range in conditions could lead to differences in surface chemistry, as noted for the high radar emissivities seen at higher elevations, as well as in geomorphic processes. For example, winds at the higher elevation might need to be stronger to entrain particles than at lower elevations because of the lower atmospheric densities.

6.5.2 Impact craters

Circular-shaped features on Venus were first detected on Earth-based radar images and were inferred to be impact scars, the existence of which was confirmed by *Venera 15* and *16* data. The higher-resolution *Magellan* images revealed fewer than 1,000 impact structures, ranging from 1.5 km to 270 km in diameter. The craters appear to be randomly distributed globally and their paucity suggests a surface age of about 750 Ma or less, as reviewed by planetologist Bill McKinnon *et al.* (1997).

Venusian impact craters 10 km to ~50 km in diameter are classified as complex, being typified by Adivar crater

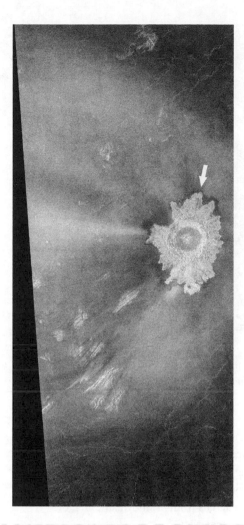

Figure 6.9. Impact crater Adivar is 30 km in diameter and has typical features for complex craters on Venus, including a scalloped outline, terraced walls, and a central peak. The continuous ejecta is radar-bright, indicative of a rough surface, while the radar-bright horseshoe-shaped zone beyond the continuous ejecta and the "tail" of bright ejecta extending to the left both indicate the influence of the prevailing winds toward the west (left) at the time of the impact (NASA *Magellan* P-38387).

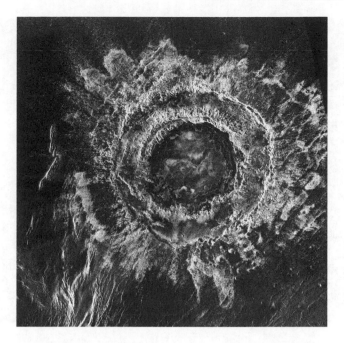

Figure 6.10. A *Magellan* image of the 150 km in diameter Meitner crater, a typical multi-ring impact structure on Venus (NASA *Magellan* C1 MIDRP60S319).

(Fig. 6.9), which has a scalloped rim, central peak, and flat floor. Except for the floor deposits, Adivar's rim and inner wall are radar-bright, a consequence of the rugged, irregular topography. The continuous ejecta deposits are sharply defined by a distinct outer boundary and are radar-bright, which is indicative of rocky debris. The "tail" and the horseshoe bright zone extending toward the west are ejecta deposits that were caught by prevailing winds from the east at the time of the impact and carried westward for deposition.

Craters larger than 50–60 km in diameter transition from complex features to multi-ring structures, such as the 150 km in diameter Meitner in the southern hemisphere **(Fig. 6.10)**. Radar-dark flat surfaces are seen on the floor and between the inner ring and the main rim, while the continuous ejecta deposits extend only about 50 km beyond the rim, evidently a result of the high surface gravity on Venus and dense atmosphere which would retard the trajectories of the ejecta.

Even before radar data were available from spacecraft, there was speculation that small craters would not be found on Venus. The reasoning was that the dense atmosphere would fracture small incoming objects into small pieces, or perhaps even consume them. The smallest obvious impact crater seen in *Magellan* data is about 1.5 km across; and, as speculated, it appears that incoming objects do break apart into smaller pieces that reach the surface to form a cluster of craters, much like the blast from a shotgun **(Fig. 6.11)**. Thus, craters smaller than 10–20 km in diameter have irregular outlines and floors, unlike the simple bowl-shaped craters of this size seen on other terrestrial planets.

For some inferred impact features, there are no obvious topographic depressions, but only a radar-dark zone on the surface **(Fig. 6.12)**. It is thought that as smaller bolides pass through the dense atmosphere they are so fragmented that only the shock wave and, perhaps, fine material reaches the surface, leaving the blast-scars visible on images.

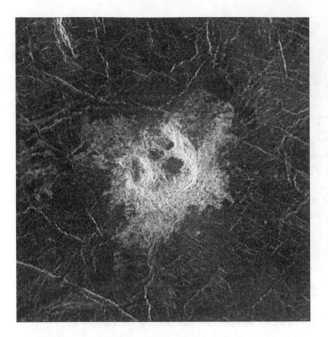

Figure 6.11. This impact structure, named Lilian, is in the Guinevere region and consists of four smaller crater-forms formed from the breakup of a bolide as it passed through the dense Venus atmosphere (NASA *Magellan* P-38290).

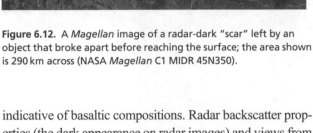

Figure 6.12. A *Magellan* image of a radar-dark "scar" left by an object that broke apart before reaching the surface; the area shown is 290 km across (NASA *Magellan* C1 MIDR 45N350).

One consequence of the high surface temperature is the formation of impact melt deposits as ejecta **(Fig. 6.13)**. These have the appearance of lava flows and are found in association with many impact craters on Venus.

6.5.3 Volcanic features

The surface of Venus is dominated by volcanic and tectonic features reflecting an extensive history of internal processes. Volcanic features include vast lava plains, cones, domes, and huge shield structures (Head *et al.*, 1992). Major volcanic provinces include Western Eistla, Atla Regio, Beta Regio, and Bell Regio, all of which are topographically high and have been modified by extensional tectonism. While most of the volcanic features are probably basaltic, given their morphology and compositional information, some volcanoes might include more silica-rich lavas.

Many planetary geologists have studied and mapped the venusian plains. Unfortunately, some of the terms and descriptions are not used consistently, and the different categories overlap. For simplicity, the volcanic plains form flat, smooth surfaces that are radar-dark and are found mostly in the lowlands and rolling uplands. *Venera 9*, *10* and *13* and *Vega 1* and *2* all landed on plains deposits and returned data

indicative of basaltic compositions. Radar backscatter properties (the dark appearance on radar images) and views from the landers **(Fig. 6.2)** suggest that some plains are mantled with clastic materials, which could be windblown sediments, ejecta deposits, basaltic materials from *in situ* weathering, or some combination of these materials.

Most of the volcanic plains have no obvious vents, but a few distinctive depressions with sinuous channels are thought to be eruption sites **(Fig. 6.14)**. Plains units show distinctive lobate flows of volcanic origin and display differences in radar brightness **(Fig. 6.15)** that are attributed to differences in surface textures, such as pahoehoe, aa, or block flows.

How were the vast lava plains emplaced? Although hints of volcanic channels on Venus were seen on *Venera 15* and *16* images, the *Magellan* mission showed these features in detail **(Figs. 6.14** and **6.16)**. More than 200 channels have been identified, some of which are as long as 6,800 km. Such a great length requires that the lava remained molten for a substantial length of time. At first, we might think that the high surface temperatures would allow the flows to retain heat, but, as discussed in the analysis by Head and Wilson (1986), this effect would be offset by the dense atmosphere that would conduct heat efficiently away from the surface. In attempts to resolve the issue, lavas other than basalt have been considered,

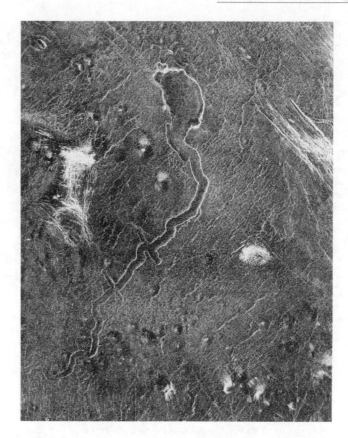

Figure 6.14. This volcanic sinuous rille is near Alpha Regio and appears to have originated from an irregular caldera-like depression; similar features are seen on the lunar maria; the area shown is 130 km by 160 km (NASA *Magellan*, F 25S345).

Figure 6.13. This 90 km in diameter impact crater, Addams, shows a massive "outflow" lobe of ejecta melt extending 550 km to the bottom of the image; such morphologic features are not seen on any planetary surface other than Venus (NASA *Magellan* F-MIDR 55S097).

including compositions that are high in magnesium and iron (called "ultramafic"), such as **komatiite**, which is thought to erupt at high temperatures and is of low viscosity. Even more exotic lavas, such as liquid sulfur and carbonatite, have also been considered. Carbonatite is molten carbonate of magmatic origin and has been observed erupting at the Oldoinyo Lengai volcano in Tanzania, where its physical properties were measured, including its very low viscosity. In any event, it is likely that lava tubes, or crusted surfaces on flows through lava channels, were important to retain heat and enable continued flow over long distances.

Fields of small shield volcanoes are found in many plains regions. Some fields cover areas a few hundred kilometers across, while the individual shields are 5–15 km in diameter. Although the resolution is not sufficient for detailed analysis, the shield volcanoes are similar to small basaltic shields in the Snake River Plain of Idaho and elsewhere on Earth. Other venusian plains are characterized by fields of small cones, suggestive of cinder cones, some of which include distinctive radar-dark halos that have been interpreted as being pyroclastic deposits **(Fig. 6.17)**.

Venusian volcanoes are among the largest and most complex seen in the Solar System. Hundreds of volcanic constructs have been identified, many of which are more than 100 km across. One of the largest, Maat Mons **(Fig. 6.18)**, reaches 9 km above planetary datum. The main construct is more than 500 km wide at its base and displays flows that can be traced as far as 600 km from their source to form gentle flank slopes. Some venusian shields either lack, or have only small, summit craters, while others display enormous complex calderas **(Fig. 6.19)**.

Evidence of more viscous lava flows than typical for basalt is seen in the form of dome-shaped volcanoes **(Fig. 6.20)**. These structures are as large as 25 km across

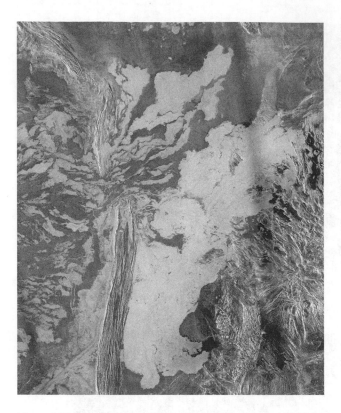

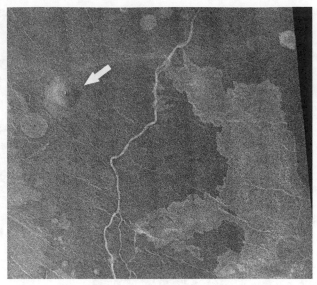

Figure 6.16. A segment of a venusian channel ~2 km wide in Guinevere Planitia and the apparent spillover of radar-bright material to form a massive flow-lobe; also visible to the left of the channel is a volcano (arrow) about 20 km across with a small summit crater (NASA *Magellan* MRPS 41387).

Figure 6.15. Abundant lava flows are seen in the lowland plains and in some upland regions of Venus. These flows in the Lada Terra region breached a large north–south ridge belt and can be traced more than 600 km from their source; differences in radar brightness indicate lava surface texture roughness; the area shown is 550 km by 630 km (NASA *Magellan* MRPS 37755).

and are hundreds of meters high. They are generally circular in plan-form and have flat, upper surfaces, leading to the informal term "pancake" volcanoes. Many of the constructs have small craters on their surfaces, suggestive of volcanic vents or explosive blasts. One field of domes is in the general vicinity of the *Venera 8* landing site, where compositional data suggest more silicic rocks, supporting the suggestion that the domes were formed by more viscous lavas.

6.5.4 Tectonic features

In the early period of exploration in the 1960s, there was speculation that Venus might experience plate tectonics in the style of Earth (i.e., spreading centers and zones of subduction) because of the similarities in planetary size and density. However, the global data sets from *Venera 15* and *16* and *Magellan* clearly show that this is not the case and that a different style of tectonism must be responsible for the loss of heat from the planet's interior. One clue to this different style comes from surface features called **coronae** (from the Latin for crowns). These are circular features 200–600 km across consisting of ridges and grooves arranged in concentric patterns **(Figs. 6.21–6.23)**. The ridges and grooves are often found on the flanks of gentle domes or surrounding shallow depressions, some of

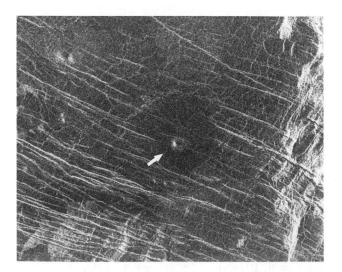

Figure 6.17. This small cinder cone (arrow) is surrounded by a radar-dark zone, which has been interpreted as pyroclastic deposits superposed on plains that have been extensively fractured by tectonism; the area shown is 110 km by 90 km (NASA *Magellan* F-MIDRP 50S345).

Figure 6.18. A computer-generated oblique view of Maat Mons, a 500 km diameter by 6 km high volcano in Atla Regio, showing extensive lavas that flowed from the flanks; the vertical scale has been exaggerated ×23 to enhance detail in topography. A 23 km in diameter impact crater and its ejecta melt flow-lobe are visible in the foreground (NASA *Magellan* P-40175).

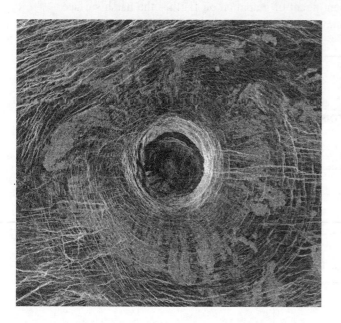

Figure 6.19. This ~20 km in diameter caldera in the Aphrodite Terra area displays a series of concentric fractures outlining the zone of subsidence and radial lava flows that extend beyond the depression. Note also the widely spaced concentric fractures beyond the lava flows (NASA *Magellan* F-MIDR O6N227).

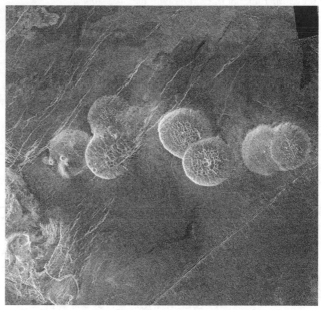

Figure 6.20. This series of domes is found east of Alpha Regio; the domes average 25 km across and their steep margins represent emplacement of viscous lavas, possibly of silicic composition (NASA *Magellan* P-37125).

which are probably calderas. Coronae are considered to represent crustal deformation resulting from mantle plumes or upwelling "hot spots."

In addition to coronae, tectonism indicative of both extension and compression is common on Venus. Sets of parallel rugged ridge-belts, as in Lavinia Planitia, and mare ridges in plains reflect regional compression **(Fig. 6.24)**, while troughs, broad grabens, and complex chasmata **(Fig. 6.25)** indicate extensional tectonism.

Complex ridged terrain formed by crustal deformation under both extensional and compressional processes **(Fig. 6.26)**. This terrain, found mostly in highland regions, makes up some 30% of the surface and is characterized by features called **tesserae** (from the Greek referring to four corners). Tesserae consists of tile-like blocks of crust cut by ridges and troughs 10–20 km apart, which are often transected by younger, broad grabens **(Fig. 6.26)**. Embayment and superposition of plains units show

Figure 6.21. Thouris Corona is about 190 km across and is identified by the concentric set of fractures surrounding a gentle dome. The structure is set within regional fractured plains (NASA *Magellan* mosaic ASU-IPF).

that some complex ridged terrain is among the oldest recognizable geologic units on Venus.

Mare-type ridges are common on nearly half of the venusian plains (**Fig. 6.24**), mostly at lower elevations. The ridges are several hundred meters high, as wide as a kilometer or two across, and can be hundreds of kilometers long. More than 65,000 mare ridges have been identified, and they are tectonic compression features. Detailed mapping reveals the global lithospheric stress field in which lowland plains were in compression at the time of ridge formation, while some uplands were in extension, exemplified by chasma formation.

6.5.5 Gradation features

Even in the absence of water and fluvial streams – the dominant agent of gradation on Earth – the harsh surface environment of Venus can be expected to lead to weathering

CONCENTRIC 	1. SIZE RANGE -- 75–2600 km 2. VOLCANISM – MODERATE TO HIGH; SMOOTH PLAINS, DOMES, FLOWS, RILLES, EDIFICES 3. TECTONISM -- OBLIQUE, RADIAL CONCENTRIC FAULTS
CONCENTRIC -- DOUBLE RING 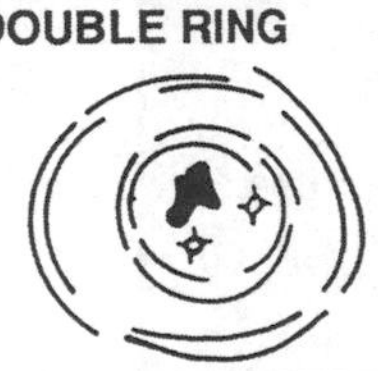	1. SIZE RANGE 125–870 km 2. VOLCANISM -- MODERATE TO HIGH, SMOOTH PLAINS, DOMES, FLOWS 3. TECTONISM -- OBLIQUE, RADIAL FAULTS
RADIAL/ CONCENTRIC 	1. SIZE RANGE – 175--500 km 2. VOLCANISM -- GENERALLY LOW; SMOOTH PLAINS DEPOSITS 3. TECTONISM – RADIAL, CONCENTRIC FAULTS
ASYMMETRIC 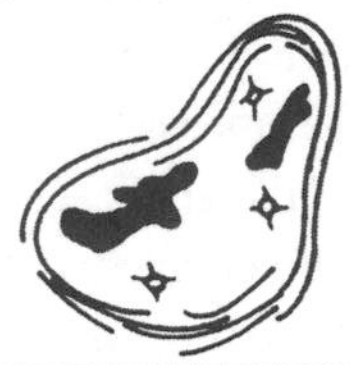	1. SIZE RANGE -- 125--500 km 2. VOLCANISM -- MODERATE TO HIGH, SMOOTH PLAINS, DOMES, FLOWS, EDIFICES, RILLES 3. TECTONISM – OBLIQUE FAULTS, FAN GRABEN
MULTIPLE 	1. SIZE RANGE 125--500 km 2. VOLCANISM -- LOW TO HIGH; SMOOTH PLAINS, DOMES, FLOWS 3. TECTONISM – OBLIQUE FAULTS

Figure 6.22. Classification of venusian coronae (from Stofan *et al.*, 1992).

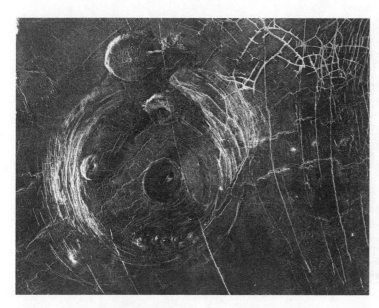

Figure 6.23. Aine Corona is 250 km in diameter and is located in plains south of Aphrodite Terra. It has several flat-topped volcanic domes, some of which are superposed on the ring fractures, showing that volcanism and tectonic deformation were occurring at essentially the same time; the area shown is 300 km by 235 km (NASA *Magellan* P-38340).

Figure 6.24. Mare ridges in Rusalka Planitia were formed by compression. Their superposition both on the radar-dark and on the radar-bright volcanic plains indicates that their formation occurred after the lavas had been emplaced in response to regional compression (from Solomon *et al.*, 1992).

Figure 6.25. This *Magellan* image shows part of Devana Chasma, a rift system more than 2,500 km long between Beta Regio and Phoebe Regio. It represents some 20 km of horizontal extension and is comparable to major continental rift systems on Earth; the area shown is 670 km by 845 km (NASA *Magellan* P-41294).

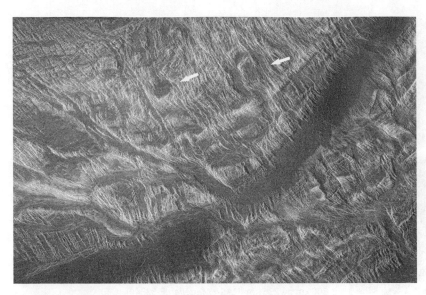

Figure 6.26. Complex-ridged terrain in Ovda Regio, showing the square blocks of tesserae (arrows) that are highly fractured. The dark band cutting through the terrain is a younger graben partly filled with smooth-surfaced lava; the area shown is 225 km by 150 km (NASA *Magellan* P-37788).

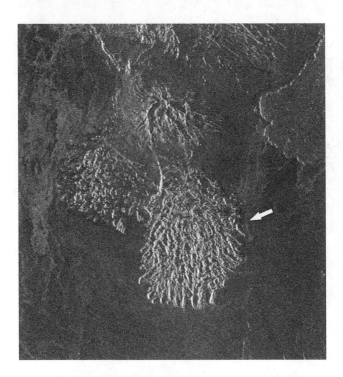

Figure 6.27. This fan-shaped landslide mass (arrow) is about 25 km long and originated from a volcano. The radar-bright mass indicates rugged, blocky topography (NASA *Magellan* F-MIDR 75N327).

fractured terrain, clastic materials produced from pyroclastic eruptions and tectonic "grinding" of crustal rocks also generate smaller fragments.

As clastic materials are generated and reduced in size, additional weathering of materials would occur in the presence of the high temperatures and sulfuric acid environment. Coupled with *in situ* weathering of bedrock, the resulting fine-grained debris could account for the generally low radar backscatter signatures seen in many areas of Venus.

Once produced, clastic materials generated either by primary processes or from weathering are subject to agents of transportation through wind and gravity. Mass movement in the form of landslides is seen in several areas where steep slopes occur **(Fig. 6.27)**. These occur on a wide range of scales and involve mostly rock avalanches, some of which are surrounded by radar-dark surfaces interpreted to be fine-grained materials generated from the mass movement.

In the present surface environment, wind appears to be the primary means for redistributing surface materials on Venus. Surface winds are very sluggish, due in part to the high density of the atmosphere; they were measured by the Soviet *Vega* landers and NASA *Pioneer Venus* probes to be about 0.5–2 m/s. But, because the atmosphere is so dense, winds need not be moving very fast to entrain sand and dust **(Fig. 3.36)**, and the measured winds are well within the range predicted for particle transport by the wind.

Fields of sand dunes were identified in *Magellan* radar images in two areas. The Aglaonice field covers 1,300 km^2 and is associated with ejecta deposits from the impact crater

and the formation of gradation features. Physical weathering to produce loose rocks and small grains **(Fig. 6.2)** results from a variety of processes, including impacts **(Fig. 4.7)**. For example, NASA scientist Jim Garvin estimates that the amount of impact-generated debris on Venus would produce a layer 0.3–1 m thick if spread evenly over the entire planet. Given the extensive volcanic features and

Aglaonice. Individual transverse dunes are several hundred meters wide and have crests oriented generally north–south, while wind streaks within the field suggest formative winds from the east. The Fortuna–Meskhent dune field was found in a valley between Ishtar Terra and Meskhent Tessera and also has numerous wind streaks within the field **(Fig. 6.28)**. The formative wind patterns suggest flow from the southeast to the northwest, with a shift toward the west in the northern part of the field. Although it is a

puzzle why more dunes were not found, experiments using the NASA Venus Wind Tunnel show that, in the sluggish winds on Venus, relatively small bedforms develop, including so-called "micro-dunes" that are only centimeters in size, which is far too small to be detected on the radar images. They might, however, generate a rougher surface than smooth plains, and it is possible that some of the radar-bright plains could have sets of micro-dunes.

By far the most abundant aeolian features on Venus are wind streaks, with more than 6,000 having been mapped on *Magellan* images. These occur in a wide variety of forms, with most being associated with topographic features, such as small hills **(Fig. 6.29)** or ridges, and include both radar-bright and radar-dark signatures **(Fig. 6.30)**. Wind streaks are thought to result from the interaction of surface winds with landforms, perhaps generating zones of preferential erosion and deposition. For example, bright streaks could form by the removal of fine-grained clastic

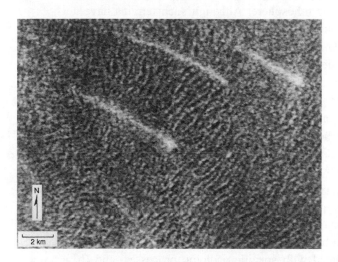

Figure 6.28. Part of the Fortuna–Meskhent dune field, showing sets of transverse dunes and a few radar-bright streaks, indicating prevailing winds from the southeast (lower right); the area shown is about 55 km by 55 km (NASA *Magellan* MRPS 39824).

Figure 6.29. A radar-bright wind streak 26 km long formed in association with a 5 km wide hill; wind was blowing from the lower left toward the right at the time of streak formation (NASA *Magellan* P-38810).

Figure 6.30. Radar-dark wind streaks associated with ridges; winds were blowing from left to right at the time these features formed; the area shown is 44 km by 64 km (NASA *Magellan* MRPS 3883).

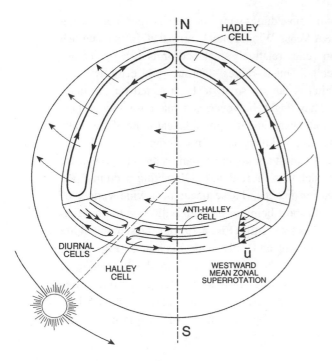

Figure 6.31. A diagram showing the global circulation of Venus' atmosphere as modeled and confirmed by near-surface wind patterns derived from the orientations of wind streaks. This pattern reflects Hadley cell circulation that takes into account factors such as solar heating and planetary rotation.

materials to expose an underlying rougher surface. Wind streaks "point" in downwind directions and are used to map global near-surface winds. The results show that global winds blow toward the west **(Fig. 6.31)**.

6.6 Geologic history

Comparison of the geologic histories for the terrestrial planets poses an intriguing problem. As shown in **Fig. 4.50**, the rock records for Mercury, the Moon, and Mars can be traced back to the final stages of terminal accretion, marked by an abundance of large impact craters. Except for Earth, the geologic time scales of the terrestrial planets are based mostly on the numbers of superposed impact craters on mappable units and the stratigraphic horizons provided by widespread ejecta from major impacts, such as the Fra Mauro Formation on the Moon. The paucity of large craters on Venus precludes this approach and suggests a surface age of 0.75 Ga or younger. This leads to a controversy regarding the styles of resurfacing, with two competing ideas: (a) rapid, catastrophic overturning of the lithosphere; and (b) slower, local resurfacing by volcanism. Of the craters

on Venus, 62% are relatively pristine and only 4% are partly embayed by lava flows, with the remainder being partly degraded by tectonic processes. If resurfacing took place by prolonged volcanism, one would expect a greater number of craters displaying various stages of burial; thus, their absence supports the catastrophic overturning model. On the other hand, if substantial blocks of terrain were found in which remnants of large craters were preserved, it would show that the earlier rock record is preserved in some places and argue against complete global overturning of the lithosphere. Although such terrains have not been found, it is possible that the low resolution of the *Magellan* data does not allow one to discern remnants of highly deformed craters.

Most geophysical models of the interior suggest that the current heat flow from the surface is less than the heat being generated, even taking into account the large number of hot spots. The result is that with time heat is accumulating below the lithosphere, which could lead to a catastrophic overturning, and it is thought that such cycling could occur every 0.5 Ga.

The history subsequent to the resurfacing at 0.75 Ga (regardless of mechanism) is also debated, with two general hypotheses in contention: (a) an episodic history punctuated with specific geologic processes, and (b) a non-episodic history characterized by uniform geologic processes spread throughout time. In the first case, it is assumed that all geologic features of a specific type formed at the same time, such as the coronae, regardless of geographic location. This inference is based partly on stratigraphic relations with other terrains, such as plains, and tectonic features. In the second case, detailed mapping suggests that the formation and evolution of many features, such as the large volcanoes, spans a long period of time, with volcanic materials "inter-fingering" with tectonic deformation in a patch-work pattern of local resurfacing. In both models, there is agreement that complex-ridged terrain and the formation of tesserae represent some of the earliest visible rock record, but that, in the non-episodic model, some younger tesserae can be identified.

What processes might be taking place today on Venus? There are no observations of changes on the surface with available data; i.e., no new lava flows, new landslides, or changes in wind-related features. But it must be remembered that the available data are extremely limited for detecting such changes, and variations in the abundance of sulfur dioxide in the atmosphere led Larry Esposito (1984) to suggest that volcanoes might at present be erupting, an idea supported by some *Venus Express* data.

Assignments

1. Geophysical models of out-gassing by magma reaching the surface of Venus suggest that the high atmospheric surface pressures would retard explosive eruptions. Yet, *Magellan* images show the presence of cones and likely pyroclastic deposits. Suggest some possible explanations for this apparent discrepancy.

2. Briefly discuss three ways in which Venus and Earth are similar and three ways in which they are different.

3. Outline the primary data sources for deriving information on the surface compositions for Venus.

4. Explain two methods used to determine the topography of Venus.

5. Discuss at least three consequences of the high surface temperatures and atmospheric pressures on Venus in relation to the morphologic features seen.

6. Briefly explain the ideas as to why the venusian surface is geologically young compared with those of the Moon and Mercury.

CHAPTER 7

Mars

7.1 Introduction

Few objects in the sky hold the fascination in the public mind as much as Mars. Easily seen with the naked eye, the "Red Planet" has been linked with various gods of war through the ages. The late 1800s and early 1900s saw both serious and not-so-serious writings on martian life, including the presence of advanced civilizations, and culminating in the infamous radio broadcast of H. G. Wells's fictional *War of the Worlds*, in which martian spacecraft land on Earth. This broadcast filled many a family with terror as the story unfolded with the destruction of whole cities.

Building on public support for the exploration of Mars, the Red Planet has been visited by more spacecraft than any other object except Earth's Moon. Along with Europa and possibly Titan, Mars is a favorable planet in the exploration for possible present-day or past life.

With a diameter of 6,779 km and a mass of 6.4×10^{23} kg, Mars gravity is 0.37 that of Earth. The total surface area of Mars is just about equal to the land surface on Earth above sea level. One Mars year is 686.98 Earth days, while one Mars day is 24 hr, 39 m, 35.2 s. Its present-day spin axis is inclined 25.19° (slightly more than Earth), which leads to distinctive seasons. The seasons are defined by Mars' position in orbit and described by **aerocentric longitudes** (L_s) of the Sun in degrees. L_s is the angle between the Mars–Sun line and the line of equinoxes. L_s of 0° is set at the martian equinox for the beginning of winter in the northern hemisphere (L_s 0° to 90°), with northern spring (L_s 90° to 180°), northern summer (L_s 180° to 270°), and northern autumn (L_s 270° to 360°).

Mars wobbles on its spin axis, which means that at times the poles receive substantially more sunlight in their respective summers and are proportionally much colder in the winters than at other times. This is thought to lead to martian ice ages on time scales of 50–100 Ma as a function of the wobble and could result in large climate changes.

Mars has an atmosphere composed mostly of carbon dioxide (95%), with minor amounts of nitrogen, argon, oxygen, carbon monoxide, and water vapor. But with an average surface pressure of only 6.5 mbar, the atmosphere is very "thin" in comparison with Earth's atmosphere – about the equivalent of Earth's atmosphere at an altitude of about 130,000 feet (nearly 40 km) above the surface, or four times higher than commercial jet flights. While the average surface temperature is −63 °C, the range of temperatures is −133 °C to +27 °C.

Mars is the most Earth-like of all the objects in the Solar System and exhibits evidence of processes related to running water, glaciers, wind, volcanism, and tectonic deformation. However, unlike Earth, Mars preserves extensive parts of its ancient record of impact cratering, while the current cold, dry climate militates against there being extensive liquid water on the surface. Despite the low-density atmosphere, winds and associated aeolian processes dominate the current environment. Although Mars appears to lack apparent plate-style tectonics, as noted below, there are suggestions that such activity might have occurred early in its history.

Given its size, rocky composition, and the presence of an atmosphere, Mars exhibits surface features that are recognizable to most geologists and affords the opportunity to study familiar processes under non-Earthly conditions.

7.2 Exploration

Although even the earliest telescopes were trained on Mars, it was not until the late 1800s that attempts were made to identify surface markings and relate them to processes. The famous "*canali*" of Italian astronomer Schiaparelli were loosely translated to "canals" and spurred interest on the part of Percival Lowell. This wealthy Bostonian became fascinated with Mars and built an

Table 7.1. **Selected successful missions to Mars (all NASA expect where indicated otherwise)**

Spacecraft	Encounter date	Mission	Geosciences data
Mariner 4	Jul. 1965	Flyby	Imaging; closest approach 9,912 km
Mariner 6	Jul. 1969	Flyby	Imaging; closest approach 3,330 km
Mariner 7	Aug. 1969	Flyby	Imaging; closest approach 3,518 km
Mariner 9	Nov. 1971	Orbiter	Imaging; ultraviolet and infrared spectrometers; infrared radiometer
Viking 1	Jul. 1976	Lander and orbiter	Lander imaging; wind speeds, temperatures and directions; chemical and physical properties of surface; orbiter imaging; gravity; atmospheric water levels, thermal mapping
Viking 2	Sep. 1976	Lander and orbiter	Lander imaging; wind speeds, temperatures and directions; chemical and physical properties of surface; orbiter imaging; gravity; atmospheric water levels, thermal mapping
Mars Pathfinder	Jul. 1997	Lander and rover	Imaging; surface composition, meteorology
Mars Global Surveyor	Sep. 1997	Orbiter	Imaging; altimetry; spectroscopy; magnetometer
Mars Odyssey	Oct. 2001	Orbiter	Thermal emission; imaging; gamma-ray spectrometer; radiation detector
Mars Express[a]	Dec. 2003	Orbiter	Imaging; spectroscopy; atmospheric monitoring; radar sounding
Mars Exploration Rovers			
Spirit	Jan. 2004	Rover	Imaging; spectroscopy; robotic arm; magnets; composition
Opportunity	Jan. 2004	Rover	Imaging; spectroscopy; robotic arm; magnets; composition
Mars Reconnaissance Orbiter	Mar. 2006	Orbiter	Imaging; radar sounding; spectroscopy
Phoenix	May 2008	Lander	Imaging; composition; weather, surface properties

[a] European Space Agency.

observatory now bearing his name in Flagstaff, Arizona. Lowell Observatory was originally dedicated to the study of the Red Planet by mapping its linear features and following its seasonal changes. Lowell Observatory continues today to be a key center for planetary studies.

Among the first geologically oriented observations were those by Dean McLaughlin, who speculated in 1954 that surface changes indicated vast dust storms and volcanic eruptions. Today we recognize that dust storms and other aeolian processes are active today, but the question of active volcanism remains open.

During the heyday of the space race, NASA and the Soviet Union sent numerous spacecraft to Mars. For a variety of reasons, most of the Soviet missions were failures, but, as noted earlier, the Soviets engaged in a highly successful exploration of Venus in the same period. In contrast, the United States saw successes for Mars missions in the mid to late 1960s **(Table 7.1)**, first with the *Mariner 4* flyby, quickly followed by *Mariners 6* and *7*. By happenstance, all three flybys observed the same heavily cratered part of Mars **(Fig. 7.1)**, leading much of the scientific community to speculate that Mars was simply another Moon-like object. Fortunately, in the early 1970s, the *Mariner 9* orbiter showed the true diversity of Mars, revealing the huge volcanoes, the tectonic rift system of Valles Marineris, and the extensive now-dry river channels. This mission was followed a few years later (1976) by the *Viking* Project, which involved two orbiters and two landers for the first successful surface operations on Mars. All four spacecraft operated concurrently, and for many years this was the most complicated robotic mission to have been flown in deep space.

The last of the *Viking* data were sent to Earth in 1982 from *Viking Lander 2*. It was 15 years before the next successful mission to Mars, marked by the landing of *Mars Pathfinder* in 1997 and the operation of its little rover, *Sojourner*. Coupled with the recognition that some meteorites were blasted from Mars by impacts and sent on trajectories carrying them to Earth, there was

(a)

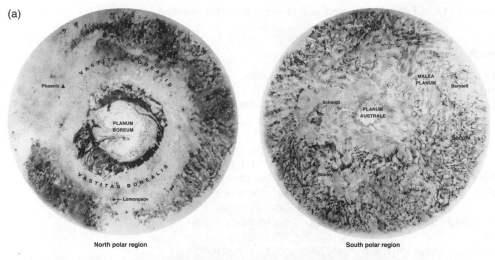

North polar region   South polar region

(b)

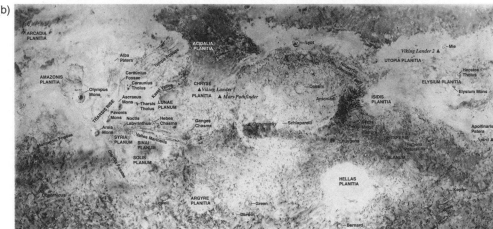

Figure 7.1. Shaded relief maps of Mars showing albedo patterns for the polar regions (a) and the equatorial regions (b) with selected place names and the locations of the successful landings. By happenstance, the first three NASA missions to Mars (*Mariners 4, 6,* and *7*) were all flybys that imaged only the heavily cratered southern highlands, leading many to think that Mars was much like Earth's Moon.

renewed interest in Mars, especially with the proposal (now largely rejected) that one of the Mars meteorites contained evidence of life. NASA implemented an expansive program for Mars exploration involving landers, orbiters, and roving vehicles, which continue to provide an incredible wealth of data, despite notable spacecraft failures by NASA, Russia, Great Britain, and Japan.

Key recent missions include the NASA orbiters *Mars Global Surveyor* (*MGS*) with its Mars Orbiter Camera (MOC), Thermal Emission Spectrometer (TES), Mars Orbiter Laser Altimeter (MOLA), and instruments for measurements of the remnant magnetic field; *Mars Odyssey* with its Thermal Emission Imaging System (THEMIS); and the *Mars Reconnaissance Orbiter* (*MRO*) with its High Resolution Imaging Science Experiment (HiRISE) and the Compact Reconnaissance Imaging Spectrometer for Mars (CRISM); and the ESA's orbiter *Mars Express* with its High Resolution Stereo Camera (HRSC) and near-IR spectrometer, OMEGA. In addition to the images from these missions, the

topographic information from MOLA, the HRSC, and HiRISE is particularly important for geomorphic studies of Mars. The global altimetry from MOLA enables broad views of the planet to be generated from its digital elevation model (DEM) and assessment of regional slopes. The local DEMs derived from stereoscopic images enable topographic resolutions as good as a few meters for selected areas.

Successful landings in the same period include the *Mars Exploration Rovers* (*MER*) *Spirit* and *Opportunity* (Squyres *et al.*, 2003), operating concurrently in two different regions of Mars and returning *in situ* data on surface compositions, physical properties, and active surface processes, and *Phoenix*, which landed in the north polar region, returning definitive evidence of water-ice on and near the surface. All of these spacecraft include instruments in addition to those noted here (such as the radar sounding systems on *Mars Express* and on the *MRO*) and have provided unparalleled information on Mars, much of which has yet to be "mined" by the science community.

7.3 Interior

As with the other terrestrial planets, the interior of Mars is partitioned into a core, mantle, and crust/lithosphere, reflecting differentiation following its formation. Although the thicknesses of each zone are not well known, our knowledge of the size and density of Mars enables some estimates to be made. The overall density of Mars is lower than that of Earth, even taking into account the difference in diameters and the corresponding adjustments for interior properties (i.e., **gravitational compensation**). If the core is mostly iron with some oxygen and sulfur, geophysical models suggest a core diameter of ~4,400 km. On the other hand, if the core is composed of nickel–iron, it would be only ~2,600 km in diameter. Thus, depending on the core size, the mantle would be 1,500–2,100 km thick with a density of 3.41–3.52 g/cm^3. By analogy with Earth, the martian mantle is likely to include various sub-zones, but this idea cannot be tested in the absence of seismic data.

The surface of Mars' crust (the compositionally distinct zone of a planet) is mostly mafic, according to remote sensing data, laboratory analyses of martian meteorites, and data obtained from landed spacecraft. The crustal thickness can be derived from gravity data determined from tracking perturbations in the paths of spacecraft in orbit around Mars, as was done for the discovery of mascons for the Moon. These data suggest that the crust is 70–80 km thick in the older terrain of the southern highlands and thins northward to 35–40 km in the lowland plains. It is also thin beneath large impact structures, such as Hellas (**Fig. 7.2**).

In contrast to the crust, the lithosphere (the physically rigid outer shell of planets) appears to be relatively thin in the older terrains, such as the cratered highlands, where it is modeled to be <20 km thick, and to be as thick as 100 km in the youngest terrains. This relationship is thought to reflect the thermal history of Mars, which in early times involved high heat flow from the interior, leading to a thin rigid zone at the surface. Progressively lower heat flow as the planet cooled enabled a thicker rigid zone to develop.

As with Venus, there is debate regarding the style of tectonics on Mars. Following global imaging of Mars by *Mariner 9*, terrains were searched for evidence of plate tectonics, but none was found. Nonetheless, geophysists modeled the thermal evolution of Mars and debated the possibility, but there was little agreement on the results.

One of the most exciting discoveries of the *MGS* mission was the presence and distribution of a remnant magnetic field in the ancient terrain of Mars. Because the terrains associated with the large impact structures

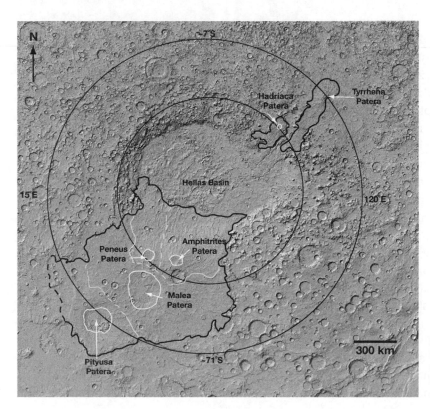

Figure 7.2. A shaded relief map of the Hellas impact basin, showing its inferred structural rings (concentric circles) and the Circum-Hellas Volcanic Province; white outlines show the positions of calderas and the associated volcanic deposits.

Hellas and Argyre do not show the remnant field, the conditions leading to the magnetic field must have been lost prior to the impacts, or before ~4 Ga.

The remnant magnetic field patterns are very similar to the bilateral symmetry of Earth's paleomagnetic stripes on the sea floor that are indicative of sea-floor spreading. Detailed analysis of the martian patterns even suggests the presence of transform faults to some planetary scientists. Although the *MGS* data are of low resolution, the discovery of the remnant field could mean that Mars had early in its history a magnetic field that has since been lost and that the stripping pattern would be consistent with a form of plate tectonics.

7.4 Surface composition

Information on Mars' surface and near-surface compositions comes mostly from the martian meteorites, remote sensing, and *in situ* measurements made from landers and rovers. More than four dozen meteorites have been found on Earth that bear the distinctive geochemical "fingerprints" of Mars when matched against measurements made by landers. Although these rocks display a wide range of crystallization ages (4.5 Ga to less than a couple of hundred million years) and histories, most are of basaltic or similar mafic compositions, confirming earlier interpretations of multispectral data from Earth-based telescopes.

Global or near-global remote sensing data are available from orbiters that flew in the 1990s to early 2000s. The TES instrument on the *MGS* (Christensen *et al.*, 2001) mapped mineralogies using the mid-IR part of the electromagnetic spectrum **(Fig. 2.14)** and showed that plagioclase feldspar, pyroxene, and olivine are common, along with some limited exposures of high-silica materials, all consistent with basalts as the primary rocks on the surface. As noted by TES designer Phil Christensen, the data also suggested the presence of andesitic materials, although the data can also be explained by invoking the presence of primary igneous glass and/or weathered basalts. The Gamma Ray Spectrometer (GRS) on *Mars Odyssey* mapped elemental compositions and showed that iron concentrations are high (as expected for basalts) and relatively uniform across the surface, while chlorine values vary substantially.

The near-IR multispectral mappers OMEGA and CRISM have returned compositional data that can be correlated with rock units of different ages. As outlined by French planetary scientist Pierre Bibring and his colleagues (Bibring *et al.* 2006), phyllosilicates characterize early Mars history, reflecting the presence of slightly alkaline liquid water, while younger rocks contain hydrated sulfates, and the youngest materials consist of anhydrous iron oxides.

Some of the remote sensing data have been "ground-truthed" by measurements from landed spacecraft and rovers. For example, TES data suggested the occurrence of crystalline hematite in some areas of Mars. Because this mineral forms predominantly in the presence of liquid water, these areas became of high priority for surface exploration and this resulted in the selection of the Meridiani Planum landing site for the rover *Opportunity*. Subsequent rover data confirmed the TES observations through analysis of the hematite "blueberries" **(Fig. 7.3)**. Similarly, *in situ* surface data show the dominance of basaltic compositions and the presence of a suite of chemically weathered minerals, including sulfates and halides. Surprisingly, though, there is a general absence of abundant carbonates in the remote sensing and lander data. While carbonates have been found in some martian meteorites and at some sites, they were expected to be rather common on Mars, given the carbon dioxide atmosphere, and their relative absence remains a puzzle. The calcium carbonate detected by *Phoenix* is thought to have resulted

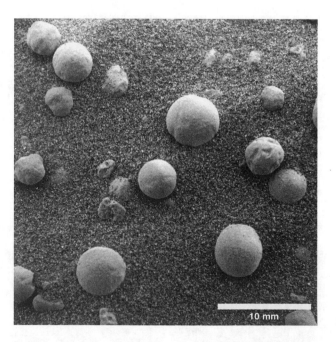

Figure 7.3. A Microscopic Imager view of spherical hematite grains, called "blueberries" because of their spectral properties, as seen at the rover *Opportunity* landing site (the image covers an area of 31 mm by 31 mm).

from reactions of atmospheric carbon dioxide with films of liquid water on soil particles.

7.5 Geomorphology

7.5.1 Physiography

Superficially, Mars can be divided into two primary terrains somewhat similar to the Moon, the *northern lowland plains* and the *southern cratered uplands*, constituting a so-called global crustal dichotomy (**Fig. 7.4**). The northern lowland plains are generally below the 0 km reference elevation. Lacking a sea level, elevations on Mars were first defined following the *Mariner 9* mission as the elevation at which atmospheric carbon dioxide pressure is 6.1 mbar, the triple point of water on Mars. Subsequently, MOLA data were used to redefine elevations referenced to the center of mass for Mars. Both reference systems are in the literature, and caution must be exercised in using the data uniformly, since there are significant differences between the two systems on the planet. Note also that both east and west longitudes are used in papers and on Mars maps as geographic coordinates; it is best to check for the system used in any given case.

The southern cratered uplands represent the final stages of early heavy bombardment and generally stand at elevations higher than 1 km above datum. As discussed below, this terrain has been heavily modified by surface processes, some of which are probably active today. The relative lack of impact craters exposed on the surface of the lowland plains suggests its relative youth. However, radar sounder data reveal the presence of dozens of circular features beneath the lowland plains surface that are thought to be large (100 to ~450 km) buried impact craters that formed in Mars' early history. Although there is debate as to whether the dichotomy formed as a result of endogenic processes (such as mantle convection) or represents one or more mega-impacts (the basins which would form the lowlands), it would appear that the dichotomy was present very soon after Mars' crust developed and that the lowlands were resurfaced.

Superposed on the border of the northern lowland plains and the southern cratered uplands is the volcanic Tharsis rise, which includes four enormous shield volcanoes and numerous other structures. The shields, dome volcanoes, and associated lava flows in this area constitute the Tharsis Volcanic Province. Other volcanic regions include Elysium, Syrtis Major, and the Circum-Hellas Volcanic Province.

The Hellas basin is found in the southern cratered uplands and, at more than 1,800 km across, is one of the largest impact structures in the Solar System. A similar but smaller impact basin, Argyre, is also found in the southern hemisphere. The floors of both structures have been partly filled with a variety of materials, while their ejecta deposits have been mantled or so heavily modified that they are not clearly recognized.

Valles Marineris is the "Grand Canyon" of Mars, which stretches more than 3,800 km eastward from the Tharsis rise across the northern equatorial region. This tectonic feature has multiple canyon systems, with some floors being more than 3 km below the surrounding terrain.

Mars has distinctive north and south polar regions that are easily seen by noting the presence of white deposits of frozen carbon dioxide (**Fig. 7.1**). These polar frosts

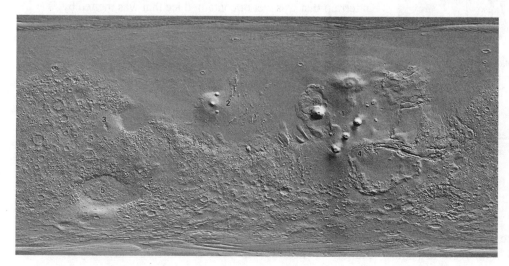

Figure 7.4. An image of Mars generated from Mars Orbiter Laser Altimeter data showing the "global dichotomy" between the sparsely cratered northern lowlands and the heavily cratered southern uplands. The Tharsis volcanic province (1) is superposed on the boundary between the two prominent terrains; also shown are the Elysium (2), Syrtis Major (3), and Syria Planum (4) volcanic provinces and the Hellas impact basin (5).

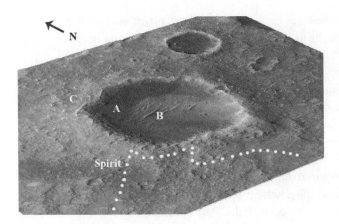

Figure 7.5. An oblique view of the 210 m in diameter Bonneville crater and the path of the *Spirit* rover; A, B, and C show the locations of various ripples and dunes, some of which fill many of the small craters on Mars, partly accounting for the shallow depths of the craters (NASA HiRISE frame PSP_00151_1655, part).

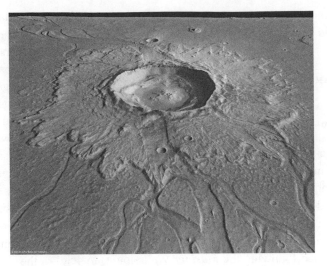

Figure 7.7. An oblique view of a complex impact crater 20 km in diameter in the northern hemisphere of Mars, showing terraced inner walls, a small central peak, and flow-like ejecta deposits; also shown are fluvial channels on which the impact was superposed (ESA HRSC image #435).

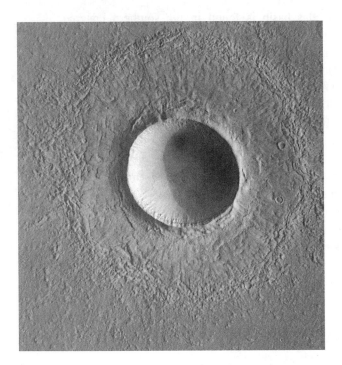

Figure 7.6. Small fresh impact craters on Mars are typically bowl-shaped, exemplified by this crater 2.6 km in diameter on northern Elysium Planitia (NASA PIA02084).

expand and shrink with the martian seasons and are underlain by more permanent water-ice, all interlayered with windblown dust and other deposits.

7.5.2 Impact craters

Martian impact craters range in size from the 1,800 km Hellas basin down to meter-size depressions seen at lander sites, many of which are thought to be secondary craters highly modified by erosion and in fill by windblown sediments **(Fig. 7.5)**. In general, primary martian craters show a progression in morphology with size from simple bowl-shaped features **(Fig. 7.6)**, through complex craters with terraced walls and central peaks **(Fig. 7.7)**, to central peak–ring structures, such as Lowell crater **(Fig. 7.8)**. This progression is similar to that on the Moon, but the transitions in morphology occur at smaller sizes on Mars, due in part to the higher gravity.

Ejecta deposits of many martian impact craters exhibit flow-like patterns **(Fig. 7.7)** distinct from those seen on the Moon and Mercury and suggest that the impacts occurred in terrain that was wet or contained ice that was melted by the impact. The mixture of rock, soil, and water is thought to have formed a slurry-like mass that was ballistically emplaced but then slid outward across the surface from the crater. If this idea is correct, then the distribution and timing of these impacts could provide insight into the nature of the surface through Mars' history, and numerous groups have been mapping the craters showing the flow-like patterns for this purpose (Barlow *et al.*, 2000). An alternative suggestion by Pete Schultz of Brown University is that flow-like ejecta could result from interaction of the ejecta with the atmosphere.

Craters in some regions show extensive modification, mostly by processes of gradation. Many craters in the higher latitudes are described as "pedestal" craters because they are

Figure 7.8. Lowell crater, west of the Argyre basin, is 201 km in diameter and exhibits the multi-ring structure typical of larger craters on Mars (NASA PIA02836).

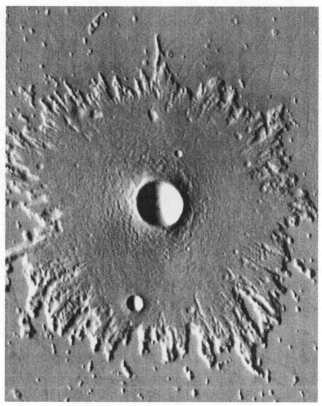

Figure 7.9. "Pedestal" craters, such as this structure 5 km in diameter in the northern hemisphere, are typified by a surrounding mesa-like platform (the pedestal), which corresponds approximately to the extent of ejecta deposits (NASA THEMIS V02215005).

on a mesa-like platform that corresponds roughly to their ejecta deposits. This effect was first seen on *Mariner 9* images, and it was suggested that impacts occurred on materials that could be deflated but that the mesas were rendered more resistant to erosion by the presence of the ejecta that armored the surface. Although this idea still has merit, it is difficult to imagine how fine-grained ejecta at the distances seen in **Fig. 7.9** would be effective in resisting deflation.

Impact crater size–frequency distributions (i.e., crater counts) provide a means for assessing surface ages in the absence of radiogenic age determinations. As discussed in **Chapter 2**, crater count ages can be derived assuming that (1) only primary impact craters are counted (or that secondary craters are taken into account), (2) there has been no "erasure" of craters, as from erosion, and (3) a valid method for calibrating the data exists. Mars is particularly challenging on all three issues. First, non-impact processes, such as volcanism, can produce circular depressions, and much of Mars involves volcanic surfaces, as described below. Moreover, planetary geologist Alfred McEwen and colleagues have noted the abundance of secondary craters on Mars, which could strongly influence the distributions of craters, especially in the smaller sizes. Second, much of the surface has been highly modified by wind, water, and (in places) glaciation, leading to erosion of impact craters.

Third, in the absence of radiogenically dated surfaces on Mars, there is no direct calibration for the crater counts, and the data from the Moon (**Fig. 2.10**) must be extrapolated to Mars. This requires making adjustments to account for the difference in factors such as gravity (higher **g** results in smaller craters on Mars than on the Moon for the same impact event) and Mars' proximity to the asteroid belt, which could lead to more impacts as a function of time in comparison with the Moon.

The issues surrounding crater counts for Mars have been partly addressed, employing certain assumptions and models. For example, volcanic craters are assumed to be recognizable from their morphologies (e.g., they tend to be non-circular), while the issue of secondary craters was analyzed statistically and shown to be resolvable (Werner *et al.*, 2009). The physics of gravity scaling of impacts is relatively well understood, and crater sizes can be adjusted accordingly. Extrapolating the crater count curve from the Moon to Mars, however, is problematic, keeping in mind that the curve for the Moon itself has

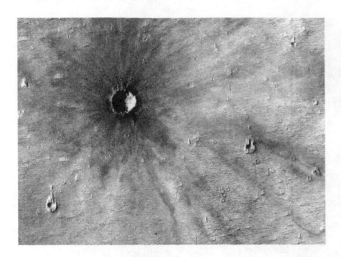

Figure 7.10. Repeated imaging of the same areas reveals newly formed features; this small (54.5 m) impact crater formed between February and July 2005; the dark ejecta is fragmented basalt excavated from beneath a lighter mantle of dust (NASA HiRISE image ESP_011425_1775).

Figure 7.11. The Olympus Mons shield volcano in the Tharsis province is more than 600 km across, making it one of the largest volcanoes in the Solar System (NASA *Viking Orbiter* 646A28).

built-in uncertainties, especially for the very early and the very young ages. For Mars, various astrophysical models are used to account for factors such as the proximity to the asteroid belt and potential contributions from comets. In addition, the discovery of recently formed small impact craters **(Fig. 7.10)** helps establish the current flux of incoming objects on Mars. Despite the uncertainties and assumptions, in the absence of any other dates for Mars, crater counts remain as the sole technique for assessing surface dates.

7.5.3 Volcanic features

Mars has some of the most impressive volcanoes in the Solar System, with the Olympus Mons shield **(Fig. 7.11)** and similar features in the Tharsis region. Mars' volcanoes can be classified into constructs formed by central vents (i.e., "point-source" eruptions), such as Olympus Mons, and volcanic plains emplaced from fissures or inferred vents lacking discernible surface features **(Table 7.2)**. The total areal coverage of martian volcanic materials is more than half the planet's surface. While most attention has been focused on the shield volcanoes, various plains units of volcanic origin represent most of the volcanism on Mars.

The central volcanoes include shields, domes, highland patera, and a unique structure, Alba Patera. The shield volcanoes have all the attributes of classic Hawaiian shields in that they are composed of countless flows of basaltic composition, commonly emplaced by lava tubes and channels erupted from central vents, or calderas, or from parasitic flank eruptions that built gently sloping structures. The martian shields are remarkable on account of their great size; Olympus Mons is more than 550 km across, stands 21 km above the 0 km reference datum, and contains a summit caldera some 80 km across. Some volcanoes, such as Hecates Tholus in the Elysium province, appear to be blanketed with fine-grained basaltic material inferred to be pyroclastics. The steeper slopes of some shields suggest a more evolved magma and a transition from effusive to mild explosive activity.

Tharsis Tholus exemplifies dome volcanoes **(Fig. 7.12)**, which are characterized by flank slopes that are steeper than the shields. They are inferred to represent either more silicic lavas or lower rates of effusion that would produce short flows that piled on top of each other, rather than spreading great distances from their sources. Some dome volcanoes could simply be the steeper summit regions of shield volcanoes for which more gentle flanks have been buried. Unfortunately, many of the central-vent volcanoes on Mars are covered with a mantle of windblown dust that precludes obtaining compositional information by remote sensing. TES data, however, suggest the presence of silica-rich materials in limited areas, and it is reasonable to expect evolved magmas to be present, given the extensive volcanism. For example, the Syrtis Major volcano has remote

Table 7.2.   **Classification and extent of Martian volcanic features (from Greeley and Spudis, 1981)**

Type	Example	Characteristics	Extent $(10^6 km^2)$	Percentage of Mars' surface
Central volcanoes				
Alba	Alba Patera	Unique, extremely low-relief shield-like volcano; sheet and tube-fed flows	1.13	0.78
Highland paterae	Tyrrhena Patera	Low-relief, degraded, radially textured volcanoes	0.23	0.16
Shields	Olympus Mons	Broad, moderate-relief central vent volcanoes; tube-fed flows	1.23	0.85
Domes	Tharsis Tholus	Steeper-sided central volcanoes	0.08	0.05
		Total area of central volcanoes	2.67	1.84
Volcanic plains				
Simple flows	Hesperia Planum	Regional plains; wrinkle ridges only; no-flow-lobes	42.08	29.22
Complex flows	Tharsis Plains	Complex flow units; flow-lobes abundant; rare wrinkle ridges	9.16	6.36
Undifferentiated	Northern plains	Plains of uncertain origin (flow-lobes rare), although probably volcanic	27.37	19.00
Questionable	Aureole materials	Modified units associated with volcanic units	5.92	4.11
		Total area of volcanic plains	84.53	58.69
		Total volcanic surface area on Mars	85.20	60.53

Figure 7.12. Tharsis Tholus is a dome volcano of dimensions about 110 km by 170 km and has a large complex caldera at its summit (NASA *Viking Orbiter* 858A23).

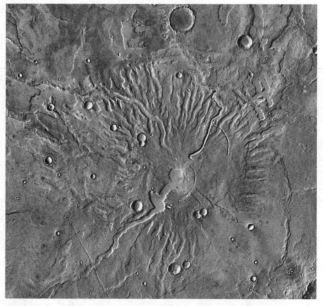

Figure 7.13. Tyrrhena Patera, northeast of the Hellas basin, is a central volcano thought to be an eroded ash shield; the radial features are probably volcanic channels that have been modified by fluvial erosion; the youngest features are lavas erupted from the central caldera, which flowed down the channel to the southwest; the area shown is about 140 km across (NASA THEMIS mosaic).

sensing signatures suggestive of dacite, which could indicate a more evolved magma than basalt.

Highland paterae consist of very low-profile volcanoes that have calderas, radial flows, and channels **(Fig. 7.13)**.

These represent the oldest volcanic constructs on Mars, dating from 3.9 Ga. As the name implies, they occur primarily in the cratered uplands of Mars with most paterae in the Circum-Hellas Volcanic Province. The characteristics of the patera flanks suggest erosion by relatively "soft" material, and it has been suggested that these volcanoes are ash shields from explosive eruptions. Some paterae appear to have evolved to more effusive eruptions, as evidenced by superposed lava flows.

Alba Patera is a huge, unique, very-low-profile feature in the northern part of the Tharsis region. Radial flows extend more than 400 km from its central caldera, with many of the flows having been emplaced through lava tubes. Their great length suggests very fluid lavas, probably of basaltic or ultra-mafic compositions. Alba Patera covers some 1.13×10^6 km^2, nearly equal to all of the shield volcanoes on Mars combined. Were it not for its very low relief, many planetary scientists might consider it the largest single volcano in the Solar System.

Ridged plains are found in many parts of Mars (**Fig. 7.14**) and are very similar to those seen on the Moon, Mercury, and Venus. Although such ridges are primarily tectonic features, they are thought to be characteristic of deformed thick basalt flows. Mare-type ridges are also seen in materials filling the floors of the large calderas on Mars. Coupled with compositional data from remote sensing showing basaltic compositions for the martian ridged plains, these considerations lend support to the interpretation that such plains are volcanic. Referred to as simple flows in **Table 7.2**, the ridged plains are considered to represent flood eruptions that buried their eruptive fissures. In contrast, complex flows are composed of thin, multiple flows that form distinctive lobes (**Fig. 7.15**), many of which were emplaced through lava tubes and channels. As such, they reflect lower volumes of lava and slower rates of effusion than the more massive simple flows of the ridged plains. Transitional between the large shield volcanoes and the lava flows are small shield volcanoes (**Fig. 7.16**).

Undifferentiated flows (**Table 7.2**) constitute much of the northern lowland plains. Although these plains involve a wide variety of materials resulting from aeolian,

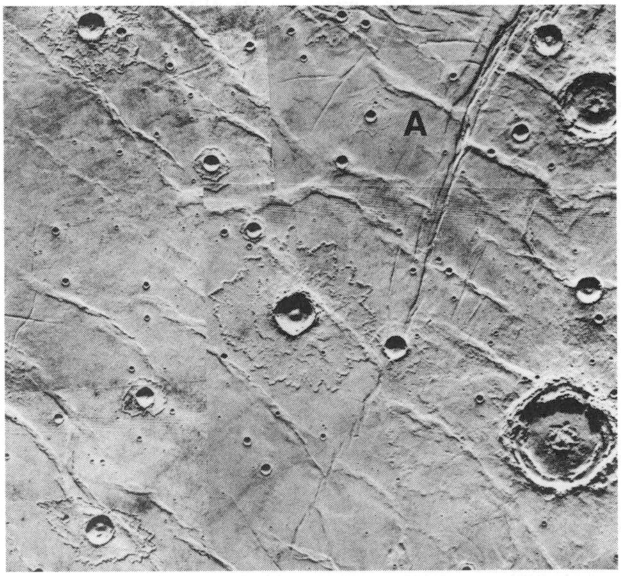

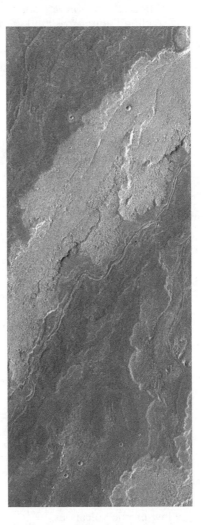

Figure 7.14. An image mosaic of ridged plains in the southern hemisphere; the ridges resemble those seen on the Moon, Mercury, and Venus; the ridge at "A" appears to have been offset by right-lateral displacement by the fracture; most of the craters display ejecta flow-lobes (NASA *Viking Orbiter* 610A01–3, 608A09).

Figure 7.15. Complex lavas in the Arsia Mons area, showing multiple flow-lobes and lava channels; the area shown is 20 km by 47 km (NASA THEMIS orbit 35460).

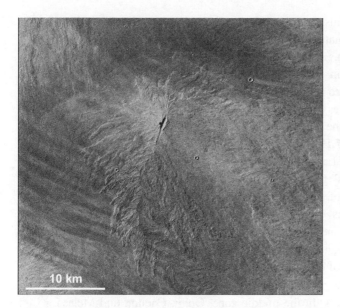

Figure 7.16. A small shield volcano on the plains south of Pavonis Mons, showing multiple thin lava flows erupted from an elongate central vent (ESA HRSC frame 0891).

Figure 7.17. The Amenthes Rupes lobate scarp stands higher than 1 km and is thought to be a low-angle thrust fault; the area shown is about 500 km by 500 km (NASA *Viking Orbiter* mosaic).

periglacial, and possibly fluvial and lacustrine processes, the bedrock is considered to be volcanic because flow lobes are seen in some areas and remote sensing data are indicative of basaltic compositions. The "questionable" category of volcanic material **(Table 7.2)** refers to deposits found around some of the large shield volcanoes and mapped as aureole material. These deposits are interpreted as being debris mass-wasted from the flanks of the volcanoes, perhaps comparable to the **lahars** seen on Earth which consist of volcanic mudflows triggered by eruptions.

The presence of water and ice on and near the surface, at least in the past, suggests that magma–water interactions were common. Such interactions would explain the formation of small cones in many areas, as well as the explosive eruptions to form the ash shields in the Circum-Hellas Volcanic Province. In addition, the Medusae Fossae Formation in the equatorial region has been suggested to be a vast volcanic ash deposit that could have involved volatile-rich explosions.

7.5.4 Tectonic features

Surface features resulting from tectonic deformation on Mars include both compressional and extensional structures. These features are found in association with volcanoes, impact structures, local deformation, and regional-scale deep-seated deformation of the lithosphere.

Extensional structures include joints, grabens that are as large as several kilometers wide, hundreds of meters deep, and tens of kilometers long, and troughs (*chasmata*) of Valles Marineris that are 1,000 km long and as wide as 100 km. The smaller extensional features are thought to involve high-angle faults that, when projected to the subsurface, might converge at depths as great as several kilometers. The larger troughs of Valles Marineris could reflect fracturing completely through the lithosphere to greater depths. On the other hand, some extensional features could reflect more shallow deformation resulting from intrusion by dikes, or swarms of dikes, including parts of the Valles Marineris rift system.

Compressional features include ridges and lobate scarps. The ridges are comparable to mare ridges on the Moon and are linear-to-arcuate features as large as 100 m high, several hundred kilometers long, and more than 10 km wide. As on the Moon, they often consist of a broad, gentle "swelling" surmounted by a narrow crenulated ridge and are considered to reflect folded layers of rocks, such as lava flows, underlain by thrust faults. The Hesperian ridged plains in the southern hemisphere and Lunae Planum just north of the equator are characterized by ridges and are inferred to be thick basalt flows that were subjected to regional compression.

Lobate scarps on Mars **(Fig. 7.17)** are comparable to those on Mercury and can reach heights of several kilometers. Most lobate scarps are found in the ancient heavily cratered terrains and resulted from large thrust faults.

The Tharsis rise (or "bulge") imposes the greatest influence on the tectonic patterns on Mars. The rise stands some 10 km above datum and has related tectonic patterns covering one-fourth of the martian surface. Following its discovery during *Mariner 9*, two general ideas emerged for the origin of the Tharsis rise: it could reflect dynamic uplift and support by one or more mantle plumes, or it could be an enormous pile of volcanic materials loaded on an elastic lithosphere. Gravity data and precise topography from the *MGS* have enabled partial testing of these ideas. Coupled with refined geochemical models of the interior, it now appears that the time-frame needed to develop dynamic plumes of the size required to support Tharsis probably exceeds the very age of Mars. Consequently, the loading model is more widely favored, as reviewed by planetary scientists Matt Golombek and Roger Phillips (2010). It is also important to note that work by Solomon and Head (1982) suggested that magma fed from plume(s) and erupted onto the surface could provide a type of "feedback" in which the fractures providing the conduits to the surface could be further opened by loading of the erupted materials onto the surface.

Mapping the Tharsis radial fracture systems **(Fig. 7.18)** and their stratigraphic relations to the lavas and other materials in the region enables the centers and timing deformation to be derived. As summarized by Carr (2006), most of the faults associated with Tharsis formed early in Mars' history and the basic structure has not changed in the last 3 Ga.

Valles Marineris is a vast rift system stretching eastward from the Tharsis rise, with which it might be associated **(Fig. 7.19)**. This system can be divided into three parts, Noctis Labyrinthus in the west, the central rift zone, and the "chaos" terrain in the east. Noctis Labyrinthus is on the flank of the Tharsis rise and is marked by intersecting grabens and elongate pits, suggestive of extension and subsidence likely associated with the Tharsis uplift. The main canyon reflects extensional rifting that has formed a series of parallel chasmata, some of which are deeper than 7 km, exposing thousands of meters of layered deposits in the walls. In the middle zone, Ophir, Candor, and Melas chasmata merge to form a central depression 600 km wide. Some of the canyons, such as Hebes Chasma, form closed depressions, suggestive of substantial collapse. Local volcanism within the canyonlands along the bases of some canyon walls was identified by geologist Baerbel Lucchitta and is of basaltic composition as indicated by high-resolution multispectral data from the CRISM instrument.

The eastern part of Valles Marineris grades into "chaos" terrain, characterized by extensive mass wasting that forms a jumbled relief. The chaos terrain appears to "feed" an extensive set of huge channels that drain east and northward into the lowland plains. Although the chaos terrain includes vestiges of possible extension,

Figure 7.18. An oblique view showing grabens and other fractures associated with the Tharsis rise in the southern hemisphere where lavas have flooded into the heavily cratered terrain; the higher density of fractures in the cratered terrain suggests that most deformation occurred before the latest emplacement of the lavas, although some lavas are also fractured, indicating continued tectonism (ESA HRSC #095).

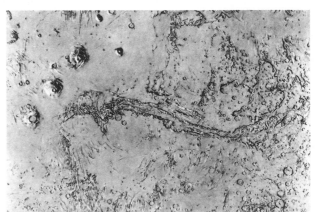

Figure 7.19. A shaded relief map of the Valles Marineris region. This area, known collectively as the canyonlands, is composed of three elements: Noctis Labyrinthus in the west, the central canyon system, and an eastern area characterized by chaotic terrain and outflow channels. The area shown is 3,500 km by 6,000 km (courtesy of the US Geological Survey, Flagstaff).

Figure 7.20. This small landslide found on the inner wall of an impact crater is typical of many such features where slopes are steep; the area shown is 17 km by 20 km (NASA THEMIS orbit 27216).

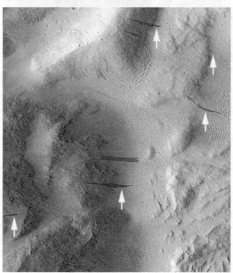

suggested by linear scarps roughly parallel with the fractures in the main canyon system, the primary formational process was mass wasting on a grand scale.

7.5.5 Gradation features

Like Earth, Mars exhibits a rich history of gradation involving wind, water, mass wasting, and periglacial processes. Mass wasting has enlarged the canyonlands, forming huge scarps and landslides along the walls, and generating the chaos terrain. The aureole deposits noted above in association with the large shield volcanoes likely involved mass wasting to form lahars.

Small landslides are ubiquitous on Mars, especially along over-steepened slopes, as on impact crater walls **(Fig. 7.20)**. An active landslide was even "captured" in an image in early 2008 that showed dust and huge blocks of rock falling down a 700 m scarp. Some active mass wasting features, called "dark slope streaks," continue to form today **(Fig. 7.21)**. They represent exposure of dark materials when bright dust or bring water slides downslope. These

Figure 7.21. MOC images of "dark slope streaks" (arrows) that formed between August 1999 (top image) and April 2001 (bottom image) in terrain north of Olympus Mons; the area shown is 3 km by 3.5 km (NASA PIA03226).

features fade with time, as dust settles from the atmosphere and mantles the dark substrate or as water evaporates.

Other flow-like features seen on images are likely to be glaciers **(Fig. 7.22)**, as suggested by Jim Head *et al.* (2010) from their analysis of *Mars Express* data. Results from the radar system on the MRO reveal the presence of ice beneath many of the putative glacial features, even in the lower latitudes closer to the equator.

The search for evidence of past and present water on Mars, either as ice or in liquid form, is a fascinating story. When the first spacecraft images of Mars were returned in the 1960s, Mars was thought to be a dry, Moon-like planet

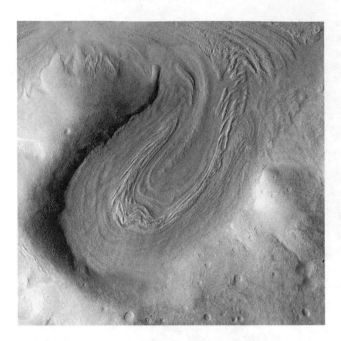

Figure 7.22. An oblique view of an inferred glacier 5 km wide in the Protonilus Mensae area, showing moraine-like ridges (NASA MRO CTX frame).

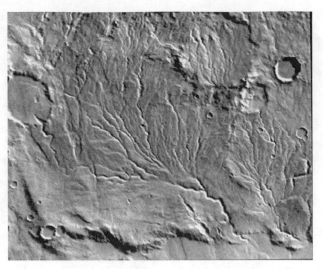

Figure 7.23. These integrated valley networks in the southern cratered highlands suggest surface run-off from precipitation; the area shown is about 200 km across (from Mars Digital Image Map, image processing by Brian Fessler, Lunar and Planetary Institute).

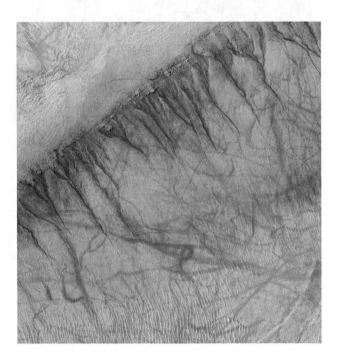

Figure 7.24. Gullies seen in Kaiser crater in the southern hemisphere; the gullies are thought to form from the seepage of water derived from a distinctive rock layer cropping out on the inner crater wall; also visible are numerous aeolian ripples (lower part of image) and squiggly dark streaks left by the passage of active dust devils; the area shown in this MOC image is 3 km by 3 km (NASA PIA03753).

dominated by impact cratering. The first clues as to the possibility of water came with the *Mariner 9* images showing channels interpreted to have been cut by fluvial activity in the past when the climate was sufficiently warm to allow liquid water on the surface. The largest features are the so-called outflow channels, some of which are tens of kilometers across and more than 1,000 km long. Detailed images of these features reveal complex terraces and inter-braided channels. Some channels have large impact craters "interleaved" with channel elements, suggesting either prolonged flow or episodic flow. The sources for the large fluvial features include the chaos terrain and fractures that appear to have released great quantities of ground water. The presence of giant ripples along some channels suggests that the water was released as enormous, catastrophic floods, similar to the channeled scablands in eastern Washington that were formed by the release of water from the bursting of glacial lakes.

In addition to outflow channels, other water-carved features are integrated valley networks **(Fig. 7.23)**. These are thought by many planetary geologists to represent run-off from precipitation, although others suggest that they form by seepage from multiple springs. For example, gullies on some scarps were discovered in MOC images and were later found to be actively forming today. Their morphology and geologic relations to the

terrain in which they occur suggest local springs from distinctive rock layers **(Fig. 7.24)**. One of the most striking features associated with water is the delta seen in Eberswalde crater **(Fig. 7.25)**. While surface conditions

2 km

Figure 7.25. This complex delta consists of alluvium deposited from a channel that flowed into Eberswalde crater; overlapping channel segments and deposits in the delta indicate repeated flooding events (NASA PIA04293).

on Mars today are borderline for liquid water, the probable presence of salts would lower the freezing temperature and enhance its probability.

Each new mission to Mars has not only provided a growing body of evidence for the presence of water but also facilitated the making of estimates of the amount that might be present. The total amount of water is commonly expressed as the depth of water that would cover the entire globe if Mars were a smooth, spherical object. These estimates range from a depth of few meters to nearly 100 m.

The *Mars Exploration Rovers*, *Spirit* and *Opportunity*, revealed evidence for groundwater modification of surface materials from the identification of minerals that commonly form in association with water, such as the hydrated iron oxide goethite. The *Spirit* rover analyzed outcrops within the Columbia Hills of Gusev crater and revealed loose, clastic materials and indurated deposits rich in sulfate and silica. Although several diverse interpretations have been proposed to explain the origins of these materials, including volcanic pyroclastic emplacement and local fumarolic activity, most ideas require alteration in the presence of water as discussed by Ray Arvidson and *MER* colleagues (Arvidson *et al.*, 2008). After a journey of more than 30 km from its landing site,

Opportunity reached crater Endeavor in 2011 and began exploration of this 22 km impact structure. Analysis of a small boulder on the crater rim revealed the presence of high levels of zinc and bromine, which are considered to be good indicators of deposition in warm water.

The *Phoenix* landing site was selected in the north polar region, Vastitas Borealis. Shallow trenches dug by the lander's mechanical arm revealed the presence of hard-as-rock water-ice 5–18 cm beneath the surface. Coupled with the discovery of various salts and minerals, and observations of active snowfall and frost formation in late summer, this shows that water clearly plays an important role in the chemistry and geology of the polar region. Data from the *Mars Reconnaissance Orbiter* show that ice and water processes also occur in the lower latitudes. For example, Shane Byrne of the University of Arizona examined newly formed impact craters and found water-ice that "faded" with time, the inference being that the impact penetrated several meters into a zone of relatively clean ice, which subsequently sublimated as it was exposed to the atmosphere.

Mars' polar regions have long been recognized as distinctive from the rest of the planet. Sparsely cratered layered terrains include alternating deposits of dark and light materials thought to be ice (both carbon dioxide and water) and dust settled from the atmosphere. Some layers are less than 1 m thick and could represent seasonal or longer-term cycles perhaps associated with variations in the position of Mars' spin axis. The total thickness of the south polar ice cap is ~3 km and consists of about 85% carbon dioxide ice and 15% water-ice. Shallow-depth radar mapping from the ESA's *Mars Express* orbiter reveals the presence of massive buried deposits of carbon dioxide ice in the south polar layered terrain, as reported by Roger Phillips *et al.* (2011). If released to the atmosphere during times of Mars' high obliquity, the atmospheric surface pressure could nearly double, enhancing aeolian activity and the possibility of there being liquid water on the surface.

In addition to the polar ice caps, various periglacial features are seen. For example, **Fig. 7.26** shows irregularly shaped pits in the residual southern carbon dioxide cap; similar features imaged repeatedly changed shape with time, which is attributed to the sublimation of ice. So-called "spider terrain" is also seen in the south polar regions. This terrain consists of radial fractures **(Fig. 7.27)**, some of which have associated dark streaks **(Fig. 7.28)** that are interpreted to result from carbon dioxide geysers that eject dark rocky material. These are

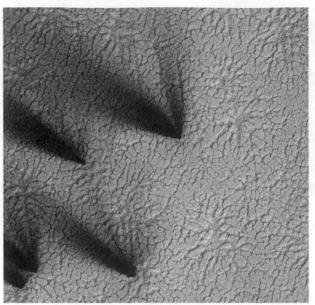

Figure 7.28. Dark deposits thought to result from geysers generated by heating of the surface and release of volatiles from shallow subsurface reservoirs (NASA HiRISE PSP_003364_0945, part).

Figure 7.26. "Cottage cheese terrain," dubbed thus for its pitted morphology, occurs in the south polar region and is thought to result from the sublimation of ice, leaving irregular-shaped depressions. Repeated imaging shows that the features enlarge with time. This image covers an area 2.5 km by 3.5 km; the pits are about 4 m deep (NASA MOC M0306646, part).

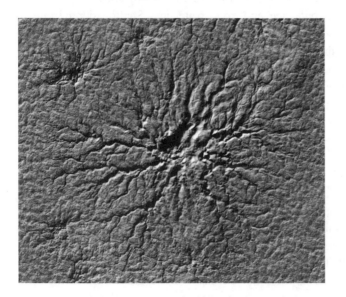

Figure 7.27. "Spider terrain" is characterized by radial fractures, some of which leave fan-shaped deposits of dark material; the area shown is 540 km by 470 m (NASA HiRISE ESP_014413_0930 part).

thought to result from a type of "solid state" greenhouse in which sunlight penetrates the ice, heats the contact zone with the underlying rocky materials, and sublimates the ice to form a high-pressure gas that ruptures the ice explosively.

While recently formed craters and the presence of water on and near the surface show that impact, hydrologic, and periglacial processes are active today, by far the most prevalent surface modifications are from aeolian activity. Even before the first mission flew to Mars, Earth-based telescopic observations showed seasonal patterns of brightening, some of which were attributed to dust storms. Subsequent data show that dust is lifted by the wind locally and regionally, with some storms completely enshrouding the globe in as little as one month. One such global storm was raging across the planet when the *Mariner 9* orbiter arrived, and imaging of the surface had to await the settling of dust from the atmosphere. In contrast to common windblown dust on Earth, which is 10–20 microns in diameter, martian dust is extremely fine, of diameter a few microns. Very strong winds are needed to move such fine grains, yet measurements from landers suggest that such winds are rare. However, the discovery of dust devils in *Viking Orbiter* images suggests an effective mechanism for lifting dust into the atmosphere. Dust devils are local vortexes formed by heating of the surface, which act like vacuum cleaners in that they suck dust from

Figure 7.30. Barchan sand dunes in Chasma Boreale; slip faces on the northeast sides of the dunes (toward the upper right) indicate prevailing winds in that direction; the area shown is about 1.5 km wide (NASA MOC 2–147).

Figure 7.29. The active dust devil on the horizon over Gusev crater is one of more than 800 features observed from the rover *Spirit* during three complete dust devil seasons. Each season began in late martian spring when the surface began to heat with the approaching summer. Also visible in this view is a ripple composed of sand, showing that both sand and dust are present on Mars (NASA *Spirit* image).

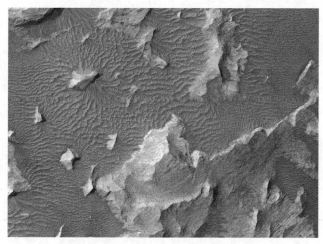

Figure 7.31. Sand dunes are seen in many parts of Mars and, just as on Earth, their location and orientations are often controlled by local topography; this image was taken in the Iani Chaos region and covers an area of about 1.0 km by 0.8 km (NASA HiRISE PSP_008100_170).

the surface **(Fig. 7.29)**. Once the dust has been lifted, it is easily carried aloft and transported long distances in suspension, even in the thin martian atmosphere.

More than 800 active dust devils were recorded by the rover *Spirit*, and dozens more have been imaged from orbit. The passage of dust devils leaves distinctive dark tracks, which result from the removal of bright dust to expose a darker substrate. It is thought that as much as half of all the dust in the atmosphere has been lifted by dust devils.

Each subsequent mission to Mars has revealed even more wind-related features like those first discovered on *Mariner 9* images, including dunes, ripples, wind-sculpted hills (yardangs), and wind-eroded rocks (ventifacts). Most of the dunes are transverse features (i.e., their axis is normal to the formative wind direction) and have similar sizes and shapes **(Fig. 7.30)** to dunes on Earth. High-resolution images of nearly all regions on Mars show dune fields and large patches of ripples, the orientations of which are controlled by surrounding topography **(Fig. 7.31)**. The *Opportunity* rover provided the first robotic views of outcrops on another planet. As the rover continued its exploration, it traveled to the rim of an impact crater and imaged well-formed cross beds exposed in the crater rim, which

have been interpreted as representing fossil sand dunes **(Fig. 7.32)**.

Yardangs on Mars have been imaged in exquisite detail by the HRSC **(Fig. 7.33)**. These features show that relatively easily eroded materials, such as moderately indurated fine-grained ash or dust, are common in some deposits such as the Medusae Fossae Formation. Evidence of wind erosion is also common at many landing sites. Ventifacts **(Fig. 7.34)** show the results of blasting by sand, perhaps derived from nearby sandy ripples **(Fig. 7.29)**. Because very strong winds are needed to set particles into motion in the low-density martian

Figure 7.34. Ventifacts are wind-abraded rocks, such as these imaged by the rover *Spirit* in Gusev crater; the basaltic rocks show elongated grooves that were sand-blasted by winds coming from the right side of the image (NASA *Spirit* Pancam image).

Figure 7.32. An outcrop imaged by the *Opportunity* rover on the rim of Victoria, an 800 m impact crater. The outcrop is about 12 m high and exhibits cross-bedding that is thought to reflect paleo-sand dune deposits (NASA PIA10210).

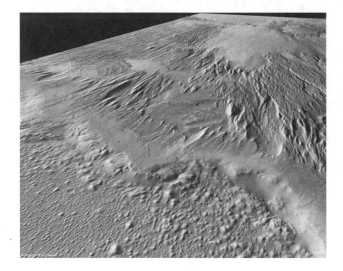

Figure 7.33. An oblique view of the Medusa Fossae Formation in the Eumenides Dorsum region, showing yardangs, the elongate streamlined-shaped hills cut into the formation (ESA HRSC image).

Figure 7.35. Bright (high-albedo) wind streaks in the Cerberus region; these and similar features indicate the prevailing wind direction at the time of their formation (in this case, winds blew from the upper right to the lower left). The area shown is 156 km by 184 km (NASA *Viking Orbiter* 545A54).

atmosphere **(Fig. 3.33)**, sand grains are accelerated to high speeds and are effective agents of abrasion.

Contrasting albedo patterns that appear, disappear, or change their size, shape, or orientation are very common and are called **variable features (Fig. 7.35)**. These include dark (low-albedo) wind streaks, bright (high-albedo) wind streaks, and dust devil tracks left by the passage of dust devils, noted above. The albedo contrasts are thought to result from removal or deposition of wind-blown particles. Thus, keeping in mind the general correlation with particle size, high albedo can be indicative of fine-grained material such as dust, whereas lower albedo can result from larger grains, such as sand. Although composition must also be considered, bright wind streak features are considered to be dust deposits, while dark wind streaks are sand deposits or exposures of a dark substrate from which dust has been eroded. Thousands of variable features have been mapped, and their orientations are used to assess the prevailing wind directions at the time of their formation.

7.6   Geologic history

The first global geologic maps of Mars were made following the *Mariner 9* mission and included a geologic time scale. Derived by Mike Carr and the late Dave Scott, both of the US Geological Survey, the time scale includes three periods, the Noachian (oldest), Hesperian, and Amazonian (youngest). Although this time scale is relative, impact crater size–frequency distributions for the major units associated with these periods enable estimates of "absolute" dates. As discussed in **Section 7.5.2**, the dates are based on extrapolations from lunar crater counts that are calibrated against radiogenic ages from samples returned from the Moon. The ages of various events have been coupled with the wealth of new data for Mars since *Mariner 9* to refine the general geologic sequence, as discussed by Mike Carr and Jim Head (2010).

The Noachian Period (4.6 to 3.5 Ga) began with the solidification of the martian crust from an inferred magma ocean. In this time, the global crustal dichotomy formed and the uplift of the Tharsis rise was initiated. The southern uplands preserve impact craters that represent the final stages of heavy bombardment in the inner Solar System and include the Hellas and Argyre basins. This record is suggested to underlie the northern lowlands, where vestiges of large circular structures are seen in the subsurface by radar sounding from orbit. The discovery of the remnant magnetic field in Noachian-age rocks (but not in younger materials) suggests that the interior of Mars was able to generate a magnetic field.

Noachian-age surfaces have been heavily modified by a wide variety of processes, including volcanism, tectonic modification, and gradation by wind and water, but the exact timing of these modifications is poorly constrained. Jean-Pierre Bibring, the lead on the French OMEGA near-infrared spectrometer on *Mars Express*, found that phyllosilicates are common in Noachian materials; because these clay minerals form in the presence of alkaline water, he and his colleagues proposed that the early climate on Mars was much warmer and wetter than present-day conditions. The integrated valley networks suggest the presence of surface run-off in this period of Mars' history, which would be consistent with Bibring's interpretation.

The Hesperian Period (3.5 to 1.8 Ga) is marked by the eruptions of vast sheets of lava that formed the ridged plains, including the type example, Hesperian Planum. Central volcanism included the formation of the highland paterae, Syria Planum, and early eruptions in the Elysium and Tharsis provinces, coupled with the continued development of the Tharsis rise. In addition to extensive volcanism, most of the large outflow channels were cut in this period.

The extensive Hesperian volcanism that occurred in the Hesperian Period could account for the formation of sulfate minerals and related materials revealed in remote sensing data obtained from orbit. These minerals might have formed from volcanic sulfur dioxide combined with acidic water. The transition from Noachian to Hesperian times also saw the end of the martian magnetic field. It is thought that the loss of the magnetic field would have exposed Mars to the solar wind, which would have stripped away Mars' atmosphere, leading to a colder, drier environment. Thus, the outflow channels would reflect release of water from the subsurface rather than from surface runoff, as suggested by their association with the chaos terrain.

The Amazonian Period (1.8 Ga to the present) is the youngest subdivision of time on Mars. However, one must remember that this time corresponds to all of the Phanerozoic Eon and much of the Proterozoic Eon of the Precambrian on Earth. Tectonic deformation and volcanism continued from the Hesperian Period but were much less pronounced. Very young lava flows of limited areal extent are seen in many areas, and active volcanism cannot be ruled out with currently available data. Gradation of all types modified the surface through the agents of wind, ice, gravity, and local liquid water; many of these processes continue today, along with the formation of small impact craters. In the absence of abundant volcanism and liquid water on the surface in this period, slow chemical weathering takes place through oxidation of the iron-rich basaltic materials by atmospheric peroxides (which have been detected by the *Phoenix* lander), which leads to the formation of iron oxides (i.e., "rust") that gives Mars its red color.

Assignments

1. Name the three primary regions of volcanic activity on Mars and describe the key differences in the style(s), ages, and areal extents of volcanism among them.

2. What is the evidence for any Earth-like plate tectonics on Mars in its geologic history?

3. Discuss the primary ways in which *gradation* operates on Mars in the current environment in comparison to Earth.

4. Go to the relevant NASA website(s) and find images for the floor of Mars' Gusev crater; describe the different types of surface features in terms of the geologic processes responsible for their formation and note whether any of these processes are active today (be specific and provide the evidence for your answer).

5. Examine **Fig. 7.7** showing a 20 km impact crater on Mars. Find images of impact craters of comparable size on the Moon and Mars and describe the differences in the morphology of the ejecta deposits; discuss the reason(s) for the differences, if any, among the craters.

6. *Viking Lander 1*, *Mars Pathfinder*, *Spirit*, and the *Mars Science Laboratory* all landed in different areas on Mars. Compare and contrast each site with regard to the potential for astrobiology.

CHAPTER 8

The Jupiter system

8.1 Introduction

Jupiter, one of the brightest objects in the sky, was named after the mightiest of the Roman gods because of its dominance. More massive than the other planets combined, Jupiter with its rings and satellites has been likened to a "miniature Solar System." More than 400 years of telescopic observations and, more importantly, flights of the *Pioneer, Voyager, Galileo, Cassini,* and *New Horizons* spacecraft have yielded images and other data for the Jovian system that are among the most spectacular in the Solar System.

8.2 Exploration

Scientific exploration of the Jupiter system was begun in 1610 by Galileo Galilei. He had been waiting many days for the night-time skies to clear so that he could try out a new, technically advanced instrument. But it was January in the town of Padua in northern Italy where he worked and winter skies were frequently cloudy. Then, on January 7, a break in the weather brought a sparkling clear night, and Galileo was able to use the new invention to discover a fascinating set of worlds. These discoveries not only brought Galileo much acclaim but also led to a series of military contracts to put the invention to other uses. The invention was the telescope, and, although Galileo did not invent it, he was probably the first to use the telescope to study the heavens, leading to his discovery of the four large moons of Jupiter.

Galileo's discovery was pretty heady stuff and not without substantial controversy. The late 1500s and early 1600s saw the emergence of ideas that the Earth might not be the center of the Universe. Egocentric humans naturally assumed that planets, as well as stars, circled around the all-important Earth. But careful tracking of the planets and stars in the 1500s by Nicolaus Copernicus suggested that this might not be true. Care had to be taken, however, because such views were contrary to popular beliefs and were considered heretical. In fact, the Italian, Giordino Bruno, was too outspoken in expressing his negative views of a Sun-centered system. In 1600, he was tried at an Inquisition in Rome, found guilty of heresy, and was burned at the stake.

So here was Galileo, announcing to the world that he had found four objects orbiting something other than the Earth. Could this also be the heretical rantings of another trouble maker? Clearly, Galileo had to walk a fine line between maintaining scientific integrity (he knew what he had seen) and not upsetting the establishment. But Galileo also had some very powerful friends, the Medici family, who controlled much of central Italy at the time. Galileo's work was directly or indirectly supported by the Medici, and, when he discovered the moons of Jupiter, he had the good political sense to refer to them as the Medician "stars."

Improvements in telescopes following Galileo's observations enabled refined views of Jupiter and definition of the moons' orbits and other general characteristics. Surprisingly, the innermost moons, Io and Europa, were found to have densities suggesting rocky materials, while data for the outer two moons, Ganymede and Callisto, suggested a large proportion of water or water-ice.

The Space Age ushered in the first visits by spacecraft to the outer planets. In 1973 and 1974, two spacecraft, *Pioneers 10* and *11*, respectively, streaked past Jupiter and obtained the first close-up observations, including the first spacecraft images of Jupiter's moons. The picture quality was rather limited, however, because the spacecraft had to spin like bullets to keep on course. This was something like trying to take pictures while turning somersaults – not an easy task, even in the Space Age. The first high-quality pictures were taken by the *Voyager 1* and *2* spacecraft in 1979. Unlike the spinning *Pioneer* spacecraft, the *Voyagers* maintained their course using small rockets and had a stable platform for taking pictures.

The *Pioneers* and the *Voyagers* were flyby scouts to provide the reconnaissance for the next stage of exploration, the *Galileo* project. The *Galileo* mission involved an orbiter and a probe designed to plunge deep into Jupiter's atmosphere. The probe was outfitted with instruments to measure the temperature, water content, composition, and other properties of the atmosphere. The orbiter carried ten instruments, including those used to map the composition of Jupiter's clouds and the surfaces of the moons, an imaging system, and various instruments to study small particles and Jupiter's electromagnetic environment.

Getting the *Galileo* spacecraft to Jupiter was not easy. Although the project began in 1977, it was 18 years before the spacecraft arrived at Jupiter. The project was to be the first planetary mission using NASA's new shuttle as a means for leaving the Earth. However, shuttle delays and the tragic explosion of the shuttle *Challenger* deferred the launch of *Galileo* until 1989. Moreover, it was found that the launch capability was less than originally planned, and a direct path to Jupiter was not possible.

Fortunately, talented engineers at the Jet Propulsion Laboratory found that by flying past Venus, looping around the Sun twice, and flying past the Earth sufficient momentum could be gained to "sling-shot" onward to Jupiter. Although this saved the mission, there was also a price to pay; during the period before leaving the inner Solar System, the fragile high-gain antenna on the spacecraft was folded up like an umbrella. As the spacecraft left the inner Solar System and the command to unfurl the antenna was sent, it opened only part way and, despite months of trying to shake it free, it remained stuck. This meant that all communication would have to be funneled through the inefficient low-gain antenna, which required a complete redesign of data acquisition for it to be as efficient as possible. Painful decisions were made to decide which observations would have to be thrown out. For example, originally more than 70,000 pictures were to be taken, but the number had to shrink to fewer than 2,000. The camera, however, had an adaptable computer system, which enabled a great deal of flexibility in taking pictures efficiently.

By happenstance, the *Galileo* spacecraft was in a unique position to view a momentous event in Solar System activity; the collision of the comet Shoemaker-Levy 9 with Jupiter in 1994. This comet had been discovered two years earlier, and, during previous passes of

Jupiter, it was ripped into ~21 fragments, some more than 2 km across. Analyses of the orbits of the fragments led to predictions of impact with Jupiter. Because of the position of the Earth with respect to Jupiter at the time of the impact, Earth-based telescopes could not view the actual collisions but only the impact scars left in Jupiter's clouds. The *Galileo* spacecraft, however, was in a position to capture the events as they occurred, enabling the first observations of active, large impacts in the Solar System.

In the final stages of its journey to Jupiter, in 1995 a beachball-size probe from *Galileo* plunged into Jupiter's atmosphere where its instruments operated for nearly an hour. This marked another first in Solar System exploration – measurements made directly within the atmosphere of a giant planet. As the probe relayed its data back to Earth, the *Galileo* spacecraft began a series of maneuvers to place it in orbit around Jupiter. This involved a close flyby of Io, using gravity to slow the spacecraft and to put it in a trajectory to begin orbiting Jupiter in 1996. The mission lasted until 2003, completed 35 orbits around the planet, and enabled observations of the entire Jupiter system.

In December 2000, the *Cassini* spacecraft flew past Jupiter on its way to Saturn. This provided the unprecedented opportunity to make simultaneous observations from two spacecraft of a planet in the outer Solar System. Then, in February 2007, the *New Horizons* spacecraft flew through the Jupiter system on its way to rendezvous with Pluto in 2015. During the flyby, it was possible to obtain new data to assess changes that had occurred on Io since the *Galileo* mission, monitor cloud patterns on Jupiter, and collect other data.

8.3 Jupiter

Views of Jupiter (**Fig. 8.1**) show intricate clouds in a great variety of colors. The colors result from trace amounts of sulfur and organic materials, some of which were probably implanted by impacts. What lies below Jupiter's clouds cannot be determined directly with today's technology, but models provide insight into the characteristics of the deep interior. Because of Jupiter's massive size, the density of the gasses must increase tremendously, reaching stages in which the hydrogen would first be liquid and then transition into a metallic state. At the very center of Jupiter, it is thought that a rocky core about the size of Earth might be found.

Figure 8.1. Jupiter is the most massive object in the Solar System, after the Sun. This mosaic was assembled from *Cassini* images during its flyby of Jupiter in 2000. The Great Red Spot is the prominent oval-shaped storm system seen to the lower right (NASA PIA02873).

Despite its large size, Jupiter spins rapidly on its axis, making one complete rotation in less than 10 hours. This causes the equator to bulge outward and contributes to the intricate patterns seen in the clouds. The distinctive horizontal bands visible in global views of Jupiter's atmosphere relate to major zonal "jets" having speeds of 50 m/s. Some jets travel westward, while others are in eastward motion, leading to fantastic shear zones between the jets. Such shear results in eddies, curls, spirals, and feathery clouds.

The best-known storm system is the Great Red Spot **(Fig. 8.1)**. This feature is of size more than 40,000 km by 14,000 km, several times larger than the Earth. Rising 24 km above the surrounding clouds, it is a tremendous up-welling of the atmosphere. Although the reason for its red color is not well understood, it could result from phosphorus, and perhaps compounds of that element, including phosphine (PH_3).

Jupiter, like its cousin Saturn, releases more energy than it receives from the Sun. When Jupiter first formed, it was probably ten times larger than it is now. It began to collapse under its own weight, and, in the process, heat from friction in the gasses was generated, raising the temperature to 50,000 °C or more. That heat is still "leaking" into space and is responsible for the convective cells in the atmosphere which produce some of the cloud patterns.

Jupiter has an enormous magnetic field. Although the exact mechanism responsible for producing magnetic fields around planets, including Earth, is not well understood, it is thought to operate something like a dynamo. In the case of Jupiter, the zone of metallic hydrogen within the fast-spinning planet generates a magnetic field that extends thousands of kilometers from Jupiter, making it the largest feature in the Solar System, even larger than the Sun. The field is extremely intense, creating a zone of radiation harmful not only to life but also to sensitive spacecraft components and resulting in a type of "space weathering" on satellite surfaces.

Jupiter has at least 63 moons, including the four **Galilean satellites** (Io, Europa, Ganymede, and Callisto) that are in the same size class as Mercury and Earth's Moon. All four are in synchronous rotation, keeping the same side facing Jupiter, resulting in the terms **jovian hemisphere** (facing Jupiter), **anti-jovian hemisphere**, **leading hemisphere** (for the side facing in the direction of travel in orbit around Jupiter), and **trailing hemisphere**. In addition, Io, Europa, and Ganymede are in **Laplace resonance**, in which the orbital periods are in a ratio of 4:2:1; in the time taken for Io to orbit Jupiter four times, Europa orbits twice, and Ganymede orbits once.

The Galilean satellites display a wide spectrum of geologic histories, from extremely active Io, the innermost large moon, to Callisto, which displays little evidence of internal geologic activity since its formation some 4.6 Ga ago. Europa and Ganymede appear to be intermediate in terms of activity. There is the possibility that Europa could also be active today, albeit not to the same level as Io. Given their orbital geometry (i.e., Laplace resonance), Io, Europa, and Ganymede are in an intricate "push–pull" tug of war with Jupiter and each other **(Fig. 8.2)**. This results in tidal stressing of the satellites and the generation of internal frictional heat, the magnitude of which decreases with distance from Jupiter. As portrayed in **Fig 8.3**, this could explain the differences in the geologic appearances among the satellites and their interior configurations **(Fig. 8.4)**.

8.4 Io

One of the most spectacular discoveries in the exploration of the Solar System was the existence of active volcanoes on Io. This discovery was made in 1979 during the flyby of the *Voyager 1* spacecraft. A series of images had been taken for navigation and some of the images included the limb, or edge, of Io in the field of view. A young navigator-engineer,

Linda Morabito, was analyzing the images and noticed what appeared to be a smear extending from Io into the dark sky **(Fig. 8.5)**. Her first impulse was to dismiss the feature as a flaw in the image or an artifact of some sort introduced during computer processing. A check with the

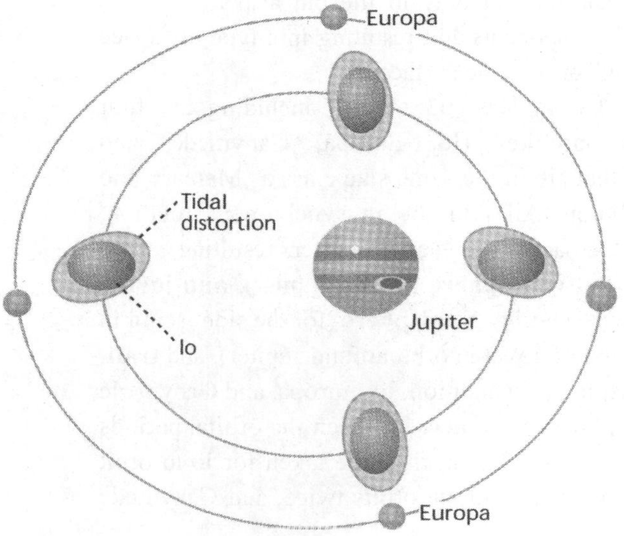

Figure 8.2. The orbits of Io and Europa are non-circular around Jupiter (not drawn to scale). At times, Io draws close to Jupiter, then pulls farther away. Coupled with interactions with Europa, this results in a constant tug of war within Io, causing the rocks within the moon to be pulled closer to Jupiter during the near part of the orbit, with a relaxation during the distant part of the orbit. This tidal flexing results in internal friction and the generation of enough heat to melt the rocks below the surface, forming magma and fueling Io's active volcanoes. Similar tidal processes operate on Europa and Ganymede, but with less intensity (from Greeley and Batson, 2001).

science team showed that this was not the case. With additional processing, the feature was found to be a plume of gas and debris erupting from the volcano now named Pele.

Perhaps equally remarkable about the actual discovery of volcanism was the prediction that volcanoes would be found on Io. A team of geophysists had calculated the amount of heat generated from Io's tidal flexing in its orbit around Jupiter and determined that magma might be generated. They published their prediction of active volcanoes just two weeks before the *Voyager 1* image was snapped. Such scientific serendipity is rare and serves as an inspiration for all who seek new insight into how planets operate.

Io **(Fig. 8.6)** is about the diameter of Earth's Moon and has a similar density, suggesting that it is a rocky object with a small metallic core **(Fig. 8.4)**. It is the reddest object in the Solar System, but its surface also includes pale shades of brown, yellow, and orange. *Galileo* data and observations from the *Hubble Space Telescope* and *Cassini* provide a pretty good idea of what the various colors represent. For example, SO_2 gas is a major component of the eruptive plumes rising from Io's surface. As the gas condenses into solid ash, it rains onto the surface, leaving diffuse, light-toned deposits around the volcanic vents. Data also show that one molecular form of sulfur, S_2, is present in the eruptive plumes. However, this form of sulfur is unstable in the cold environment of Io and converts to two other molecular forms, S_3 and S_4, both of which are reddish orange. But the cycle is even more complicated because these forms last only a few years before converting to S_8, which is yellow.

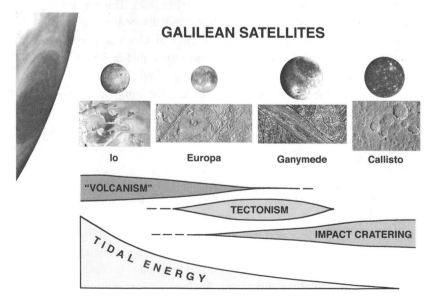

GALILEAN SATELLITES

Io   Europa   Ganymede   Callisto

"VOLCANISM"

TECTONISM

IMPACT CRATERING

TIDAL ENERGY

Figure 8.3. A stylized diagram of the Galilean satellites relating their dominant geologic processes to the degree of tidal energy generated by interactions among the moons and Jupiter. Io experiences the maximum tidal stresses, resulting in volcanism, while Callisto apparently experiences little, if any, tidal stressing; thus Callisto's surface preserves the impact cratering record from the early history of the Solar System. Europa and Ganymede are intermediate cases, with Europa showing some evidence for cryovolcanism and Ganymede being extensively modified by tectonic resurfacing.

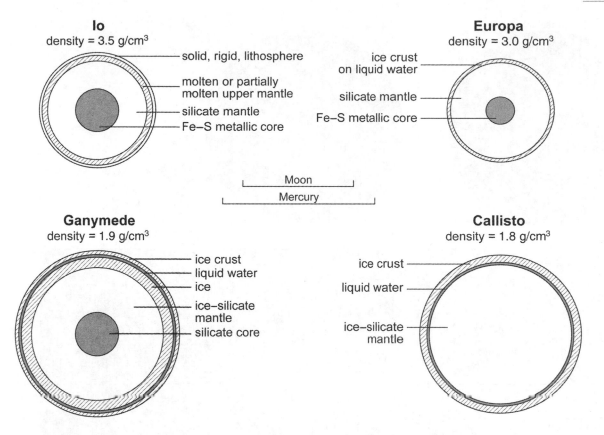

Figure 8.4. Modeled interiors of the Galilean satellites: Io and Europa are composed mostly of silicate materials; Ganymede and Callisto are mixtures of rocky materials, liquid water, and water ice. Europa has an outer shell of H$_2$O, the surface of which is frozen. Liquid saltwater is likely present on Europa, Ganymede and Callisto, as suggested by the behavior of the Jupiter-induced magnetic fields. Scale bars indicate the diameters of Mercury and Earth's Moon.

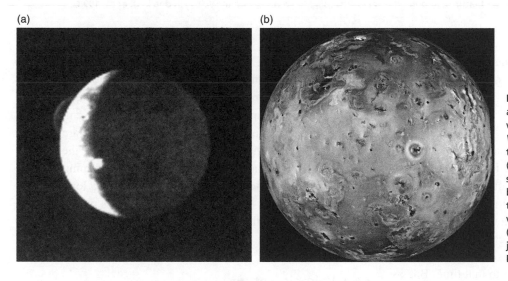

(a) (b)

Figure 8.5. (a) The discovery of active volcanoes outside Earth was made through analysis of this *Voyager 1* image which shows this 260 km high explosive plume (upper left) from the volcano subsequently named Pele. The bright spot on the terminator is the eruptive plume from the volcano, Loki (NASA PIA00379). (b) A view of Io showing the anti-jovian hemisphere (*Galileo* NASA PIA00583).

The study of high-resolution images shows that almost all of the black areas on Io are either the floors of volcanic calderas or relatively recent lava flows. Moreover, the temperatures of many of the dark deposits exceed the boiling point of sulfur, and they are thought to represent the eruptions of silicate magmas.

With these observations in mind, Io is seen as a color kaleidoscope that is constantly changing as new lava

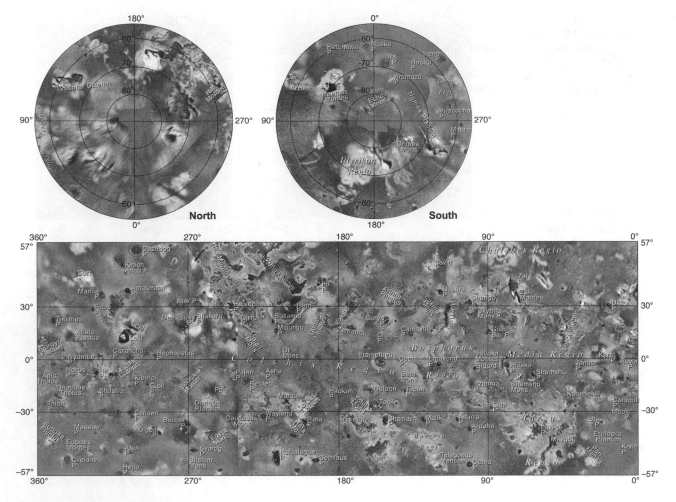

Figure 8.6. A map of Io with prominent named features. Reprinted from *Icarus*, 135, McEwen, A., Keszthelyi, L., Simonelli, D. *et al.*, 1998, 181–219, with permission from Elsevier.

flows are erupted and sulfur-rich materials blanket the surface, convert to more stable forms, and change color.

8.4.1 Impact features (none!)

Even before the volcano discovery image had been taken, Io presented an intriguing face to the *Voyager* cameras as the spacecraft drew closer to Io. From a distance, dark circular smudges set against Io's red and yellow surface were seen in the low-resolution pictures. At first, the smudges were thought to be impact craters, a reasonable assumption because craters had been seen on every solid-surface planet and moon explored up to that time. But, as *Voyager 1* drew closer to Io, the pictures became sharper, and the dark areas were resolved to be volcanic craters and calderas, many of which were seen to have lava flows extending from their rims. No impact craters were seen then or by later missions.

From estimates of the volumes of volcanic materials erupted, the flux of impacting objects in the Jupiter system, and the lack of impact craters seen in imaging data, Io is thought to be volcanically resurfaced at a rate of ~1 cm/yr; thus, impact craters probably do form, but they are buried so quickly that none are preserved within the visible record. This suggests that Io is volcanically turning itself "inside-out," and apparently has done so for a long time, as reviewed by Ashley Davies (2007). This means that Io has one of the youngest surfaces in the Solar System.

8.4.2 Volcanic features

More than 400 volcanoes are documented on Io, and they reflect both effusive and explosive eruptions. Some of the volcanoes discovered during the *Voyager* mission were seen to still be active some 28 years later during the *New*

Horizons flyby and are likely to continue erupting for years to come.

With one or two exceptions, the volcanoes on Io are not classic cone-shaped mountains like Mount Fuji. Rather, they are wide, low-profile features composed of lava flows that spread great distances from their sources and are termed paterae. In some cases individual flows can be traced more than 500 km, suggesting that they must have been extremely high-temperature, low-viscosity lavas. This suspicion was confirmed when temperatures as high as 1,340 °C were recorded, higher than those of most lavas erupted on Earth today. However, very early in Earth's history, a type of lava called komatiite was produced, and it is estimated that its eruptive temperatures were somewhat similar to those observed on Io. Moreover, komatiite is very rich in iron and magnesium, which is consistent with estimates of the compositions of some of the lava flows on Io. Planetary volcanologists conclude that we might be seeing a type of volcanic activity on Io comparable to some eruptions on Earth ~3 Ga ago.

Williams and Howell (2007) classified three styles of eruptions on Io: (1) flow-dominated (effusive) eruptions, (2) explosion-dominated eruptions, and (3) intra-patera volcanism. Flow-dominated eruptions, typified by the Amirani lavas **(Fig. 8.7)**, produce large flow fields that emanate from central vents or fissures over months to years, similar to eruptions in Hawaii. The colors of the lavas and temperature measurements suggest that they are basaltic. These eruptions produce low-profile shield volcanoes such as Maasaw Patera **(Fig. 8.8)** and pancake-shaped features such as Apis Tholus **(Fig. 8.9)**. Some flows have associated bright deposits that appear to be generated from the lavas flowing over sulfur dioxide deposits **(Fig. 8.10)**. Flow-dominated eruptions can also include explosions, producing Prometheus-type plumes. Named for their type locality on Io, such explosions produce umbrella-shaped plumes of sulfur dioxide 100–200 km high that form circular ash deposits around their vents.

Explosion-dominated eruptions produce enormous plumes that can exceed heights of 500 km, as well as large flow fields of lava, but they are short-duration, sporadic, high-energy events called "outbursts." Typified by Pele **(Fig. 8.5)**, these eruptions eject materials at speeds as high as 1 km/s, and rain huge quantities of reddish sulfur and silicate ash onto the surface **(Fig. 8.11)**. The associated lava flows involve very high rates of effusion; for example, the emplacement

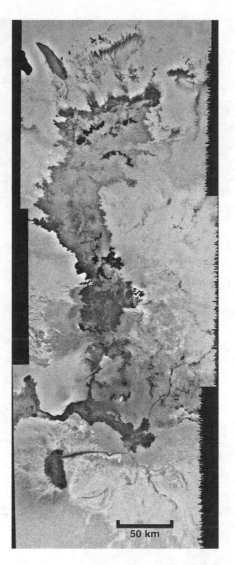

50 km

Figure 8.7. The vent for the Amirani lava flow-field is in the lower center of this picture of Io; compound lava flows extend 250 km north, toward the top of the image. The smaller flow extending west from the vent was found to have a high temperature, suggesting that it was active at the time of the observations; some flows extending to the south were tube- and channel-fed (NASA *Galileo* PIA02506).

of the Pillan lava field **(Fig. 8.11)** was estimated to involve flow rates of 1,740 to 7,450 m^3/s, comparable to the largest rates on Earth for basaltic lavas in recent time.

Intra-patera eruptions occur within calderas and can accompany some explosive eruptions. For example, Loki volcano **(Fig. 8.12)** repeatedly changes temperature and shape with time, while its caldera contains confined lava lakes with surface crusts that are disrupted by magma circulation in the lava lake.

Figure 8.8. Maasaw Patera, Io, displays dark lava flows that radiate from the caldera, forming a low-profile shield volcano. The caldera is 50 km by 25 km (NASA *Voyager 1* 199J1+000).

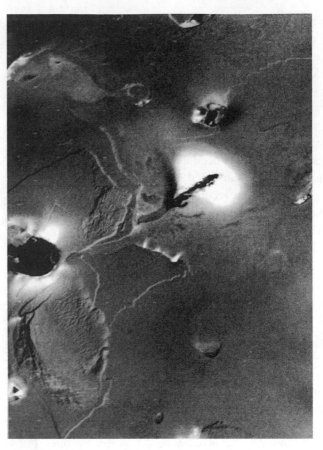

Figure 8.10. This image shows dark lava that flowed over smooth plains; the bright white zone around the flow is SO_2 "cooked out" of the plains by the hot flow and deposited as frost (NASA *Voyager 1* 75J1+000).

Figure 8.9. "Pancake"-shaped volcanoes Apis Tholus (upper right) and Inachus Tholus (center) on Io probably formed from very fluid lavas that spread evenly from the vents (bright central zones); the area shown is 600 km by 800 km (NASA *Voyager 1* 71J1+000).

Local eruptions of liquid sulfur are likely to occur in addition to the major styles of eruptions outlined above. Sulfur and sulfur compounds are common and some are probably deposited by fumarolic activity. Because sulfur has a low melting temperature, it is probably mobilized by heat from silicate eruptions to produce flows of sulfur. This process was documented on the flank of Mauna Loa volcano in Hawaii and accounts for local flows of nearly pure sulfur.

The extensive explosive eruptions lead to significant loss of mass from Io. It is estimated that some 10 tons/s of sulfur and sulfur dioxide escape from Io. Much of this material is caught by Jupiter's magnetic field, where it forms a donut-shaped "torus" enveloping the orbital path of Io.

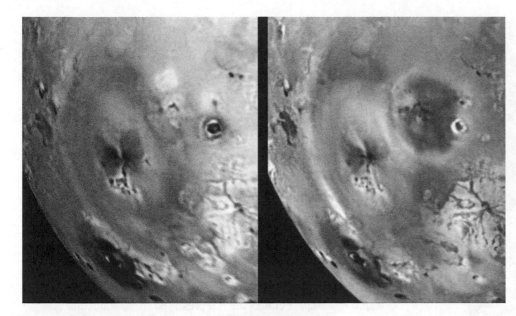

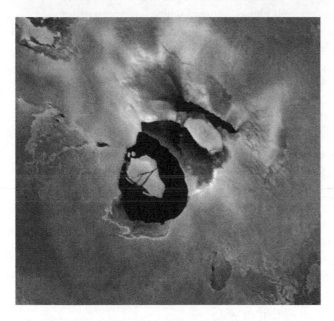

Figure 8.12. Loki Patera is a huge caldera containing a lava lake (dark areas) that has changed with time. The light areas within the lake are slabs of crust, which shifted position and shape. The diffuse areas surrounding the lake represent ash and other debris from explosive eruptions; the area shown is 894 km across (NASA *Galileo* PIA00710).

8.4.3 Tectonic features

Io displays some of the highest mountains in the Solar System, exemplified by Boosaule Montes, which rises 17.5 km above the surrounding plains. Io's mountain ranges average 100 km in length and 6 km in height. The steep slopes on the mountains and the walls of calderas indicate that the crustal compositions are mostly silicate materials rather than sulfur because sulfur has insufficient strength to support such a large mass, even in the lower gravity of Io (Clow and Carr, 1980).

Most of the mountains resulted from uplift by block-faulting, although some could be of volcanic origin. In addition, some of the ranges have been offset by slip-faults, as evidenced by possible geometric reconstructions **(Fig. 8.13)**. This poses a puzzle for Io: mapping of the distribution of the mountain ranges and the volcanic vents fails to show any clear trends or other suggestions of tectonic patterns. Consequently, researchers are currently attempting to determine the processes that lead to the locations of the eruptive centers, block-fault mountains, and strike–slip faults.

8.4.4 Gradation features

McCauley *et al.* (1979) suggested that isolated mesas, irregular surfaces, and segments of subdued scarps reflect some process of erosion **(Fig. 8.14)**. Since Io's environment precludes fluvial or aeolian processes, they evoked a **sapping** mechanism through the release of SO_2 as an erosional agent. They suggested that subsurface liquid SO_2 under hydrostatic pressure would be released through fractures; at the triple point for SO_2, part of the liquid would crystallize and part of the system would expand, forming SO_2 snow and vapor. Upon reaching the surface, the solid–fluid mixture would be released energetically at a velocity of ~350 m/s, spraying snow as far as 70 km from the fracture to form the bright

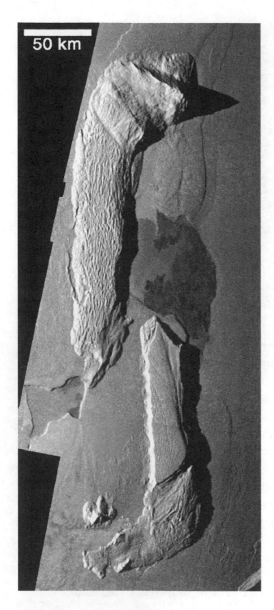

Figure 8.13. A mosaic of *Galileo* images showing two Io mountain ranges, north Hi'iaka Mons (upper left) and south Hi'iaka Mons, that could have been offset by strike–slip faulting; the two blocks appear to fit back together geometrically (NASA *Galileo* PIA02540).

Figure 8.14. An oblique *Galileo* view of a mesa near the Tvashtar volcano, Io; the irregular margin of the mesa probably resulted from sapping processes associated with the removal of sulfur dioxide (NASA *Galileo* PIA02519).

of sulfur-rich materials. Volcanic resurfacing on Io is so rapid that impact craters are not preserved. Io's mountains are among the highest in the Solar System and most appear to be uplifted fault blocks, some of which have been offset by strike–slip faults. Gradation is suggested by the mass wasting seen in a few areas and sapping processes associated with SO_2.

8.5 Europa

In many respects, Europa **(Fig. 8.15)** is unique in the Solar System. Slightly smaller than Earth's Moon, it is mostly a rocky object but has an outer shell of water some 130 km thick, the surface of which is frozen **(Fig. 8.4)**. The thickness of the ice layer above the ocean is unknown but estimates range from very thin (i.e., a few kilometers) to very thick, in which the entire mass is frozen nearly all the way to the rocky mantle. In any case, the total amount of water (liquid and frozen) is more than five times the amount of all of Earth's water.

The first clear spacecraft views of Europa were provided by the *Voyager* spacecraft in 1979. These images, however, were rather low-resolution, returning pictures that, at best, are about 2 km per pixel. Nonetheless, even these low-resolution views provided tantalizing hints that Europa is an interesting object. Seen in global views, Europa gives the appearance of a string-wrapped baseball without its cover. Many of the "strings" are actually ridges and fractures, suggesting that some tectonic process has cracked the icy crust **(Fig. 8.16)**.

The *Galileo* orbiter imaged substantial parts of Europa, some in resolutions as high as 7 m per pixel for tiny areas. Geologic analysis shows a prevalence of bright plains that

patches of SO_2. The release of SO_2 would lead to undercutting and erosion, leaving irregular scarps and mass-wasted debris **(Fig. 8.14)**. These processes would continue until the source of the SO_2 was depleted locally.

8.4.5 Io summary

The surface of Io is dominated by volcanism resulting from tidal stressing generated by interactions with Jupiter and Europa. Most of the volcanism involves silicate compositions, including high-temperature eruptions, and explosions

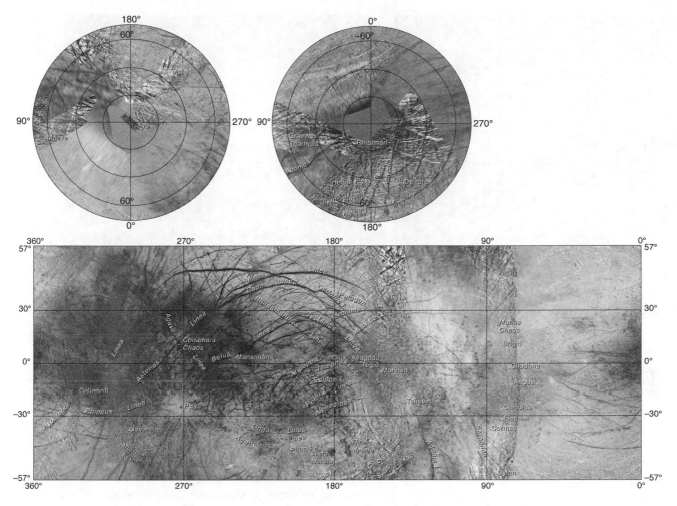

Figure 8.15. Europa, showing key place names. The trailing hemisphere (centered on 270° W) is darker, partly as a result of implantation of sodium- and sulfur-rich ions derived from volcanic eruptions on Io (from Spencer *et al.*, 2004).

have been extensively modified by tectonic processes, including the formation of structural lineaments exceeding 1000 km long, zones of crustal spreading and infill, and the formation of terrain broken into zones of "chaos." Impact craters and cryovolcanic features are also seen.

Earth-based and spacecraft data show that non-ice materials, such as magnesium salts, are present on the surface, along with organic and sulfur compounds. Surface materials are derived from three sources: (1) the rocky interior, reflecting the original compositions and differentiated products related to the origin of Europa (this material is brought to the surface by tectonic and cryovolcanic processes); (2) materials carried to Europa by comets, which would include organic compounds; and (3) implantation of high-energy particles from Io.

Europa's characteristics make it a high priority in the search for life beyond Earth. Life as we know it depends upon liquid water, an energy source to support biological processes, and the presence of the right chemistry (primarily C, H, O, N, P, S, Fe, and various trace elements). Europa meets these requirements, and, coupled with the recognition that life on Earth exists even in extreme environments, Europa is targeted for extensive future exploration.

8.5.1 Impact features

Voyager images showed relatively few large impact craters on Europa, leading to the conclusion that it, like its neighbor Io, has a geologically young surface and might even experience active volcanism. The same calculations as led to the correct prediction of active volcanoes on Io were also made for Europa. The results, however,

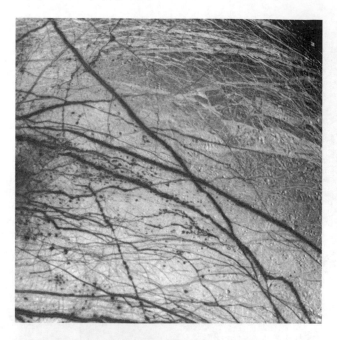

Figure 8.17. Pwyll crater is the youngest large (26 km in diameter) impact on Europa; the bright rays of length > 1,000 km consist of water-ice grains (NASA *Galileo* PIA01211).

Figure 8.16. This *Galileo* image of the northern hemisphere of Europa is 1,260 km across. The light-toned plains are oldest, while the dark linear fractures cut across the plains and are the youngest features in this area (NASA PIA00275).

were not so clear because the amount of heat predicted from tidal stresses was right on the border for the amount needed to generate magma or to keep the base of the ice crust liquid.

Only about 24 impact craters ≥10 km were identified in *Galileo* images, reflecting Europa's very young surface, estimated to be aged 40–90 Myr **(Fig. 4.50)**. This is remarkable in comparison with Earth's Moon, which is only slightly larger but far more heavily cratered. The youngest known Europan crater is Pwyll, 24 km in diameter, which displays abundant secondary craters and retains its bright rays. Pwyll probably formed less than 5 Myr ago **(Fig. 8.17)**.

Crater morphology provides insight into ice thickness at the time of impact. Morphologies vary from small bowl-shaped depressions with crisp rims formed entirely in ice, to shallow depressions with smaller depth-to-diameter ratios. Craters up to 25–30 km in diameter have morphologies consistent with formation in a warm but solid icy shell, while the two largest impacts (Tyre and Callanish, **Fig. 8.18)** might have punched through brittle ice about 20 km deep into a liquid zone.

It should be noted that, regardless of the overall thickness of the icy crust, there could be isolated pockets of salt water (brine) within the ice shell at shallow depths.

8.5.2 Tectonic features

Europa's tectonic features include ridges, bands, and fractures. Ridges are common and appear to have formed throughout the visible history of Europa. They range from 0.1 to > 500 km long, are as wide as 2 km, and can be several hundred meters high. Ridges include simple structures, double ridges separated by a trough **(Fig. 8.19)**, and intertwining ridge-complexes. Whether the variations represent different processes or different stages of the same process is unknown. Cycloidal ridges are similar to double ridges but form chains of linked arcs **(Fig. 8.20)**.

Most models for the formation of ridges include fracturing in response to processes within the ice shell, most of which are related to tidal stresses. Some models (Prockter and Patterson, 2009) suggest that ridges form by frictional heating and possible melting along fracture shear zones. Other models suggest that oceanic material or warm mobile subsurface ice squeezes through fractures to form the ridge. In this case, ridges might represent "communication" between the surface ice and the deeper ocean. Such exchanges could provide a means for surface oxidants to enter the ocean, which would make the formation of habitable zones more likely.

Bands reflect fracturing and separation of slabs of the ice crust **(Fig. 8.21)**, much like sea-floor spreading on Earth, and most display bilateral symmetry. The youngest bands tend to be dark, while older bands are bright, suggesting brightening with time. Geometric reconstruction of bands suggests that a spreading model is appropriate, indicating extension in these areas.

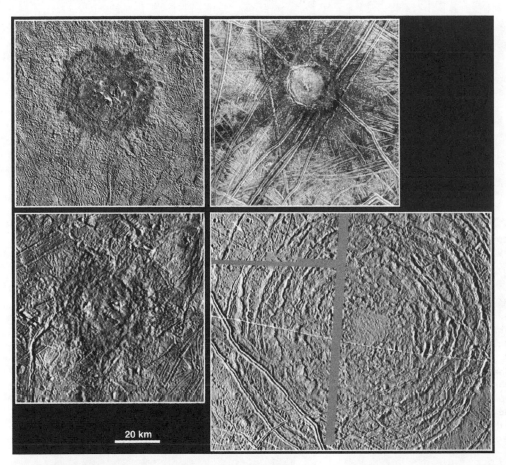

Figure 8.18. Impact structures on Europa: upper left is Pwyll crater; upper right is Cilix and its dark ejecta, central peak, and wall terraces; lower right shows concentric ring fractures of crater Tyre, the ridge system that post-dates the formation of the crater, the bright linear fracture that cuts across the crater, and the ridges (the gray bands are gaps in the image data); lower left shows crater Manannán (the scale bar is common to all four images) (NASA *Galileo* PIA01661).

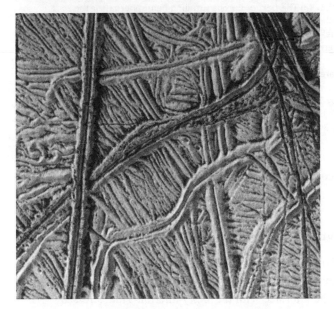

Figure 8.19. Europa's icy plains, seen here in high resolution, consist of complex, intertwining ridges, many of which occur in pairs. The youngest ridges cut across older features, tend to be dark, and stand higher than the surrounding terrain. The ridge-pairs are separated by narrow troughs. The straight, dark ridge-pair (left) is ~1.5 km wide. Ridges may be places where warm ice squeezed through fractures. The area shown is 20 km wide and is on Europa's anti-jovian hemisphere (NASA PIA01178).

Figure 8.20. Cycloidal ridges result from tidal deformation of Europa's icy shell; each arc in the "cycloid" is 50–80 km long; also shown is a central peak-crater between the cycloidal ridges (NASA *Galileo* 15ESREGMAP02).

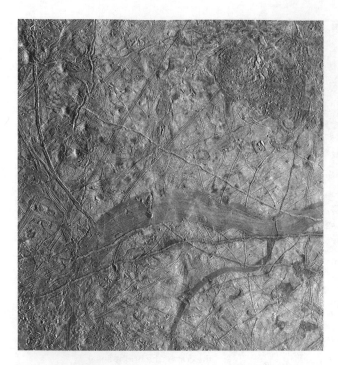

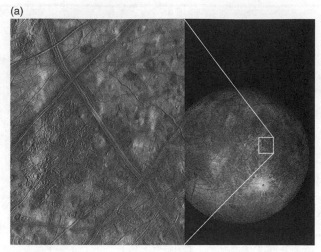

Figure 8.21. This view of Europa's trailing hemisphere is 335 km across. Bright plains mark the right side, while domes occur in a cluster in the upper left. The dark irregular patch in the upper right is a zone of chaos; cutting east–west across the upper part of the image is a gray band, or spreading center, where the crust was separated and filled in with ice or water (NASA *Galileo* PIA01125).

Figure 8.22. (a) This part of Europa's trailing hemisphere near the equator shows two large ridge systems, older plains, and dark spots. Below the "X" formed by cross-cutting ridges is a dark area of the icy crust severely disrupted into chaos terrain, named Conamara Chaos. *Galileo* data suggest that the dark ridges, spots, domes, and disrupted terrains are all of essentially the same composition, suggesting that material was brought to the surface by cryovolcanism; the area is 250 km by 300 km (NASA *Galileo* PIA03002). (b) A high-resolution view of the disrupted ice crust in the Conamara region; the area shown is 70 km by 30 km (NASA *Galileo* PIA01127).

Fractures are as small as a few meters wide and can exceed 1,000 km in length. Some fractures cut across nearly all surface features, indicating that the ice shell is subject to deformation on the most recent time scales. The youngest ridges and fractures could be active today in response to tidal flexing. Young ridges could also be places where the ocean has recently exchanged material with the surface and would be prime targets as potential habitable niches.

Chaos terrain is characterized by fractured plates of ice, many of which have shifted into new positions within a matrix **(Fig. 8.22)**. Much like a jigsaw puzzle, many plates can be fit back together. Some ice blocks appear to have disaggregated and foundered into the surrounding finer-textured matrix, while other chaos areas stand higher than the surrounding terrain. Models of chaos formation, reviewed by Collins and Nimmo (2009), suggest whole or partial melting of the ice shell, perhaps enhanced by local pockets of brine that would have lower melting temperatures. Chaos terrain commonly has associated dark reddish material thought to be derived from the subsurface, possibly from the ocean. However, these and related models are poorly constrained because

the processes within and below Europa's ice shell are not known.

8.5.3 Volcanic features

Volcanism of two forms is suggested for Europa. The first involves silicate volcanism at the base of Europa's water shell at the interface with the rocky interior. Tidal stresses of the sort that generates Io's volcanism could flex the rocky mantle to produce enough heat to melt the rock, leading to volcanism analogous to that seen on Earth's sea floor. If such activity occurs within Europa, it could maintain liquid water beneath the ice shell and lead to convection and zones of up-welling.

The second style is cryovolcanism, involving melted and partly melted ice, or slush. When the low-resolution *Voyager* images showed Europa's young terrains, it was suggested that it had been resurfaced with floods of water by cryovolcanic processes. While this might account for some of the bright plains, much of the surface is more the result of tectonic deformation. Nonetheless, small inferred "ice ponds" **(Fig. 8.23)**, chaotic terrain, infilling of spreading center bands, and features called **lenticulae** appear to involve intrusion and extrusion of melted or partly melted ice.

Lenticulae average about 10 km in diameter and include pits, dark spots **(Fig. 8.24)**, and domes where the surface is upwarped and commonly broken. These features could form by up-welling of compositionally or thermally buoyant ice diapirs through the ice shell **(Fig. 8.25)**. As

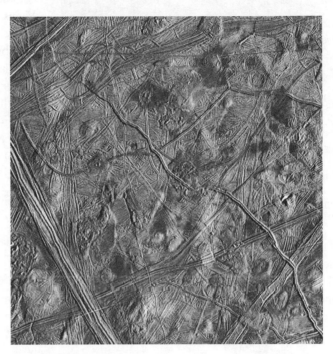

Figure 8.24. Many Europan domes are either dark or have associated dark patches, which might be non-water materials brought to the surface from below the ice crust. The area shown is 130 km across (NASA *Galileo* PIA00588).

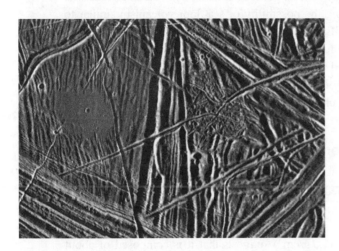

Figure 8.23. Some local areas of Europa's ridged plains appear flooded, as suggested by the smooth, flat area in the middle-left part of the picture. The smooth area is 3.2 km across and was cratered by a small impact. The area shown is 11 km by 16 km (NASA *Galileo* PIA00592).

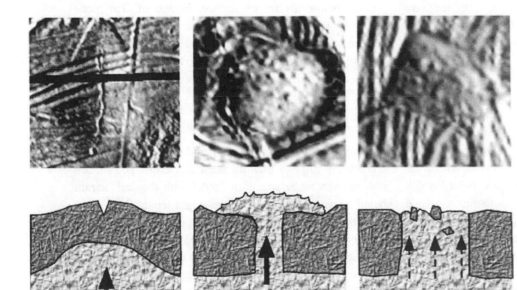

Figure 8.25. *Galileo* images and corresponding diagrams showing ways in which lenticulae might form. Left: a plume of warm ice or liquid water rises, buckling the surface; middle: a blob of low-density material breaks through the crust and flows onto the surface; right: local heating melts through the ice crust.

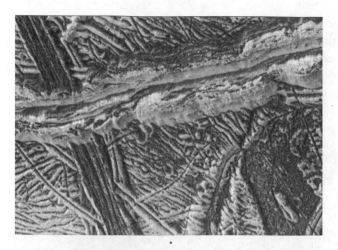

Figure 8.26. Not all young ridges are dark; the prominent ridge-pair extending east–west (right to left) cuts across the set of older dark ridges on the left. Bright ice avalanches have been shed from the west–east ridge and cover not only the dark ridge-set, but also the underlying plains. The area shown is 20 km wide (NASA *Galileo* PIA01179).

such, their size would imply that the thickness of the ice shell was at least 10–20 km at the time of formation (McKinnon, 1999). Alternatively, Rick Greenberg *et al.* (1999) suggest that the lenticulae could be small forms of chaos formed by wholesale melting of the ice crust.

8.5.4 Gradation features

Gradation on Europa occurs through mass wasting and space weathering. Mass wasting is seen along the bases of steep slopes **(Fig. 8.26)** where blocks of dirty ice have accumulated. It is likely that similar processes have degraded the walls of impact craters which have also been deformed by viscous relaxation.

On a regional scale there is a marked difference between the leading and trailing hemispheres of Europa **(Fig. 8.15)**. The trailing hemisphere is darker, tends to be red in comparison with the leading hemisphere, and probably results from the implantation of materials by Jupiter's magnetosphere. Speeds of materials within the magnetosphere are faster than Europa's orbit, which means that materials within the magnetosphere would "catch up" with Europa and impact the trailing hemisphere. Some of this material includes sulfur and sulfur-rich compounds from Io's volcanic plumes. In addition, ions impacting the trailing hemisphere would alter the molecular structure of the ice, causing it to darken with time.

8.5.5 Europa summary

In many respects, Europa is a hybrid planet. It is a rocky object similar in composition and size to Earth's Moon but contains an outer shell of water and ice, a characteristic common among the outer planet satellites. As with Io and Venus, Europa's surface is very young, leading to a compressed geologic time scale **(Fig. 4.50)**. There is a strong possibility that Europa is currently being resurfaced.

Tidal stresses in the ice shell and to some extent within the rocky interior generate tectonic and volcanic processes, including the possibility of sea-floor volcanism similar to that in Earth's oceans. These processes lead to complex surface features, as portrayed in the diagrams of **Fig. 8.27**. Coupled with the probable presence of organic compounds and key inorganic elements, Europa is a primary target in the search for habitable zones beyond the Earth.

8.6 Ganymede

Ganymede is the largest satellite in the Solar System and is the only moon known to generate its own magnetic field. It is about the same size as Mercury, but Ganymede's low density ($1.9\,g/cm^3$) suggests that it is composed of about 60% water and 40% rocky materials. Geophysical models indicate that it is differentiated into an iron and sulfide–iron core about 1,600 km across, a silicate mantle, and an outer ice zone 800–1,000 km thick **(Fig. 8.4)**. A saltwater liquid ocean is thought to be sandwiched within the ice at about 170 km below the frozen surface. Images of Ganymede taken by *Voyager* and *Galileo* cover about 80% of the surface and show two prominent terrains **(Fig. 8.28)**: dark, heavily cratered regions considered to be ancient and bright terrain characterized by sets of ridges and grooves.

8.6.1 Physiography

Overall, Ganymede is very bright and even its dark terrains are actually brighter than the lunar highlands. The surface also appears to be spectrally "red," with the dark terrain being somewhat redder than the bright terrain. Moreover, as in the case of Europa, the trailing hemisphere is darker than the leading hemisphere and could be enriched in sulfur dioxide from exogenic implantation. Dark non-ice materials at lower latitudes are possibly hydrated brines similar to those inferred for Europa while minor constituents include CO_2, SO_2, and possibly organic materials.

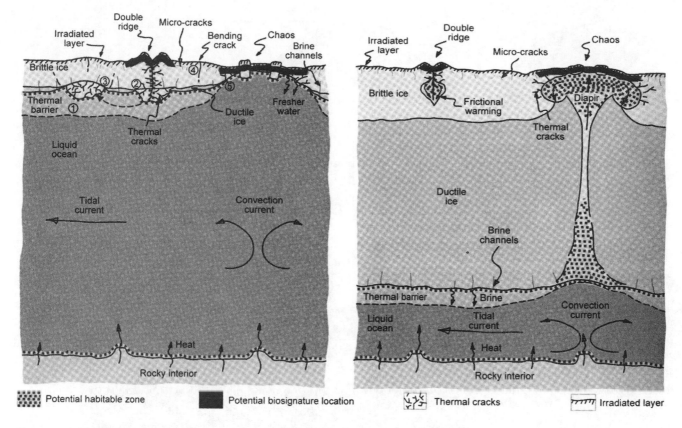

Figure 8.27. A stylized block diagram of Europa's crust showing the possible relations of surface features to processes within the icy crust for a thin ice shell (left) and a thick ice shell (right): diapirs and convection could cause thermal perturbations and partial melting in the overlying rigid ice; faulting driven by tidal stresses could result in frictional heating; and impacts could lead to refrozen central melt zones and "splash" of ejected ice and slush. Many of the zones and features shown here could be habitats for astrobiological exploration (from Figueredo *et al.*, 2003).

Dark terrain constitutes about 35% of the surface and includes large polygonal regions separated by younger grooved terrain. The most extensive region of dark terrain includes Galileo Regio, a semicircular area more than 3,200 km across on the leading hemisphere. The numerous impact craters superposed on the dark terrain reflect the period of intense bombardment in the final stages of Solar System formation. However, when craters on Ganymede are compared with those on the inner planets, relatively few craters larger than 60 km in diameter are found. Such craters might have been obliterated by some unknown process, or perhaps the cratering record is different for the outer and inner parts of the Solar System.

Bright terrain forms by the conversion of dark terrain, primarily through extensional tectonic processes (**Fig. 8.29**). For example, linear bands of grooved terrain, called sulci, dissect and convert many parts of the older terrain. Bright terrain occupies most of the polar areas and occurs as a zone hundreds of kilometers wide around Galileo Regio, as well as forming large patches in the southern hemisphere. Bright terrain is characterized by sets of ridges and grooves 100 km or more long by tens of kilometers wide. Individual ridges are several kilometers wide and may be as high as 700 m.

Although not physiographic in the sense of landforms, Ganymede exhibits distinctive characteristics in both polar regions. Above about 40° latitude, the surface shows bright diffuse deposits called "polar hoods," which are probably concentrations of ice or frost formed by the migration of volatiles from lower latitudes. High-resolution images of parts of the polar hood boundaries suggest that the deposits thin toward the equator, forming an irregular, mottled appearance.

8.6.2 Impact features

Ganymede's impact craters range in diameter from tens of meters to hundreds of kilometers, with larger craters being more shallow. The relationship between crater size and

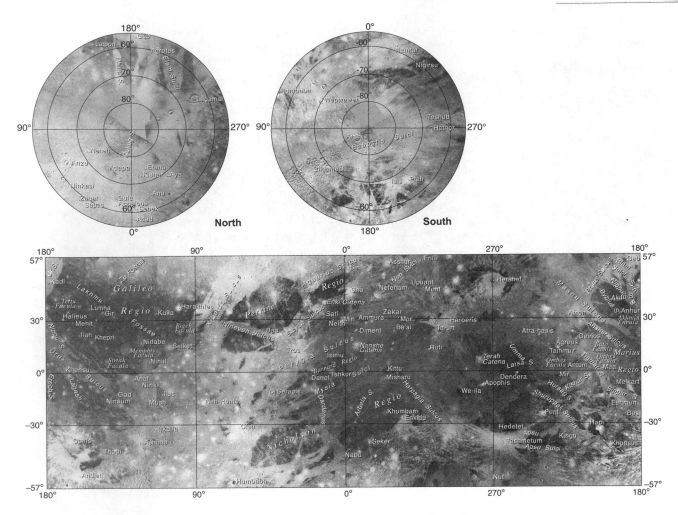

Figure 8.28. A map of Ganymede showing prominent named features (from Spencer *et al.*, 2004).

crater depth is considered to be a consequence of impacts into ice, in which crater rims "relax" by viscous flow. Given the properties of the ice and estimates of temperature as a function of depth below the surface, models show that after the impact the crater floor would first flatten and then upwarp to a convex form as the crater rim subsided. Under these conditions, long-wavelength features would be more affected and larger craters would be relatively more deformed than small craters, as seen in their in depth-to-diameter ratios (**Fig. 8.30**). However, shallow craters could also reflect conditions and properties of the target at the time of their formation in which the ice-rich crust could have been warmer in the past and more subject to deformation.

Craters smaller than ~20 km are typical bowl-shaped depressions with depth-to-diameter ratios of 1:6 to 1:12 and are not greatly affected by viscous relaxation. Thus, as models would predict, the ice-rich crust behaves essentially

as rocky material for small impacts. Craters ~20–40 km in diameter generally have flat or slightly convex floors. Central peaks occur in many craters of this size but are generally absent among the larger craters; rather, they appear to be replaced by central pits (**Fig. 8.31**), which could result from the impact punching through the ice crust into warmer, softer ice or even liquid water.

Ejecta deposits are seen around many craters on Ganymede and include flow-like patterns (**Fig. 8.32**) similar to the "ejecta flow" craters on Mars (**Fig. 7.7**). Ejecta rays extend from fresh craters and include both bright and dark forms. Some rays include mixtures of bright and dark deposits (**Fig. 8.33**), suggesting excavation into materials of different ice–rock mixtures. The floors of some craters are very dark, possibly resulting from flooding by darker subsurface materials (**Fig. 8.34**).

Among the many remarkable discoveries from the *Voyager* mission were the crater structures on Ganymede termed **palimpsests (Fig. 8.35)**. These are circular patches of bright terrain as wide as several hundred kilometers, found principally in dark terrain. The term was borrowed from historic literature in which palimpsest refers to traces of writing preserved on materials such as parchment. In a form of recycling in the Middle Ages, such material was reused by scraping the original text, but erasure was

seldom complete, and some of the writing was still visible. Similarly, palimpsests on Ganymede preserve some of the ancient landforms that have been incompletely "overwritten" by more recent processes.

Palimpsests range in size from 50 to 400 km in diameter and occur as circular-to-oval patches with little topographic expression. Most have a central bright zone composed of relatively clean ice excavated from the subsurface. From impact crater mechanics, this suggests that the icy crust was about 10 km thick at the time of impact and was underlain by water or slush. "Dome" craters **(Fig. 8.36)**, which have pronounced dome-shaped floors, are probably related to palimpsests.

Impact basins are also present on Ganymede. For example, Gilgamesh consists of a central depression 150 km wide, filled with smooth plains and surrounded by rugged, mountainous ejecta deposits. A distinct ring fracture zone found at 275 km radius probably represents the boundary for the excavation of the basin. Chains of secondary craters and individual secondaries can be traced to distances of 1,000 km.

8.6.3 Tectonic features

The oldest surface features on Ganymede are sets of semi-concentric furrows **(Fig. 8.37)**, which are considered to represent structural adjustments of the icy crust to impacts. However, because much of the dark terrain is disrupted by bright terrain, the furrows are incomplete

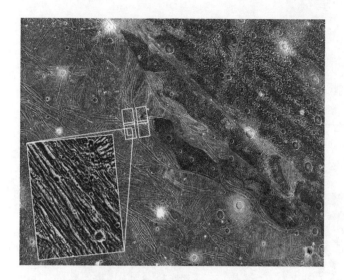

Figure 8.29. The boundary between the furrowed dark terrain of Galileo Regio (upper right) and the younger bright terrain of Uruk Sulcus (lower left): furrows of the Lakhmu Fossae system trend subparallel to the northern boundary of Uruk Sulcus, while the Zu Fossae cut them obliquely; crater forms include bright rays, central pits, and dark floors. The inset shows high-resolution *Galileo* images and details of the bright terrain (NASA PIA00705).

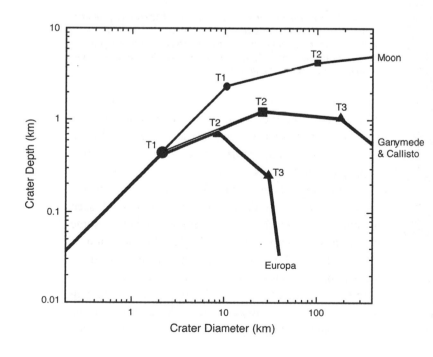

Figure 8.30. Impact crater depth–diameter data for Galilean satellites compared with the Moon. T1, T2, and T3 identify changes in morphology and shape. T1 shows simple–complex transition. T2 shows change from complex to central pit and dome morphology. T3 shows transition to large basins or multi-ring morphologies (from Schenk *et al.*, 2004).

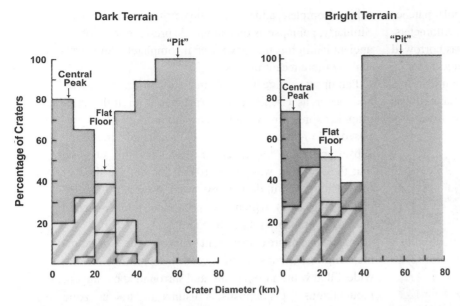

Figure 8.31. Size distributions of central-peak, flat-floor, and pit craters on Ganymede for dark and bright terrains (from Greeley *et al.*, 1982)

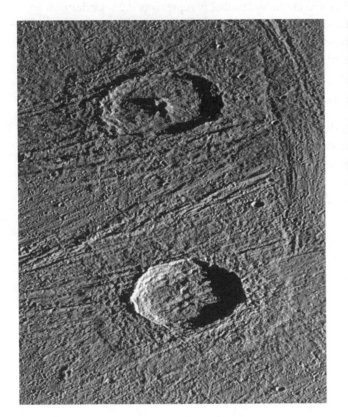

Figure 8.32. An oblique view of two fresh impact craters in bright terrain near the north pole of Ganymede. Gula crater (top, 38 km in diameter) has a central peak; Achelous crater (32 km in diameter) has an outer lobate ejecta deposit (NASA *Galileo* PIA01609).

Figure 8.33. Ganymede's Enki Catena consists of 13 craters, probably formed by the breakup of a comet. The craters cross the boundary between bright and dark terrains; the ejecta deposit surrounding the craters appears very bright on the bright terrain. The area shown is 214 km across (NASA *Galileo* PIA01610).

and in most cases the "parent" impact cannot be identified. Individual furrows are tens to hundreds of kilometers long by 5–20 km wide. Geophysical models of large impacts early in Ganymede's history (when the outer parts of the moon might have been warmer) suggest that inward flow of the asthenosphere would have been accompanied by brittle failure of the upper icy crust to produce the furrows.

Some regions of dark terrain show fracture patterns that can be correlated with the fracture and groove trends seen in younger bright terrain. This suggests that at least some of the tectonic deformation leading to the formation of bright terrain had its initiation in, and was controlled by, tectonic processes associated with the dark terrain.

Initial analyses of Ganymede's bright terrain following the *Voyager* mission suggested tectonic separation of the ice crust and infill by water–ice mixtures (slush) erupted

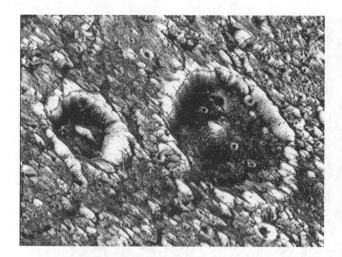

Figure 8.34. An oblique *Galileo* image of two dark-floored craters within the bright zone of Memphis Facula (Fig. 8.37). The dark floors could represent material implanted from the impact, or perhaps material brought to the surface by the impact. Crater Chrysor (left side) is 6 km in diameter; crater Aleyn is 12 km across (NASA *Galileo* PIA01607).

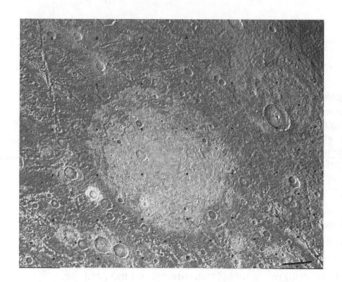

Figure 8.35. A *Voyager 2* image in Galileo Regio showing the 344 km in diameter palimpsest Memphis Facula (NASA *Voyager 2*, FDS 20637.02).

Figure 8.36. Ganymede crater Neith is about 160 km in diameter and is a dome crater; the dome is ~45 km across and is surrounded by rugged terrain. The crater rim is barely visible and is located along the outer boundary of a relatively smooth, circular area (indicated by arrows), assumed to be the crater floor, which in turn surrounds the rugged terrain (NASA *Galileo* PIA01658).

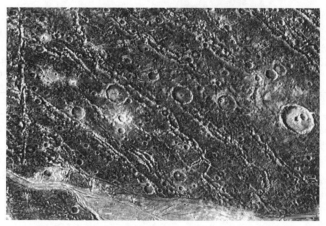

Figure 8.37. Sets of furrows on Ganymede, such as these in Galileo Regio, could be related to large impact events; also visible are several central-pit craters. The area shown is about 650 km across (NASA *Voyager 2* 546J2–001).

onto the surface. Higher-resolution *Galileo* images support this general view in some areas, as outlined in **Fig. 8.38**, but the emphasis is much more on tectonic processes than on cryovolcanism. For example, the boundary between dark and bright terrain in Uruk Sulcus **(Fig. 8.39)** shows highly fractured tilt-blocks of bright terrain cutting across the dark terrain. This style of tectonic resurfacing was diagrammed by Bob Pappalardo and colleagues in 2004 **(Fig. 8.40)**.

Most of the tectonic structures on Ganymede indicate extensional processes, as reflected in the grabens and deformed impact craters cut by faults **(Fig. 8.41)**. In addition, strike–slip deformation is seen **(Fig. 8.42)**, including *en échelon* fault patterns. Global extension of the lithosphere might have taken place as a result of liquid water freezing, much like the expansion of ice in a soda can.

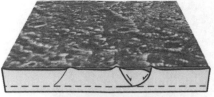

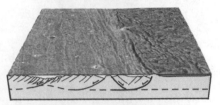

(a) dark terrain tectonism

(b) tectonic resurfacing

(c) volcanic resurfacing

(d) cross-cutting sulci

Figure 8.38. A summary model for the formation and evolution of grooved terrain: (a) normal faults inferred to define furrows in dark terrain may be reactivated to focus later grooved terrain deformation; (b) some grooved terrain may form by tectonic disruption of the preexisting surface without concurrent icy volcanism; (c) some grooved terrain may form by a combination of tectonism and icy volcanism that brightens and smooths the surface; (d) bright terrain swaths can cross-cut one another, dissecting the preexisting surface into a polygonal patchwork (from Pappalardo et al., 2004).

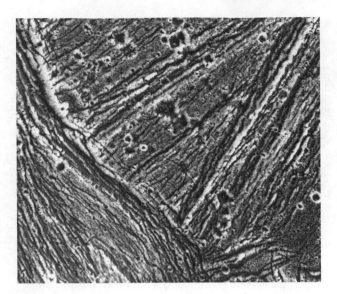

Figure 8.39. Bright grooved terrain in Uruk Sulcus, showing the older horst and graben structures in the upper right side of the image being cross-cut by younger fractures and possible tilt-block faulting in the lower left side of the image (NASA Galileo PIA00280).

Figure 8.40. An idealized model for the topography and structure of Ganymede's grooved terrain. Small ridges and troughs (of wavelength ~1–2 km), potentially formed by tilt-block normal faulting, are superposed on large swells of wavelength ~8 km. A deep trough marks the boundary between faulted and undeformed terrain (left), suggesting deformation of a hanging wall fault block above a prominent marginal fault (from Pappalardo et al., 2004).

This feature is about 50 km long, has a typical caldera-like outline, and appears to be the source for a flow. Except for these features, obvious flow lobes and channels are lacking, which could mean that (1) the images are too poor for recognition, (2) obvious features do not form in "cryolavas," (3) features deform beyond recognition quickly, or (4) such volcanism is rare or absent.

8.6.4 Volcanic features

One of the surprises from the *Galileo* mission was the near absence of clear evidence for extensive cryovolcanic features on Ganymede. Nonetheless, it is currently assumed that many of the very smooth, lightly cratered plains in some bright regions were flooded with melt water or water-rich materials that erupted onto the surface. In addition, a few features are suggestive of volcanic vents, the best candidate of which is in Sippar Sulcus **(Fig. 8.43)**.

8.6.5 Gradation features

The primary processes of gradation on Ganymede are mass wasting, sublimation, and sputtering. Slopes in the dark terrain often show V-shaped streaks trending downslope, which are interpreted as being zones of mass wasting. Aprons at the base of some streaks are probably talus deposits. Detailed analyses of the albedo show that many Sun-facing slopes are dark, while poleward-facing

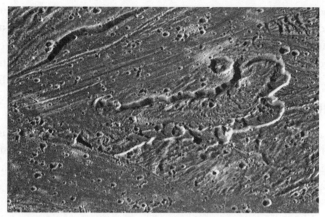

Figure 8.43. A *Galileo* image of Sippar Sulcus, a depression thought to be a cryovolcanic feature on Ganymede. The depression is about 55 km long and 17–20 km wide and has a lobate flow-like deposit 7–10 km wide on the floor (NASA PIA01614).

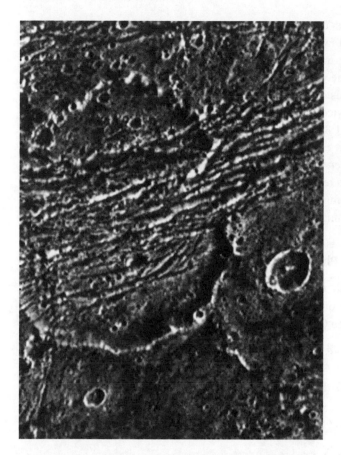

Figure 8.41. This crater in Ganymede's Nicholson Regio was fractured and spread apart for about 50% of its diameter by tectonic extension; the original crater diameter was about 25 km (NASA *Galileo* G7GSNICHOL01).

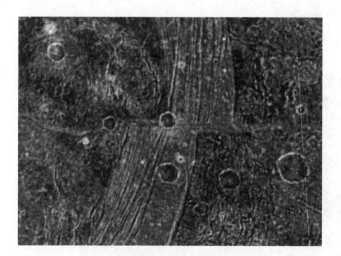

Figure 8.42. Right-lateral strike–slip faulting displaced Dardanus Sulcus on Ganymede by about 150 km (NASA *Galileo* 28GSDARDAN01).

slopes are bright. This suggests that volatiles, such as water-ice, are removed by sublimation from solar heating, leaving lag deposits of non-ice dark materials behind. **Sputtering** also changes surface albedo and occurs in the presence of radiation (in this case from Jupiter's magnetosphere), by which the molecular structure of the volatiles can be changed.

8.6.6 Ganymede summary

Ganymede grew from the collapse of the evolving gas and dust nebula that formed Jupiter and the other satellites in less than half a million years. Heat generated from accretion melted parts of the satellites to form an outer water layer, an inner silicate zone, and an iron–silicate core, forming a differentiated body. With cooling, the surface of the water froze. As reviewed by Kivelson *et al.* (2004), the presence of Ganymede's magnetic field suggests that the core is currently hot enough to be molten and to generate an internal dynamo. In addition to the intrinsic magnetic field, the field induced by Jupiter suggests that Ganymede has a deep liquid ocean sandwiched within the ice. Interior heat is still being generated, mostly from tidal interactions of the same sort as that for Io and Europa but of lower magnitude.

Strains within the icy crust have varied with time and the evolution of the water layer. Different phases of water-ice are functions of temperature and pressure that also change with depth. Some phases, such as ice I (the form familiar to ice skaters), expand as the water freezes, while other phases contract to more dense molecular structures.

Thus, the tectonic patterns exhibited in the images of Ganymede's surface reflect not only tidal interactions within the Jupiter system but also deformation associated with ice evolution.

8.7 Callisto

Callisto is the outermost of the Galilean satellites and, at 4,819 km in diameter, it is slightly smaller than Mercury. Although Callisto is not significantly affected by tidal heating, it is partly differentiated, with an ice-rich outer layer < 500 km thick, an intermediate ice–rock zone, and a rock–metal core **(Fig. 8.4)**. *Galileo* magnetometer data show that Callisto has a magnetic field induced by Jupiter, indicative of a salty ocean deep below the icy crust.

Callisto's surface composition is ice, Mg- and Fe-bearing hydrated silicates, SO_2, possibly ammonia, and various organic compounds. Carbon dioxide is found almost everywhere, but with slightly higher abundances on the trailing hemisphere and around impact features, with the youngest craters showing the largest abundance.

Callisto is the darkest of the Galilean satellites, yet it is still twice as bright as Earth's Moon. Its density is only about 1.8 g/cm^3, making it the least dense of the Galilean satellites and suggesting that it has the greatest proportion of water. Images show that its surface is heavily cratered but has relatively little relief. Except for terrains related to large impacts, the surface viewed by spacecraft is essentially uniform, consisting of dark, heavily cratered terrain.

8.7.1 Physiography

Callisto is more heavily cratered than even the oldest terrain on Ganymede **(Fig. 8.44)**. Many planetary scientists see Callisto as essentially devoid of internal activity

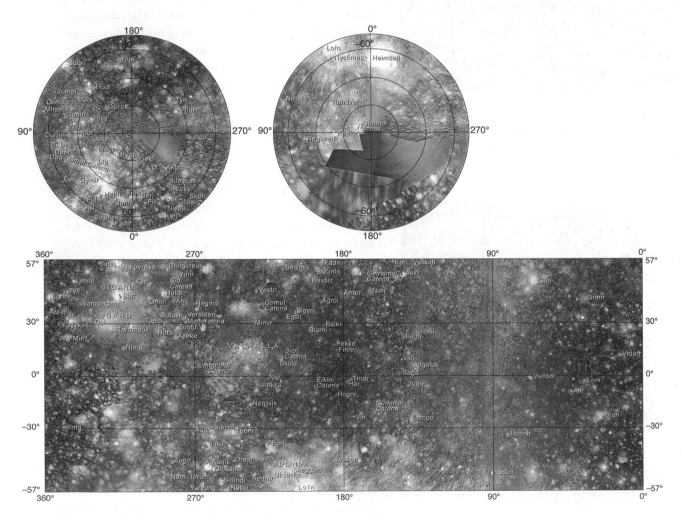

Figure 8.44. A map of Callisto with key place names (from Spencer *et al.*, 2004).

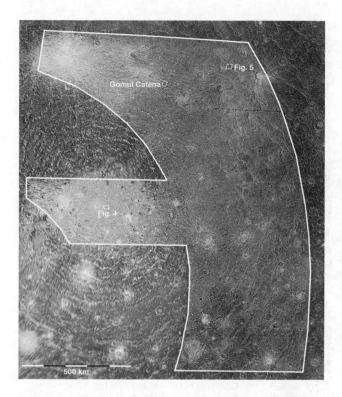

Figure 8.45. A mosaic of *Galileo* images (outlined) taken at 410 m per pixel, set in a background of *Voyager* coverage of part of the Valhalla multi-ringed structure, showing the central bright zone (left side), the inner ring zone, and the outer fracture zone (NASA ASU IPF 1148).

since its initial formation, partly owing to its lack of tidal interactions **(Fig. 8.3)**.

The most prominent feature on Callisto is a multi-ring impact structure named Valhalla **(Fig. 8.45)**. Valhalla consists of a bright central zone (palimpsest) about 600 km wide, surrounded by concentric rings extending outward for nearly 2,000 km from the center of the structure. The bright central zone probably marks the initial crater immediately following impact. At least seven other multi-ring features have been recognized on Callisto, including Asgard with its central bright zone 230 km wide. Tectonic and cryovolcanic features other than those associated with impacts have not been detected. Thus, Callisto's surface consists of ancient heavily cratered terrain composed of ice and rocky material with large, superposed multi-ring impact structures.

8.7.2 Impact features

Callisto's impact craters **(Fig. 8.46)** have morphologies similar to those on Ganymede. The smallest craters are simple bowl-shaped to flat-floored depressions that range

in diameter from about 7 km down to ~25 m, the limit of identification. The floors of small craters vary from smooth to pitted or hummocky. Central-peak craters range in size from about 5 to 40 km in diameter, and the peaks are as large as 10 km across and 0.5 km high. Many central-peak craters have scalloped rims, terraced inner walls, and hummocky floors, which is typical of similar craters seen on the terrestrial planets and the Moon.

Craters > 25 km in diameter transition from central-peak to central-pit morphologies, with central-pit diameters being 15%–20% of the crater diameter. Most central-pit craters have flat floors and, in some cases, interior terraces on their walls. For example, Tindr is a 105 km in diameter crater with a central-pit complex about 15 km across **(Fig. 8.46)**. Although the mechanism for the formation of pits remains poorly understood, a primary factor appears to be the nature of the target material (i.e., ice) rather than properties of the object that caused the impact.

Some craters > 60 km in diameter show central domes, such as Doh **(Fig. 8.46)** and Har. Doh crater is superposed on the central zone of the Asgard multi-ring structure, and has a rim composed of mountains and knobs that form a ring about 65 km across. Individual mountains are as large as 8 km across and 730 m high. Har's dome is in the center of the crater floor and is ~25 km across, or more than one-third of the crater diameter. The dome displays irregular troughs that form a crude radial pattern, giving the impression of tensional fractures that formed in response to uplift, similarly to the cracking of a bread crust. Although there are no obvious ejecta rays, a higher-albedo zone surrounding Doh probably represents ejecta deposits, including impact melt.

As crater diameters on Callisto increase above 100 km, the relative depth decreases, which may be due to viscous relaxation of the ice. Relatively young large craters are identified by concentric faults or grabens, or remnants of their rims. Two such craters, Lofn and Bran, show no evidence of a central pit or dome but have very flat floors. This morphology could have resulted from a cluster of impacts, much like a shotgun blast, or they could be a form that is transitional to multi-ring structures.

Catenae, or crater chains, are seen in many areas of Callisto **(Fig. 8.47)**. The longest found thus far is Gipul Catena, which stretches some 620 km across the surface and includes craters as large as 40 km. Two models of formation have been suggested for catenae. Initially, Callisto's catenae were thought to be secondary craters ejected from primary craters. However, except for two small chains radial to Valhalla and a few other cases,

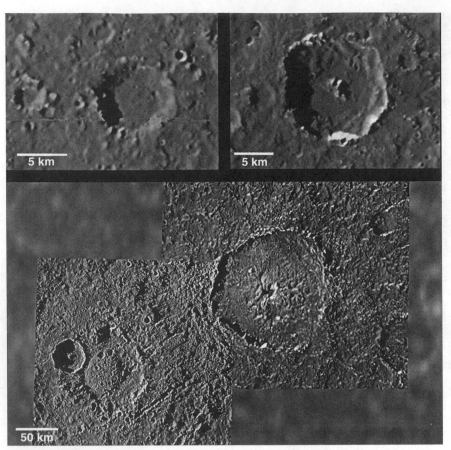

Figure 8.46. A composite of *Galileo* images showing typical crater morphologies on Callisto: upper left, a small flat-floored crater (*Galileo* frame s0401505565); upper right, a complex central-peak crater (*Galileo* frame s0401505526); and bottom, a central-pit crater (Tindr; right side) and a central-dome crater (Har; left side) (NASA *Galileo* frames s0401505300 and s0413389500).

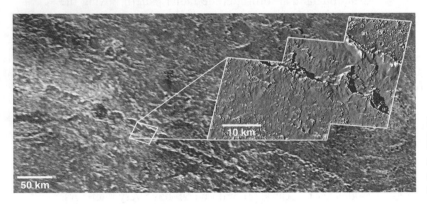

Figure 8.47. Moderate-resolution (410 m per pixel) images of Gomul Catena, north of the Valhalla multi-ring structure, and an inset of high-resolution images (40 m per pixel), showing the degraded terrain on and around the catena (NASA ASU IPF 1040).

searches for the primary craters were unsuccessful, leading to the consideration that they were formed by the impacts from fragments of a larger object. The breakup of comet Shoemaker-Levy 9 as it approached Jupiter demonstrated that tidal disruption can result in a "string" of impacting objects, which could result in the formation of some catenae. Similar chains of craters are seen on Ganymede **(Fig. 8.33)**.

Schenk (1995) classified large impact features on Callisto to include multi-ring structures, palimpsests, and "cryptic" ring structures. Multi-ring structures imaged by *Galileo* include Valhalla **(Fig. 8.45)**, Adlinda, Asgard **(Fig. 8.48)**, Heimdall, and Utgard. Most are typified by central plains, a high-albedo zone extending beyond the central area, and one or more sets of concentric scarps and troughs.

Valhalla's outer trough zone is about ~3,800 km across and is characterized by sinuous depressions **(Fig. 8.49)** considered to be grabens resulting from extension of the icy crust. Some troughs are probably related to older

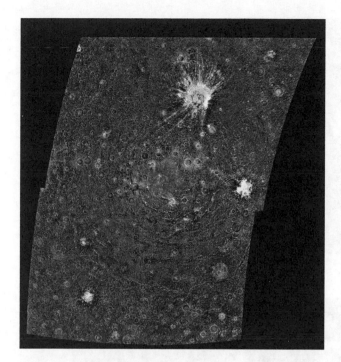

Figure 8.48. A *Galileo* mosaic showing the Asgard multi-ringed structure. Utgard is a degraded multi-ring structure partly obscured by the fresh, bright-rayed crater, Burr. Tornarsuk crater is east of the central zone of Asgard (NASA ASU IPF 1041).

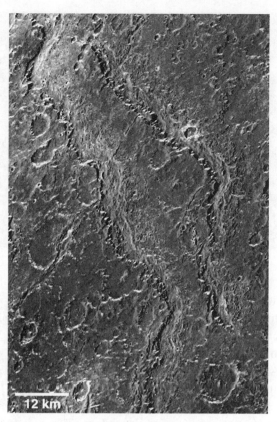

12 km

Figure 8.49. A moderate-resolution (410 m per pixel) image of the Valhalla outer ring zone centered about 27° N, 27.3° W, showing sinuous troughs, or grabens (NASA ASU IPF 1148).

impact craters, which might have weakened the ice locally and influenced the development, location, and orientation of the troughs during and following the Valhalla impact. The inner ridge and trough zone of Valhalla extends ~950 km from the center of the structure and is characterized by asymmetric ridges with the steep flanks facing outward. Many of the ridges consist of a series of bright knobs. The troughs are as long as several hundred kilometers and as wide as 20 km. They are slightly sinuous and appear to be grabens, similar to those of the outer zone. Valhalla's 360 km central zone has a mottled appearance and is moderately cratered **(Fig. 8.50)**. Numerous dark-halo craters suggest penetration into a darker substrate at depths of 1 km or less. The central zone shows remnants of degraded crater rims and small knobs of unknown origin, all set in a background of smooth, low-albedo plains. Although it has been suggested that some of the plains associated with Valhalla are cryovolcanic, there are no indications of flows or similar features. Impact may have occurred in thin, rigid ice overlying liquid water that rapidly filled the crater to form the plains.

The Asgard multi-ringed structure **(Fig. 8.48)** includes central plains surrounded by discontinuous, concentric scarps and an outer zone characterized by sinuous troughs.

The central plains are about 200 km across and are bright, partly due to ejecta from a more recent impact crater, Doh, which apparently excavated ice-rich (brighter) subsurface materials.

The Adlinda multi-ringed structure is in the southern hemisphere of Callisto, where it is defined by sets of concentric lineaments and troughs, the most prominent of which is a nearly continuous trough ~520 km long on the southwest part of the structure. The Adlinda structure has a lower frequency of superposed impact craters than the surrounding terrain, suggesting its relative youth.

8.7.3 Gradation features

One of the surprises revealed by *Galileo* images of Callisto was the paucity of small impact craters in some areas. **Figure 8.50** compares terrains within Valhalla and an area at about the same latitude west of Valhalla at the same scale and resolution. Although the viewing geometry and illumination are not quite the same, it is clear that small craters are present in some areas but absent in

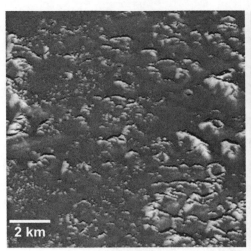 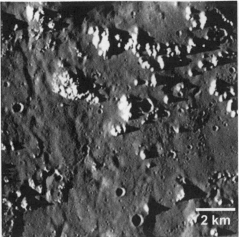

Figure 8.50. A comparison of the paucity of small craters in the Valhalla central plains (left image) and terrain (right image) not associated with a multi-ring structure. The paucity of small craters in some areas of Callisto is generally attributed to some process of crater degradation (left image, NASA *Galileo* frame s0368317401; right image, NASA *Galileo* frame s0506142927).

others. Prior to these images, the expectation had been to see increasing numbers of small craters down to the limit of resolution, as on Earth's Moon. However, some surface process appears to erase small craters and other landforms. Jeff Moore of NASA's Ames Research Center concluded that sublimation erosion could be an effective agent of gradation in which volatile components sublimate, leaving non-ice components to mass-waste downslope. The overall effect is to reduce topography and "soften" the terrain, with smaller craters being degraded or removed. Given the near lack of small craters in some areas of Callisto, the rate of degradation exceeds the rate of impact, but this process (if it occurs) does not take place uniformly over Callisto.

Degradation of Callisto's surface also includes mass wasting on the floors of several craters **(Fig. 8.51)**. These slides could result from collapse of the crater walls due to oversteepening of the slopes by sublimation erosion.

8.7.4 Callisto summary

Ganymede and Callisto are about the same size and density. Prior to the *Voyager* flybys, it was assumed that both would show surfaces reflecting similar geologic histories. However, images revealed strikingly different worlds **(Fig. 8.52)** and generated debate on the comparative evolution of the icy satellites. Nonetheless, some agreement has been reached on the general events in their histories, if not on the detailed evolution. In the early history of the Jupiter system, both objects formed from materials leading to water–silicate bodies. Heating from various sources (impact cratering, radioactive decay, tidal stresses) led to differentiation and the formation of a core. However,

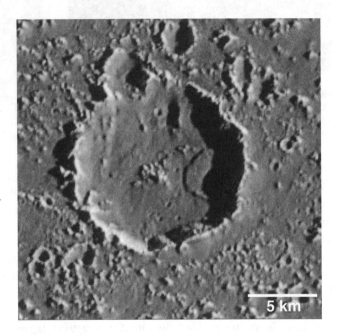

Figure 8.51. A landslide deposit inside a central-peak crater (~13 km in diameter), showing a lobate landslide extending onto the crater floor (NASA *Galileo* frame s0413382840).

differences that would result in less heat on Callisto include its smaller size (i.e., less radioactive decay) and lack of significant tidal stresses.

8.8 Small moons and rings

In addition to the four Galilean satellites, at least 59 more moons orbit Jupiter. All are small, with Amalthea, measuring 270 km by 166 km, being the largest. Four orbit Jupiter within the path traced by Io. The innermost two,

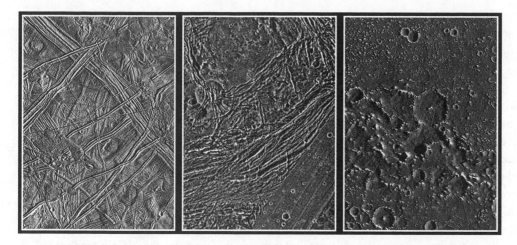

Figure 8.52. Comparisons of the surfaces of Europa, Ganymede, and Callisto at the same scale. Callisto (right) is "peppered" by impact craters but is also covered by dark material of unknown origin, as seen here in the Asgard region. Ganymede (middle) shows both craters and abundant tectonic deformation (lacking on Callisto), seen here in Nicholson Regio. Europa (left) is sparsely cratered (indicating recent geologic activity) and has abundant ridged plains, shown here around the Agave and Asterius dark lineaments. While all three moons are nearly as old as the Solar System, the ages of their surfaces are markedly different (NASA *Galileo* PIA01656).

Metis and Adrastea, were discovered from *Voyager* data in 1979, as was Thebe, whose orbit lies between those of Amalthea and Io. The remaining small moons are all found beyond the orbit of Callisto.

Before the Space Age, only Saturn was known to possess rings. We now know that all of the giant planets have ring systems, including Jupiter. Jupiter's rings were first photographed by the *Voyager* spacecraft, and later images from *Galileo* show that the rings are complex systems closely linked to some of Jupiter's small moons.

Overall, the rings include the outermost gossamer ring, a flat main ring, and an innermost donut-shaped halo ring. All are very faint and are best seen when they are backlit by the Sun. From the way in which the light is reflected from the ring particles, individual grains are estimated to be only about the size of very fine dust, something like smoke particles.

Thebe, Amalthea, Adrastea, and Metis revolve in the inner region of the Jupiter system, are mostly embedded in the ring system, and are the parent bodies of the ring material. One of the highlights of the *Galileo* mission was the discovery of how the rings form. High-resolution images show that fine dust is generated by the constant bombardment of the moons by impacts. Because the gravity of the satellites is very low, the dust escapes and spirals inward under the gravitational tug of Jupiter. This constant grinding and pulverization generates the supply of dust that makes up the ring components.

8.9 Summary

In the past few decades, more than 400 planets have been detected around stars other than our own Sun. In nearly all cases, these **exoplanets** are in the size class of Jupiter, and it is likely that smaller planets and satellites are also present but have not yet been detected. Consequently, in many regards, the Jupiter system is considered to be a prototype for these newly discovered worlds, and its study can shed light on how other Solar Systems might originate and evolve.

The Galilean moons and small regular satellites all originated from the nebular cloud that formed Jupiter. Just as higher-density materials condensed closer to the Sun to form the inner planets, higher-density materials also condensed closer to Jupiter to form Io, with less dense materials forming the outermost moon, Callisto. All four bodies are locked in synchronous rotation, showing the same "face" toward Jupiter. In addition, Io, Europa, and Ganymede are in Laplace resonance, which is unique in our Solar System for three bodies. These conditions lead to an intricate dance of the moons around Jupiter and gravitational tugs of war, resulting in tidal stresses and the generation of internal heat. It is this heat that drives the huge volcanic eruptions on Io and the tectonic deformation of the icy crusts of Europa and Ganymede.

All four Galilean satellites are differentiated, with the degree of differentiation being a function of location with respect to Jupiter. Io is highly differentiated into core, mantle, and crust, while Callisto appears to be only partly differentiated. In addition, Europa, Ganymede, and Callisto appear to have liquid water beneath icy crusts, as indicated by the behavior of the magnetic fields induced in each by Jupiter's magnetic field. Ganymede's high-density and molten core probably accounts for the generation of its own magnetic field, making it unique among the moons of the Solar System.

In addition to the four large moons, Jupiter has some 59 regular and irregular (i.e., captured) satellites, plus rings that are intimately associated with some of the smaller moons. The Jupiter system (Jupiter itself, its magnetic field, the moons, and the rings) is a high priority for future exploration. Although Europa is a central focus because of its potential for astrobiology, the entire system is of interest in consideration of the evolution of habitable worlds. Thus, Io is an end member with its lack of water and extremely harsh radiation environment, while Callisto is at the other extreme with abundant water, most of which is frozen, and its lack of significant interior heating. Ganymede is somewhat more favorable for potential habitable zones, with its internal heat and apparent liquid water at depth, but, in a sort of "Goldilocks" scenario, Europa seems to be an ideal candidate for astrobiological exploration. It generates sufficient internal heat to fuel processes that are probably geologically active today (or certainly were in the recent past), while harboring liquid water closer to the surface than Ganymede. Although it, too, is bathed in radiation from Jupiter, habitats would be shielded by ice only a meter or so below the surface.

Assignments

1. Briefly explain how internal heating is generated within Io, Europa, and Ganymede as a consequence of their orbital geometries.

2. Contrast the surface ages for Io, Europa, and Callisto and discuss the probable geologic reasons for the differences among the ages for the three moons. Include the difference between *surface ages* and ages of *satellite formation*.

3. Explain the concept of *tectonic resurfacing*, especially as related to Ganymede.

4. Discuss why Europa is a high priority for astrobiological exploration.

5. Briefly outline the differences between tectonic features on Europa and those on Callisto.

6. Describe the primary characteristics of effusive eruptions on Io in terms of compositions, styles of lava emplacement, and general morphology.

CHAPTER 9

The Saturn system

9.1 Introduction

Saturn is an enormous planet, second in diameter only to Jupiter. From the discovery of rings around Saturn nearly 400 years ago until the last three decades when rings were found around the other giant planets, Saturn was thought to be unique in the Solar System. Saturn has at least 62 satellites, including Titan, which has global clouds and an atmosphere denser than that of Earth, and the moon Enceladus with its actively spewing geysers. Titan is the only outer planet moon on which a spacecraft has landed. Consequently, the Saturn system holds a special place in our view of the Solar System.

9.2 Exploration

When Galileo viewed Saturn through his telescope for the first time in 1610, he apparently thought he was seeing three separate objects, but later observations led to his publishing a sketch in 1616 that clearly showed Saturn and its ring system. Rapid improvements in telescopes and their application to planetary observations resulted in more detailed descriptions of Saturn and prompted wide speculation on the origin and characteristics of its system of rings.

Exploration of the Saturn system by spacecraft began with the *Pioneer 11* flyby in 1979, followed by *Voyager 1* and *Voyager 2* in 1980 and 1981, respectively. *Pioneer 11* data yielded new insight into the magnetic field generated by the planet and enabled the discovery of the F Ring. The *Voyager* spacecraft returned the first clear images that revealed the great geologic diversity of Saturn's satellites.

The paths of *Voyagers 1* and *2* were planned to use a gravitational "boost" from Jupiter to speed them on their way to Saturn. *Voyager 1* flew closer to Jupiter and received a greater boost in speed that allowed it to arrive at Saturn in November 1980, nine months earlier than *Voyager 2*. Aimed to make a close pass by Saturn's giant moon, Titan, *Voyager 1* also took high-resolution pictures of Mimas, Dione, and Rhea. The path of the *Voyager 2* spacecraft was designed so that it would fly past Uranus and Neptune with a trajectory that nicely complemented that of *Voyager 1* and provided close views of Saturn's moons Iapetus, Hyperion, Enceladus, and Tethys. In addition to spacecraft observations, the *Hubble Space Telescope* has provided critical new information, especially on Saturn's clouds.

The *Cassini–Huygens* mission is a joint project that involves the NASA *Cassini* orbiter and the ESA *Huygens* probe of Titan. This mission is a "flagship-class" project and is the largest and most complex planetary spacecraft launched to the outer Solar System. Launched in 1997, it followed a trajectory similar to that of the *Galileo* Jupiter spacecraft by looping around the Sun to gain sufficient momentum to "sling-shot" into the outer Solar System. This involved two flybys of Venus (in 1998 and 1999) and one past Earth (later in 1999), during which some instruments were calibrated using Earth's Moon as a target. During its fly-through of the asteroid belt in early 2000, images were taken of asteroid 2685 Masursky to determine its diameter of 15–20 km. Later that same year, the *Cassini* spacecraft flew past Jupiter and obtained new data on the atmospheric structure of the planet and some characteristics of its rings, as well as images of Io's volcanic eruptions from which movies were made.

The *Cassini* orbiter and attached *Huygens* probe reached Saturn in mid 2004. Later that year, on December 25, the probe was released from the orbiter, and it reached Titan's surface on January 14, 2005. As it descended through Titan's atmosphere, various measurements of its temperature, pressure, and composition were made, as well as obtaining descent images. On reaching the surface, some instruments continued operation, including the cameras that returned the first images from the surface of an outer planet satellite.

The *Cassini* orbiter carries an imaging system, spectrometers in the visible, IR, and UV ranges, and a radar imaging system to map Titan's surface through the dense clouds. The payload also includes a magnetometer system, plasma spectrometer, cosmic dust analyzer, and an ion/neutral-species mass spectrometer. Following the release of the *Huygens* probe, *Cassini* was placed in orbit around Saturn in July 2004, with subsequent paths designed to make close flybys of the various satellites, similar to the sequences used by *Galileo* in orbit around Jupiter, and to make repeated observations of Saturn and the rings. Among the many exciting findings was the discovery in 2005 that Enceladus is a geologically active object with erupting plumes of volatiles from its south polar region. Particularly important for planetary geology are the observations of Titan obtained by *Cassini*'s radar system, which is capable of imaging the surface through the dense clouds. As this mission continues, swaths of radar data are slowly building a map of the surface of this intriguing moon. The orbiter is scheduled to continue operations until 2017, and, if all goes well, the result will be 155 orbits of Saturn, 53 flybys of Titan, and 11 flybys of Enceladus, thus enabling long-term study of the active plumes.

Figure 9.1. This view of Saturn is a mosaic of 30 images taken by the *Cassini* orbiter in 2008, showing its prominent rings and the relatively bland clouds (NASA PIA11141).

9.3 Saturn

Saturn's density is about 0.7 g/cm^3, which is the lowest of all the planets. Like that of Jupiter, Saturn's atmosphere consists of about 80% hydrogen, 18% helium, and amounts of oxygen, iron, neon, nitrogen, and sulfur totalling less than 2%. The interiors of Saturn and Jupiter are also similar, with both probably having mantles and rocky cores. Saturn spins on its axis extremely rapidly, taking only 10 hours to complete one rotation. The high spin rate, coupled with the interior configuration of core and mantle, probably accounts for its large magnetic field, which is only slightly weaker than that of Earth but is much weaker than Jupiter's enormous field. The rapid spin also causes a flattening of the poles and an equatorial bulge, leading to Saturn's oblate spheroid shape.

Although the atmosphere and cloud patterns on Saturn are not as spectacular as those on Jupiter, they are nonetheless interesting. Like that of Jupiter, its atmosphere is zoned, but the patterns are fainter and somewhat bland (**Fig. 9.1**). Saturn's winds have been clocked at some 500 m/s. Shear and up-welling of the atmosphere lead to some distinctive cloud patterns, including oval-shaped features that last from a few days to several years, apparently indicating the presence of storm systems. Unlike storms on Earth, which are driven primarily by heating from the Sun, saturnian storms are probably generated from heat derived from the interior of the planet. One of the most distinctive features is the Great White Spot, which is short-lived and appears and disappears about every Saturn year (30 Earth years). It is considered analogous in some ways to Jupiter's Great Red Spot; it typically begins as a small bright feature in the northern hemisphere that expands laterally into an oval feature as wide as several thousand kilometers.

9.4 Satellites

The Saturn system (**Fig. 9.2**) has about the same number of known moons as Jupiter, with ~62 satellites that range in size from less than 10 km across to the Mercury-size object Titan, which has its own atmosphere of nitrogen that exerts a surface pressure of 1.45 bar (greater than that on Earth). All of the satellites have densities suggesting that they are ice-rich objects with different amounts of rocky material in their interiors. The nine largest moons were discovered telescopically, some as early as 1655. Most of their names are taken from Greek mythology, in which Dione, Rhea, Tethys, Mimas, Enceladus, Titan, and Phoebe were giants, while Hyperion and Iapetus were brothers of Saturn.

Saturn's moons display a wide variety of orbital motions. Twenty-four are regular satellites, meaning that they are traveling in the same direction as Saturn's spin. Most of the remaining moons are irregular satellites, thought to be captured objects, which travel in a wide variety of inclined retrograde orbits.

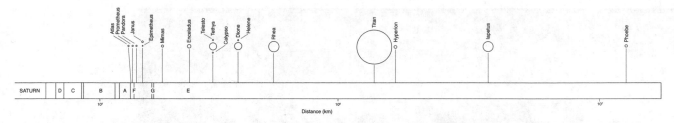

Figure 9.2. A diagram of the major and some minor satellites of Saturn and the primary ring systems. The distances shown are on a logarithmic scale from the center of Saturn.

Figure 9.3. Comparative sizes of some of Saturn's satellites; not shown are Saturn's Titan, which is half again as large as Earth's Moon, and the smaller moons of Saturn (from Moore *et al.*, 2004). Reprinted from *Icarus*, **171**, Moore, J. M., Schenk, P. M., Bruesch, L. S., Asphaug, E., and McKinnon, W. B., Large impact features on middle-sized icy satellites, 421–443, 2004, with permission from Elsevier.

All but two of the major moons are in synchronous rotation around Saturn, which, analogously to the Galilean satellites, leads to the terms **sub-Saturn, anti-Saturn, leading**, and **trailing hemispheres**. *Voyager* and *Cassini* images have been merged to generate global or near-global maps of many of the satellites, along with place names (Roatsch *et al.*, 2009). The prime meridian (zero longitude) for each moon is defined by a prominent feature, such as a crater, near the center of the Saturn-facing hemisphere, with longitudes being numbered to the west.

There are no obvious patterns relating the sizes or densities of the satellites to their distance from Saturn or to the degree of geologic activity among them. For simplicity, the moons are discussed beginning with the two satellites of high geologic interest, Titan and Enceladus; intermediate-size moons (**Fig. 9.3**) are given in the order of their orbital distance from Saturn, and then some of the small moons are discussed.

9.5 Titan

Titan is a remarkable moon. At 5,150 km in diameter, it is larger than Mercury and is in the same size class as Ganymede and Callisto. Like those satellites of Jupiter, its mass suggests that it is composed of about 45% water (mostly ice, but possibly some liquid) and 55% rocky material. What makes it remarkable is its dense atmosphere and surface conditions that allow liquids to exist. Although the satellite was the first to be discovered in the saturnian system (it was found by Christiaan Huygens in 1655), the presence of an atmosphere was not suspected

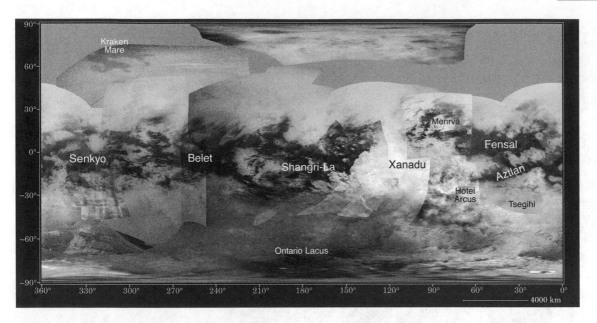

Figure 9.4. A mosaic of *Cassini* images of Titan with some of its prominent place names; most of the radar-dark patterns in the equatorial region are fields of linear dunes; the scale bar applies to the equatorial zone (NASA PIA11149).

until the twentieth century. In 1944, the Dutch-American astronomer Gerard Kuiper identified methane on Titan, which initiated speculation on the various chemical reactions that might be taking place in its atmosphere.

The presence of extensive clouds was well established by the time the *Voyager* spacecraft arrived. *Voyager* scientists hoped that the clouds would be broken and sufficiently scattered to allow glimpses of the surface. Such was not the case, and the *Voyager* pictures are disappointingly bland, showing a relatively uniform orange–brown sphere with only hints of variations in the clouds. Limb views, however, showed layers in the atmosphere that are thought to be smoglike photochemical hazes. Later, *Cassini*'s radar and infrared imaging system provided maps and moderate resolution views of Titan's surface (**Fig. 9.4**).

Nitrogen is the main component in Titan's atmosphere, along with methane, hydrogen, and other carbon–nitrogen components. It may be that, when Titan first formed, methane and ammonia ices were caught up in the accreting mass. With heating and subsequent chemical differentiation, these compounds were released from the interior, forming an early-stage atmosphere. The action of sunlight on the ammonia released the nitrogen and allowed the escape of most of the hydrogen into deep space while methane remained on the surface. Calculations show that some of the methane evaporated and was dissociated by sunlight. The products recombined to form complex organic chemicals, resulting in an orangish smog.

Following the *Voyager* flybys, there was speculation on Titan's surface processes and overall geology. All three states of methane seemed to be possible in the current environment, and it was suggested that methane gas and other molecules in the atmosphere condense to form clouds, snow, and rain, leading to rivers and lakes of liquid methane that could erode solid ice on the surface. Wind processes were thought to result from the dynamic atmosphere that could entrain small ice particles. These potential surface processes and the chemical reactions, as shown in **Fig. 9.5**, were discussed by planetary scientist Jonathan Lunine (1990).

The successful descent and landing of the *Cassini–Huygens* probe in 2005 yielded an enormous wealth of data on Titan's atmosphere and surface. Measurements of wind speeds began at 150 km above the surface and were recorded at more than 100 m/s. As the probe descended, it was buffeted by fluctuating winds resulting from vertical shear; near the surface, winds were determined to be a few meters per second, well within the range for aeolian processes. *Huygens*'s Gas Chromatograph Mass Spectrometer measured not only the chemistry of the atmosphere but also the surface materials after landing, while the Descent Imager/Spectral Radiometer obtained images of the landing site. As the probe touched down, engineering data enabled the physical properties of the surface to be evaluated; although the first descriptions of it were as a sort of "Titan mud," later analysis suggested the presence of dry,

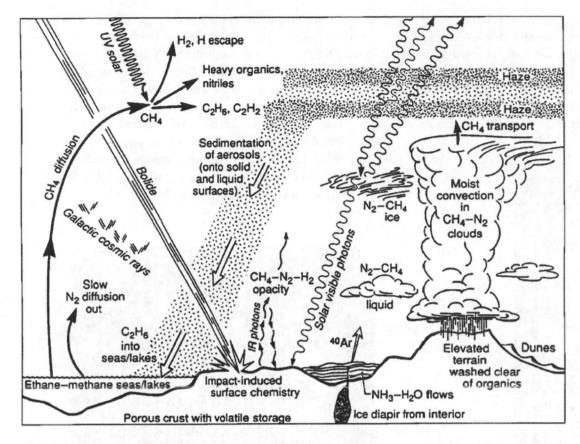

Figure 9.5. A diagram showing the inferred surface processes and cycle of methane and other components on Titan; methane can exist in gas, liquid, and solid forms. Thus, cryovolcanism and gradation in the form of wind, fluvial, lacustrine, and possibly glacial processes can occur, along with impacts and tectonic deformation, as portrayed here before the discoveries from the *Cassini* mission (from Lunine, 1990).

sand-like ice grains. Images show the surface littered with pebbles and cobbles composed of ice, and their rounded appearance suggests erosion by some fluid (**Fig. 9.6**).

Radar and infrared images from *Cassini* and the remarkable success of the *Huygens* probe amply confirm the speculation on Titan's geology and provide evidence for impact cratering, gradation, tectonism, and possibly cryovolcanic processes, some of which are active today, as reviewed by Ralf Jaumann *et al.* (2009a).

Relatively few confirmed impact structures have been identified in images of sufficient resolution for detection. This paucity indicates the relative youth of Titan's surface. Ksa crater (**Fig. 9.7**) in the equatorial region is about 29 km in diameter and has a radar-bright inner wall and a distinct raised rim. The radar-bright central peak is surrounded by radar-dark terrain, suggesting smooth deposits. Ejecta extends as far as 40 km and has hints of radial striations. The ejecta deposits are encroached on by dunes on the west and are slightly eroded by a channel on the east.

At 80 km in diameter, Sinlap crater is larger than Ksa but lacks a central peak. It does, however, also have extensive ejecta deposits with an inner, radar-dark band and an outer brighter zone, which might be comparable to the "ejecta flow" craters seen on Mars. If this is the case, the impact might have melted crustal ice to produce fluid, slurry-like ejecta.

The largest confirmed impact structure is the 440 km in diameter Menrva, found in the northern mid-latitudes. This basin consists of two rings, both of which are composed of rugged, elevated mountains (**Fig. 9.8**). Relatively smooth (radar-dark) plains are found between the rings, and there are suggestions of ejecta deposits beyond the outermost ring. In addition to the handful of other known impact craters on Titan, dozens of features of possible impact origin are seen but are not imaged in sufficient detail for confirmation.

The interior of Titan is likely to be partly differentiated. As reviewed by Sotin *et al.* (2009), Titan is modeled to have a silicate core, overlain by a zone of high-pressure ice, an ammonia-rich water ocean, and an outer solid shell consisting of organic materials and ices. Interior heating from radiogenic sources in the core, coupled with tidal heating, could lead to tectonic and volcanic processes. For example, the ridges shown in **Fig. 9.9** are likely to be tilted blocks of crust lifted by faulting. They occur in the mountainous region of Xanadu and are similar to features seen west of Shangri-La. In both regions, ridge belts are oriented east–west and are spaced about 50 km apart, suggesting tectonic compression in the north–south direction.

Various possible cryovolcanic features were imaged by *Cassini*, but the image quality is rather low. These include numerous lobate flows, small domes, and putative calderas. Some of the flows, such as Ara Fluctus, appear to originate in calderas. **Figure 9.10** shows a possible caldera about 100 km across in the southern hemisphere. Although it could be a modified impact basin, it lacks raised rims and is rather irregular in plan-form.

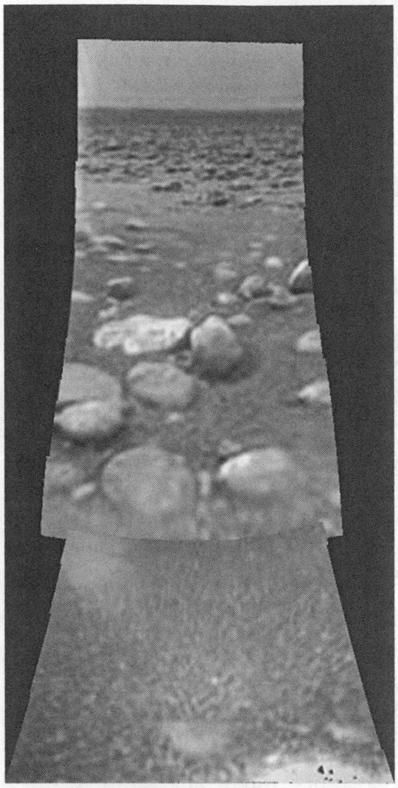

Figure 9.6. The *Huygens* camera returned these images from its landing site, viewed toward the horizon. The cobbles in the foreground are 10–15 cm across and are thought to be composed of ice; their rounded appearance is attributed to erosion by a fluid, such as liquid methane (NASA PIA06440).

Figure 9.7. A *Cassini* radar image of Titan's crater, the 30 km in diameter Ksa, showing its raised rim, central peak, smooth floor, and extended ejecta deposit. Ksa is in the terrain between the Xanadu and Fensal regions; the dark band across the image is a radar artifact (NASA PIA08737).

Figure 9.8. The multi-ring impact structure, Menrva, has an outer rim diameter of 440 km, shown on this *Cassini* radar image mosaic; the horizontal bands are mosaic artifacts (NASA PIA07365).

Figure 9.9. These three mountain chains in Xanadu stretch 100 km in an east–west direction and are tectonic features, possibly resulting from north–south crustal compression. The radar-dark circular feature south of the chains could be an impact scar; the white dotted line is a gap in the radar data (NASA PIA10654).

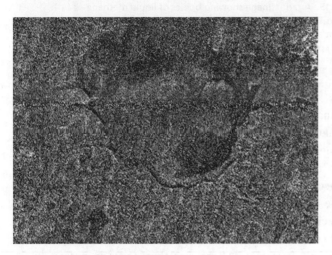

Figure 9.10. A radar image of Titan, showing an irregular-shaped basin ~100 km across, which has been interpreted as being a cryovolcanic caldera; radar-dark and radar-bright features on the basin floor represent surfaces of different roughnesses, which could be flows of different textures. The dark band through the middle of the basin is the seam between two swaths of radar data (NASA *Cassini* PIA12111).

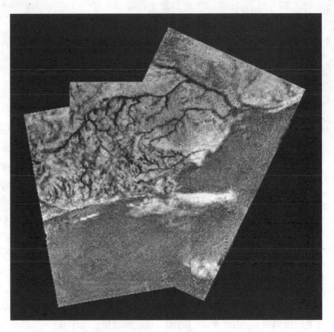

Figure 9.11. A mosaic of images from the *Huygens* descent camera showing the terrain around the probe landing site; the dark dendritic pattern is interpreted to represent fluvial run-off of liquid methane (ESA; NASA PIA07236).

By far the dominant surface process revealed by the *Cassini–Huygens* mission is gradation by wind and fluvial activity. Surface erosion and re-deposition by liquid methane is evident, and these processes are likely to be active today. The first hints of fluvial processes came from the *Huygens* descent images, which showed a quasi-dendritic pattern of dark channels (**Fig. 9.11**). Subsequent radar images from the orbiter showed that channel systems (some of which are contained within valleys) are common in many areas of Titan, with some being at least 1,000 km long. Many studies of the channel-forming liquids and the flow dynamics in the Titan environment have been conducted, as reviewed by Jaumann *et al.* (2009a). For example, methane's viscosity in the ~94 K environment of Titan would be some five times lower than that of water on Earth. Adjustments for flow conditions for Titan's lower gravity and the composition of erodible surface materials (i.e., organic compounds and water-ice) indicate that sand-to cobble-sized material (~15 cm) would be easily transported, which is consistent with materials seen in the *Huygens* images of the surface.

Some channels are radar-bright, indicating that their floors are littered with coarse material, such as boulders, while others are radar-dark, which could indicate the

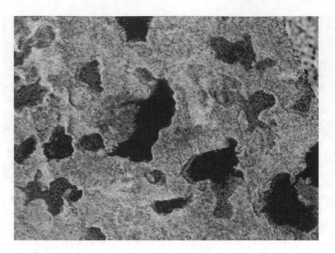

Figure 9.13. A radar image showing bodies of liquid methane–ethane in Titan's north polar area. Such lakes and larger seas are abundant and are concentrated in the higher latitudes, especially in the north polar region; repeat radar imaging shows that some lakes disappear with Titan's seasons (NASA *Cassini* PIA09183).

Figure 9.12. *Cassini* radar images of Titan's south polar area, showing complex, chaotic terrain degraded by a variety of processes, including fluvial erosion that formed the sinuous valley and channel system. The vertical black line is the seam between two images (NASA PIA10219).

presence of liquid methane or smooth ice. Many of the channels and valleys form "ordered" systems of tributaries, similar to fluvial systems on Earth, suggesting runoff from precipitation, although some could emerge from underground sources. Some models of the rate of possible rainfall on Titan suggest that "flash-floods" could occur, while other models are much more conservative and suggest that precipitation is substantially less and would be comparable to desert conditions on Earth. One possibility is that high-volume precipitation occurs rarely, leading to occasional flash-floods that recharge subsurface reservoirs, as is typical of a dry climate.

Terrain such as that shown in **Fig. 9.12** of the south polar region has been extensively modified by fluvial processes, which are evidenced by small narrow valleys, large meandering channel systems, and chaotic terrain.

Some early considerations of Titan's surface included vast oceans of liquid hydrocarbons; in fact, there was concern that the *Huygens* probe might land in such an

ocean and sink. While global oceans were not found, Titan has abundant lakes (**Fig. 9.13**), many of which are "fed" by channels and show distinctive shorelines. These features are radar-dark, indicating smooth surfaces, while the multispectral data show hydrocarbon compositions of a methane–ethane mixture. More than 1,000 such lakes have been found, ranging in size from a few kilometers across to the feature named Kraken Mare that is more than $400,000 \, \text{km}^2$ in extent. All of the lakes occur in polar regions, which is consistent with their methane–ethane compositions that require lower temperatures in order for them to be liquid. However, there are some 20 times more lakes seen in the north than in the south. This might be explained by Titan's seasons that result from its orbital inclination and its 30-year journey around Saturn each Titan year. Because the orbit of Saturn itself around the Sun is not circular, even longer-term cycles of heating and cooling are possible. Lakes have been monitored through repeat imaging and some in the south polar region disappeared toward the end of the Titan southern hemisphere summer, as a consequence of evaporation.

Aeolian processes constitute the other primary means of surface gradation. As shown in **Fig. 3.36**, wind tunnel experiments and theoretical considerations suggested that aeolian processes could occur on Titan (Greeley and Iversen, 1985). The extensive dunes were one of the most interesting *Cassini* discoveries. The dunes shown in **Fig. 9.14** are typical of a band that covers 40% of the surface between 30° north and south of the equator.

Figure 9.14. A radar image of linear dunes in the Aztlan region of Titan. The dunes are thought to consist of grains of ice a few hundred micrometers in diameter. From the pattern of the dunes and their interaction with topographic highs (the radar-bright features), winds are inferred to blow from the west (left) to the east (right). The area shown is about 275 km by 250 km (NASA *Cassini* PIA09182 part).

Figure 9.15. A *Cassini* image of Enceladus showing cratered terrain (upper right), smooth plains, and the set of parallel fractures (southern or lower part of image) referred to as "tiger-stripe" terrain (NASA PIA 06254).

>Nearly all are linear dunes, a non-genetic term that simply indicates their pattern as seen in map view. While many planetologists consider the Titan features to be longitudinal dunes, the image resolution is not adequate for this specific identification. Where the dunes interact with local hills, most terminate abruptly on the western sides; they appear to wrap around the topography and to have diffuse patterns on the east sides of the hills. These patterns suggest that formative winds are from the west to the east. Moreover, if they are longitudinal dunes, this would suggest that two wind regimes are dominant and meet at an acute angle, resulting in the duneform (**Fig. 3.39**). The puzzle, however, is that atmospheric models predict dominant winds in the opposite direction; this is a subject of current research.

Aeolian processes require a supply of small particles and winds of sufficient strength to move them. Winds measured by the *Huygens* probe are well within the range to transport grains, but the source of sand-size particles is poorly understood. In principle, such sizes are generated by volcanic, tectonic, impact, and other processes. Certainly, these processes have occurred on Titan, but how would ice behave in the presence of these agents? This, too, is a subject of current study.

In summary, the processes that have shaped Titan's surface are very familiar to terrestrial geologists, but the cold, low-gravity environment, coupled with the non-silicate composition of the crust, poses interesting challenges if we are to understand the evolution of Titan's surface. It has been argued that, in some respects, Titan is analogous to a primordial Earth, especially with regard to the origin of life, making Titan a high priority for future exploration.

9.6 Enceladus

Enceladus is an extremely bright object because of its predominantly ice surface. When it was first seen in detail in *Voyager* images of about 1 km per pixel, Enceladus posed an interesting problem: two provinces were seen globally, a heavily cratered terrain and a much younger terrain with a paucity of craters (**Fig. 9.15**). This suggested that recent resurfacing had occurred, but because the moon is only ~504 km in diameter, it seemed unlikely that such a small object could experience endogenic processes such as cryovolcanism. As is often the case in planetary science, this was a faulty assumption, as was shown by the *Cassini* discovery that Enceladus experiences actively erupting geysers. The eruptions spew water vapor, other gasses, and dust above the surface at a rate of ~100 kg/s (**Fig. 9.16**), some of which blankets the surrounding terrain. This material is also thought to inject ions into Saturn's magnetosphere and to "feed" the E Ring.

Figure 9.16. A *Cassini* image of Enceladus showing active geysers of mostly water and water-ice jetting from the "tiger-stripe" terrain in the south polar area. Despite the small size of this satellite (504 km), it experiences substantial geologic activity, apparently driven by tidal stresses in the interior (NASA PIA 12733).

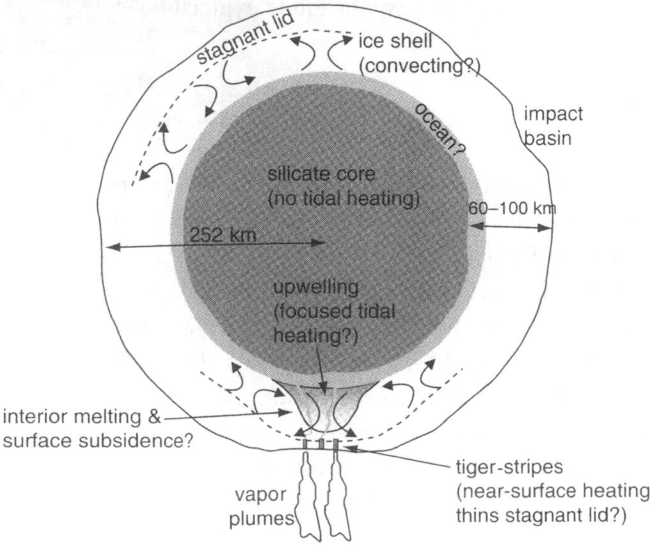

Geophysical models now suggest that the interior of Enceladus (**Fig. 9.17**) could be differentiated into a core containing silicates, overlain by a global or regional liquid water zone and an outer ice shell. Internal heat to keep the water zone liquid and to "drive" the geysers is probably generated at least partly by tidal heating associated with the orbit of neighboring Dione, as discussed by Francis Nimmo *et al.* (2007).

The global map of Enceladus (**Fig. 9.18**) shows the key terrains and named features. As reviewed by John Spencer *et al.* (2009), the principal terrains are (1) cratered plains, (2) western (leading) hemisphere fractured plains, (3) eastern (trailing) hemisphere fractured plains, and (4) the south polar regions, as well as various smooth plains.

Impact crater morphologies vary with size and apparent age. Craters of diameter <1 km are typical bowl-shaped features, while craters 1–40 km in diameter show a wide range in stages of degradation, including very shallow features that result from viscous relaxation. Degradation increases with crater size, by geographic location toward the south (especially toward the south polar terrain), and with inferred terrain age. Many of the larger craters have central peaks, upwarped central domes (**Fig. 9.19**), and lobate ejecta deposits. Moreover, many craters in the cratered plains are tectonically modified by fine fractures.

Tectonic features include scarps, troughs, ridges, and grooves (**Fig. 9.20**), giving some parts of Enceladus a superficial resemblance to Europa. However, unlike on Europa, high-standing ridges and ridge-complexes are mostly lacking. Tectonic features clearly truncate heavily cratered terrain, and the existence of the fine fractures cutting many of the impact craters on Enceladus would

Figure 9.17. A diagram showing the possible interior of Enceladus (surface topography exaggerated) and the potential convection with the generation of geysers (from Spencer *et al.*, 2009; with permission from Springer Science+Business Media B. V.: Dordrecht: Springer, *Saturn from Cassini–Huygens*, 2009, 683–724, Enceladus: An Active Cryovolcanic Satellite, M. Dougherty, L. Esposito, and S. Krimigis (eds.), Fig. #3).

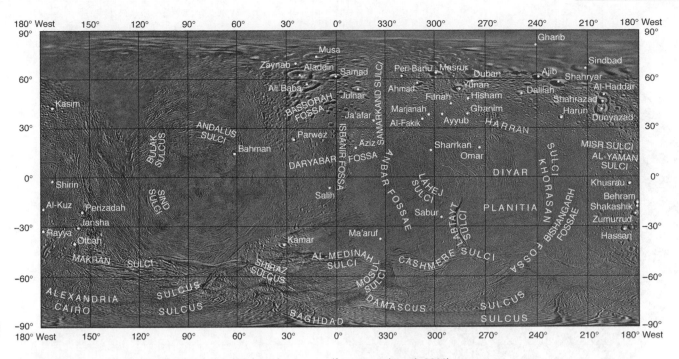

Figure 9.18. A mosaic of images for Enceladus with key place names (from Roatsch *et al.*, 2009).

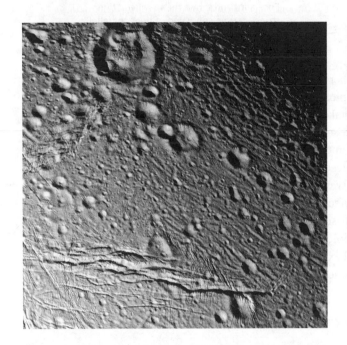

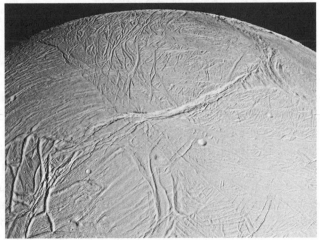

Figure 9.20. Tectonic deformation indicates the geologic activity of Enceladus; the prominent fracture-graben system seen here, named Labtayt Sulci, is as wide as 5 km and extends more than 100 km (NASA PIA06191).

Figure 9.19. A *Cassini* image of Enceladus showing cratered terrain, including a dome-floored crater (upper left) about 21 km in diameter (NASA PIA06210).

suggest that crustal deformation continued into recent geologic times and could even be active today. The largest tectonic features are canyons such as Labtayt Sulci, which is 200 km long and up to 10 km wide. Narrow linear grooves that are less than a few hundred meters wide are also seen, especially in the fractured terrains.

By far the most exciting aspect of Enceladus is the long-lived erupting plumes, making this object one of only a handful in the Solar System with confirmed geologic activity. Plumes are traced hundreds of kilometers above the surface and erupt from jets in the south polar region (**Fig. 9.21**). Repeated observations from *Cassini*

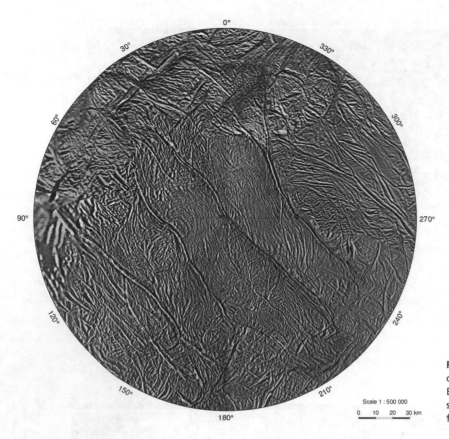

Figure 9.21. This mosaic of *Cassini* images is centered on the south polar region of Enceladus and shows the so-called "tiger-stripe" terrain consisting of huge, parallel fractures (NASA PIA12783).

and on-board compositional analysis of materials collected by the orbiter during its passage through the plumes provide critical information on the nature of the cryovolcanic eruptions. The components in the plumes are ice grains containing salt, water vapor, nitrogen, carbon dioxide, ammonia, methane, and other organic compounds.

The eruptive jets are associated with the "tiger-stripes," which consist of four parallel fractures, each of which is ~130 km long and 2 km wide and flanked by ridges as high as 0.5 km and as wide as 4 km. Remote sensing of the terrain indicates the presence of chucks of ice as large as 10 m and coarsely crystalline ice grains not seen elsewhere on Enceladus. Because coarse ice grains are short-lived in the Enceladus environment, these deposits are estimated to be younger than 1,000 years. *Cassini* measurements over the fractures show temperatures as high as 157 K, considerably elevated over the average for the surface of 68 K. Coupled with plume observations, most planetary scientists consider the eruptions (or geysers) to be derived from one or more subsurface pockets of high-pressure liquid water. A compelling argument has been made by Dennis Matson and other *Cassini* scientists that Enceladus seawater contains dissolved carbon dioxide

and other gasses that exsolve close to the surface, producing explosive plumes much like fizzy soda water once the cap is taken off.

Enceladus, along with Europa and Titan, is a future target in the exploration of life in the outer Solar System. The presence of a rich organic soup, the likelihood of liquid water (either in pockets in the icy crust or in a larger subsurface ocean), and the presence of internal heat combine to suggest an environment, or habitat, favorable to astrobiology. Regardless of the outcome of its astrobiological exploration, Enceladus remains important for geologic study.

9.7 Intermediate-size satellites

For the following moons we have sufficient data to describe their geology, and they are listed in order of distance from Saturn, beginning with Mimas.

9.7.1 Mimas

Mimas, like Enceladus, was discovered by Herschel in 1789. This heavily cratered object (**Fig. 9.22**) orbits Saturn

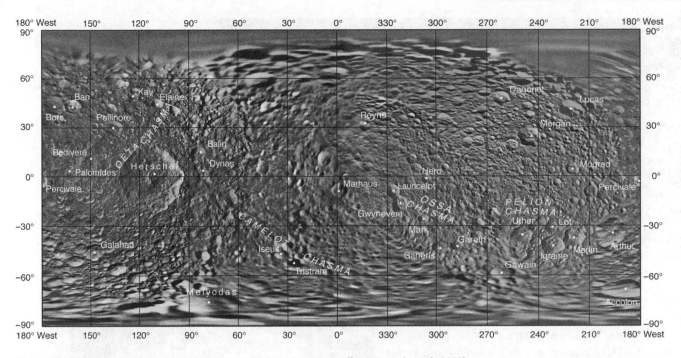

Figure 9.22. A mosaic of *Cassini* images of Mimas with key place names (from Roatsch *et al.*, 2009).

between two rings and is considered to be responsible for keeping the zone free of small particles. Mimas' density is only slightly greater than that of water (1.15 g/cm^3), suggesting that it is composed mostly of water-ice with only small amounts of rocky material. With a diameter of 396 km, it is nearly the same size as Enceladus, and these two neighbors present a remarkable contrast in geology, despite being in the same size class. While Enceladus is geologically active and displays terrains that are young, Mimas is not differentiated, shows no evidence of internal activity, and has an impact crater frequency among the highest in the Solar System.

Impact craters range in size from the limit of detection (~1 km) to the 140 km impact crater Herschel, which is nearly one-third the size of the moon (**Fig. 9.23**). Crater Herschel is 10 km deep and has a central peak that rises 6 km above its rugged floor. *Cassini* images show extensive mass wasting on the walls of Herschel and other large craters. This form of gradation includes massive slump blocks and downslope movement of debris, visible as albedo streaks. Although Mimas is heavily cratered, the size–frequency distributions vary somewhat by location, with the south polar region having a paucity of impacts of diameter >20 km. While various ideas have been proposed to explain this observation, it is plausible that larger impacting bodies are excluded from striking this area due to orbital geometries.

Figure 9.23. A *Cassini* image of Mimas and the 140 km impact crater, Herschel (NASA PIA06285).

Various grooves (chasmata), some as large as 100 km by 6 km, are visible on Mimas and are exemplified by Camelot and Ossa Chasmata. Most of the grooves are probably associated with the impact of Herschel crater, in which deformation approximately radial to the crater

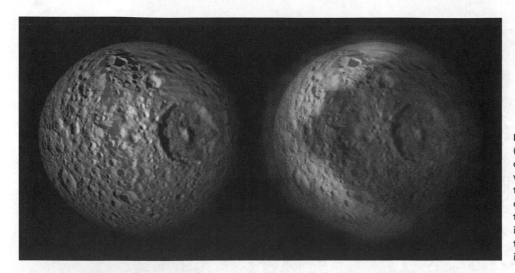

Figure 9.24. A visible image (left) of Mimas compared with a combined thermal map and visible image (right) showing the puzzling "Pac-man" pattern of temperatures in which high temperatures (>90 K) are shown in white and average surface temperatures (~77 K) are shown in gray (NASA PIA12867).

severely fractured the terrain. Oeta Chasma, however, is tangential to Hershel and would seem unrelated to this or any other obvious impact structure seen on Mimas. In addition to the large chasmata, small grooves and hints of fine fractures are seen, which are also attributable to impacts.

A bizarre discovery for Mimas was made when the *Cassini* infrared spectrometer mapped the surface temperatures. Highest temperatures were expected over the terrain warmed by the Sun in the afternoon, but maps showed a peculiar V-shaped pattern like the video game "Pac-man" eating a dot, in this case Herschel crater (**Fig. 9.24**), in which the highest temperatures were found in the morning. In addition to the "V," a warmer spot was mapped just to the west of Herschel's rim. A wide variety of explanations has been offered, including possible relationships to Hershel's ejecta deposits, but none is entirely satisfactory. This discovery demonstrates that the Solar System continues to hold new and unanticipated findings, especially when data from new instruments and higher resolutions are obtained.

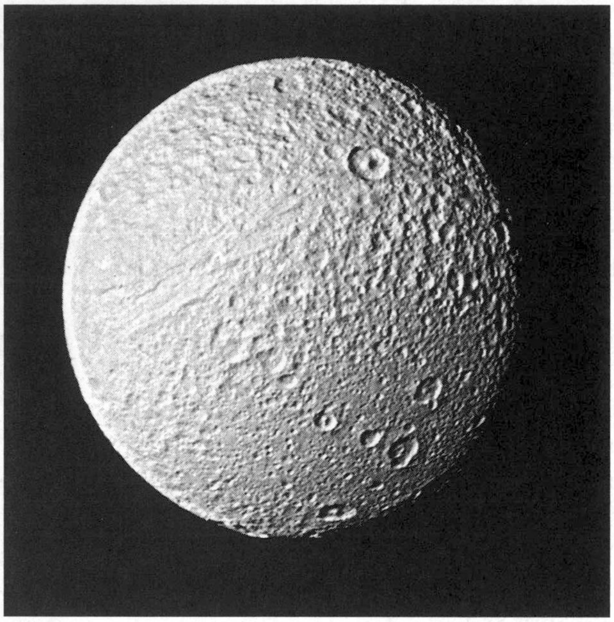

Figure 9.25. A *Voyager 2* image of Tethys showing heavily cratered terrain in the upper left of this view and the smooth plains to the lower right. The central-peak impact crater toward the top of the view is Telemachus, which is about 92 km in diameter (NASA PIA01397).

9.7.2 Tethys

Tethys is 1,048 km in diameter, more than twice the size of Enceladus, and also displays evidence of crustal deformation and resurfacing. Its density is only 0.97 g/cm^3, meaning that it is composed mostly of water-ice. Ridges, grooves, and smooth plains are present but are not as extensive as those on Enceladus. The general geology of Tethys was mapped from *Voyager* images as older, heavily cratered terrain and smooth plains (**Figs. 9.25** and **9.26**) which represent resurfacing. The cratered terrains have rugged topography and a high frequency of degraded craters >20 km in diameter. The plains are more sparsely cratered and are approximately centered on the trailing hemisphere.

The two most striking features on Tethys are Odysseus, an impact crater 445 km in diameter (about 40% of the satellite's diameter), and an enormous canyonland named Ithaca Chasma, which wraps at least three-quarters of the way around the globe. Odysseus is a shallow crater with a

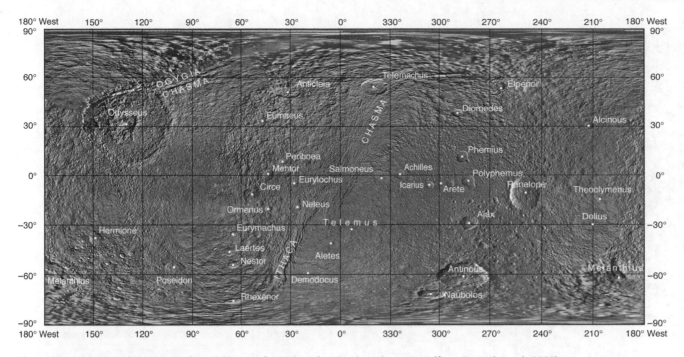

Figure 9.26. A mosaic of *Voyager* and *Cassini* images for Tethys, showing key place names (from Roatsch *et al.*, 2009).

Figure 9.27. An oblique view of Odysseus crater. This 445 km impact structure has a central peak complex and a rugged floor. The shallow depth of Odysseus suggests viscous relaxation of the rim (NASA PIA07693).

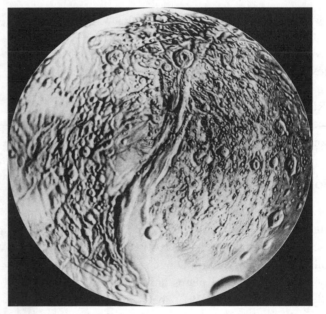

Figure 9.28. A shaded relief map of Tethys prepared from *Voyager* images centered on the sub-Saturn hemisphere at 0° longitude, showing Ithaca Chasma (courtesy of the US Geological Survey, Flagstaff).

ring-like central peak complex (**Fig. 9.27**). Its low relief may have resulted from viscous flow of the icy lithosphere. The impact that formed Odysseus could have occurred when Tethys was partly liquid and the ice in its interior was soft and plastic. Although most of the larger impact craters, such as Telemachus and Melanthius, have distinctive central peaks, others lack this structure.

Ithaca Chasma is a branching canyon system nearly 2,000 km long, 100 km wide, and ~4 km deep (**Fig. 9.28**) with wall terraces that were probably formed by enormous landslides. The canyon system is roughly antipodal to Odysseus and may have been created as some sort of crustal response to the impact. Alternatively, the canyon might have formed in response to interior expansion as

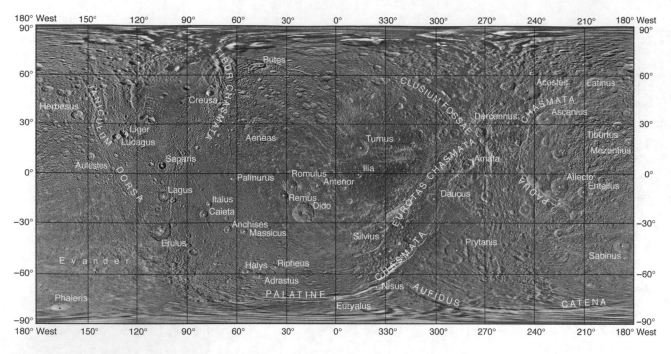

Figure 9.29. A mosaic of images for Dione with prominent place names (from Roatsch *et al.*, 2009).

liquid water froze early in Tethy's evolution, causing tensional fractures in the outer ice crust.

As on Mimas and other icy satellites, slopes on Tethys display downslope movement of debris by mass wasting. Its surface is further modified by the proximity of Tethys to the E Ring, which is fed water-ice grains by eruptions from Enceladus. These grains are thought to impact Tethys, contributing to its very high albedo.

9.7.3 Dione

Dione is 1,120 km in diameter, or about the same size as Tethys, but is considerably more dense. At 1.43 g/cm³, it is composed of water and about 46% rocky material, more than the other satellites of Saturn. Dione's surface consists of heavily to lightly cratered plains and terrains that are tectonically modified (**Fig. 9.29**). The heavily cratered terrain is on the trailing hemisphere, where the largest craters are found, while the leading hemisphere is more sparsely cratered. Prior to the imaging of Dione, some planetary scientists had suggested that the leading hemisphere should be more heavily cratered. When just the opposite was found, it was suggested that Dione might have been locked with Saturn in a different position, and that large impacts could have re-oriented the moon. The discovery of many large craters on Dione tends to support this hypothesis. Craters >80 km typically have central

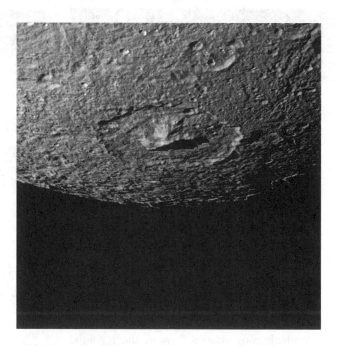

Figure 9.30. An oblique view of the southern hemisphere of Dione and Erulus crater. This impact structure is 120 km across and has a large central peak (NASA *Cassini* PIA12743).

peaks (**Fig. 9.30**). Seen in detail (**Fig. 9.31**), the steep slopes on the crater walls and central peaks are generally bright, indicative of exposures of fresh ice by mass wasting, while the crater floors tend to be darker.

Figure 9.31. A high-resolution view of the interior of a 60 km impact crater on Dione, showing its inner wall (left) and central peak (lower right); the bright slopes are fresh ice exposed by mass wasting. Also seen are fine fractures cutting across the crater floor (*Cassini* NASA PIA07748).

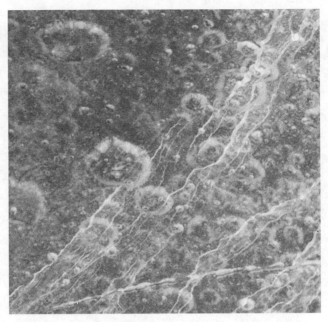

Figure 9.33. Cratered terrain on Dione; the walls and fractures exhibit bright slopes that result from mass wasting and the exposure of fresh ice, as seen here in the Carthage Linea region (NASA *Cassini* PIA07638).

Figure 9.32. This view of Dione is centered approximately on the sub-Saturn hemisphere, showing heavily cratered terrain (left) and smoother plains that have a lower frequency of superposed craters, reflecting the relative youth of the surface; this area shows bright wispy albedo patterns, some of which are bright rays of ejecta but most of which are scarps along fractures (NASA *Cassini* PIA01366).

The trailing hemisphere of Dione shows a network of bright, wispy streaks on a dark background (**Fig. 9.32**). The streaks, however, do not resemble the typical bright rays found with impact craters but follow irregular paths.

First seen on *Voyager* images, these markings were thought to be frost, perhaps deposited by explosive releases of gas from the interior through fractures in the crust. *Cassini* images now show that they are tectonic fractures (**Fig. 9.33**), the cliffs of which are very bright, indicative of exposures of fresh ice.

9.7.4 Rhea

At 1,528 km in diameter, Rhea (**Fig. 9.34**) is the largest of the intermediate-size satellites of Saturn, second in size only to Titan. Like Dione, its trailing hemisphere is dark and has bright wispy markings that are fresh ice exposed on fracture walls (**Fig. 9.35**). Parts of Rhea's surface are dominated by large, degraded craters and resemble the lunar highlands. Many of the craters show the "flattening" resulting from viscous relation, but not to the same degree as those of Ganymede.

Rhea's density suggests that it is about three-quarters water and one-quarter rocky material, but there is substantial debate as to whether Rhea is differentiated because there is disagreement on the available geophysical data. However, Peter Thomas *et al*. (2007) argued that the triaxial shape of this moon probably means that it is a homogeneous (undifferentiated) body. Thus, the observed fractures would have resulted from tectonic deformation

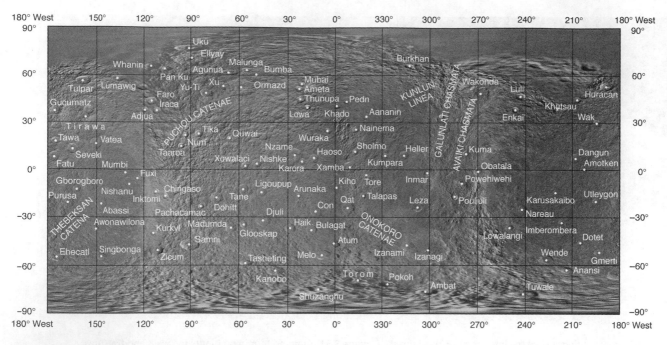

Figure 9.34. An image mosaic of Rhea with key place names (from Roatsch *et al.*, 2009).

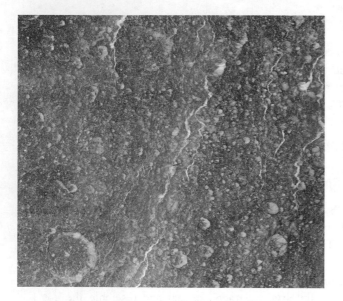

Figure 9.35. A *Cassini* image of Rhea's terrain in the sub-Saturn hemisphere showing that the bright wispy features are steep slopes where fresh ice is exposed; in the lower left part of the image is Inmar, a central-peak crater 55 km in diameter (NASA PIA12809).

Figure 9.36. A view of a heavily modified central-peak crater on Rhea; the area shown is about 130 km across (NASA *Cassini* PIA12742).

from impacts or changes in global volume related to the formation of different phases of ice.

Much of Rhea's cratered terrain is in equilibrium, and most of the larger craters display central peaks, scalloped walls, and degradation by superposed impacts (**Fig. 9.36**). At least two basin-sized impacts are present, including the 360 km Tirawa (**Fig. 9.37**) which has been heavily battered by superposed impacts (**Fig. 9.38**).

9.7.5 Iapetus

Iapetus (**Fig. 9.39**) is 1,436 km in diameter, nearly the twin of Rhea in size, but its lower density of ~1.2 g/cm^3 suggests

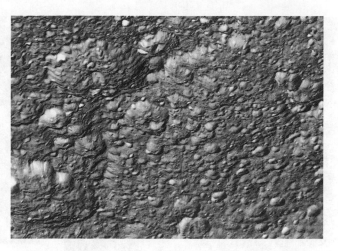

Figure 9.38. An oblique view of the western rim (arcuate ridge) of Rhea's Tirawa basin and its floor to the right of the rim; the number of superposed impacts on the rim and floor indicates a surface that is in impact equilibrium in which craters are destroyed at the same rate as that of their formation (NASA *Cassini* PIA12856).

Figure 9.37. A *Cassini* image of Rhea's leading hemisphere and the 360 km in diameter Tirawa basin near the terminator (NASA PIA08976).

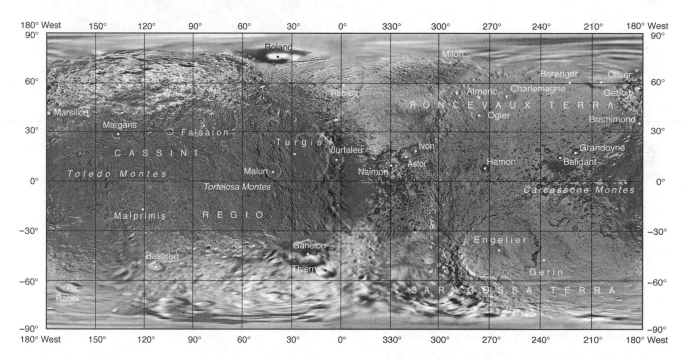

Figure 9.39. A mosaic of Iapetus showing prominent place names (from Roatsch *et al.*, 2009).

that it is composed of more ice than that proposed for Rhea. Since the satellite's discovery in the seventeenth century, the contrasts in surface brightness on Iapetus have been recognized as distinctive in the Solar System. While the leading hemisphere is extremely dark, the trailing hemisphere and polar regions are very bright. Surface features are difficult to see in the dark area (named Cassini Regio), but craters are abundant, the largest of which is the 580 km

Figure 9.41. This huge landslide on Iapetus was derived from a 15 km high wall of a 600 km impact basin and slid half-way across the floor of the 120 km crater seen in the middle of the image (NASA *Cassini* PIA06171).

Figure 9.40. A mosaic of 60 *Cassini* images of Iapetus showing the dark terrain of the leading hemisphere and the bright terrain of the trailing hemisphere; the impact structure toward the south is Engelier, a 450 km basin, one of many seen on this moon (NASA PIA08384).

in diameter Thurgis. Large impact structures are also clearly visible in the north polar area and in the trailing hemisphere bright terrain, including the 504 km in diameter Engelier (**Fig. 9.40**). Most of the large craters have central peaks and well-defined rims. The lack of relaxation of the larger craters suggests that a thick (50–100 km) ice lithosphere was present early in Iapetus' history. Gradation in the form of mass wasting is seen on some of the steep slopes (**Fig. 9.41**).

The boundary between the bright and the dark terrain is transitional. As seen in high-resolution *Cassini* images (**Fig. 9.42**), crater floors and other low-lying areas are dark, giving the appearance of "pooled" material. The frequency of the dark material increases toward the dark terrain until the entire surface has the distinctive low albedo.

Since the discovery of Iapetus by Giovanni Cassini in 1671 and study of its two-toned appearance telescopically, there has been speculation on the cause of the albedo dichotomy. Particularly controversial is the source of the dark material, which is about as black as asphalt and is thinner than ~0.5 m. The generally accepted model is that it is a lag deposit resulting from the sublimation of the icy matrix that contains carbonaceous debris mixed with metallic iron and possibly dust derived from Phoebe, which

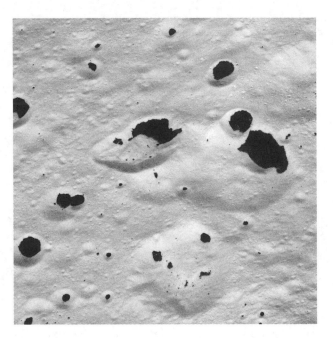

Figure 9.42. A *Cassini* image of the transition zone between the bright and dark terrains of Iapetus, showing the concentration of dark material in craters and other topographically low areas of the surface; the area shown is 37 km across (NASA PIA08374).

would be preferentially implanted on the leading hemisphere. In addition, because the polar regions are colder than lower latitudes, the volatiles would migrate toward the poles to form frost; with time, there would be a positive feedback for the formation of the lag deposits in which the dark areas would absorb more energy (causing additional sublimation and concentration of dark lag), while the frost areas would reflect more energy and retard sublimation.

A particularly fascinating discovery is the curious ridge that girdles much of Iapetus through Cassini Regio, nearly coincident with the equator (**Fig. 9.43**). The ridge is some 1,200 km long, 20 km wide, and stands as high as 10 km. It includes continuous segments >100 km long, separate peaks, and parallel ridge-sets and might extend into the bright terrain as a series of isolated peaks. Jaumann *et al.* (2009b) note that the ridge might have resulted from a disk of impact debris striking Iapetus, from tectonic deformation related to "despinning" of Iapetus, or from internal convection and up-welling; however, all of these ideas have problems and the ridge's origin remains a mystery!

9.7.6 Small satellites

"Small" satellites range in size from 25 to 220 km in diameter and include a wide variety of objects, as discussed by Carolyn Porco and her *Cassini* colleagues (Porco *et al.*, 2007). Phoebe (**Fig. 9.44**), the outermost saturnian satellite, is in a retrograde orbit inclined to the equatorial plane of Saturn. This unusual orbit suggests that Phoebe was captured by Saturn. Estimates of its mass were derived from *Cassini* flybys, which give a density of ~1.63 g/cm^3 and suggest that it has the highest proportion of rocky material of the satellites discussed above. Phoebe is roughly spherical (**Fig. 9.45**) and is the darkest object in the saturnian system, its surface being composed of carbonaceous material.

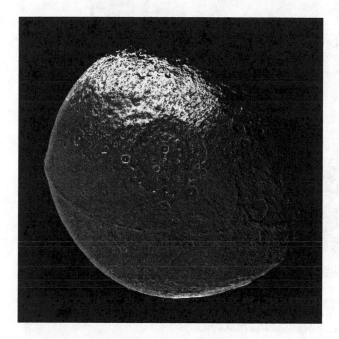

Figure 9.43. Iapetus viewed by *Cassini* showing the dark terrain of Cassini Regio and the unusual ridge that coincides with the equator; the ridge is 1,200 km long and as high as 10 km; the origin of the ridge remains to be explained (NASA PIA06166).

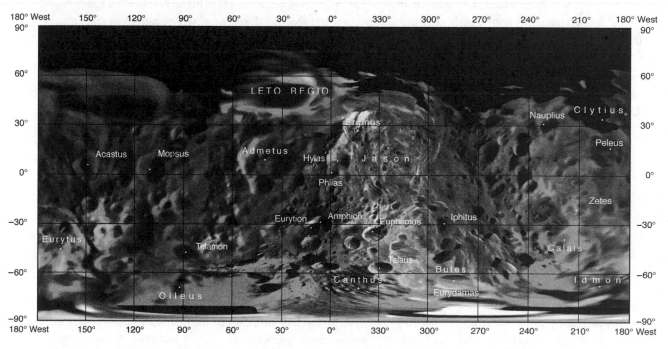

Figure 9.44. A mosaic of images for Phoebe with key place names (from Roatsch *et al.*, 2009).

Figure 9.45. A *Cassini* image of Phoebe showing the heavily crater surface (NASA PIA06064).

Figure 9.46. A mosaic of six *Cassini* images of Hyperion; the dark material in the floors of many of the craters is likely a lag deposit left from the sublimation of ice (NASA PIA07761).

Five of Saturn's small satellites are found inside the orbit of Mimas. Two, Janus and Epimetheus, are called co-orbital satellites because their orbits are within 50 km of each other. They are thought to be remnants of a larger moon broken apart by impact collision. Three additional small moons, Calypso, Telesto, and Helene, are Lagrangian satellites. It has long been known that a small object can have the same orbit and speed as a larger object, so long as it maintains a 60° arc ahead of or behind the larger object in positions called Lagrangian points. Tethys has two such small objects in its orbit; Dione has one.

Finally, Hyperion is a small, irregular-shaped object (**Fig. 9.46**) whose elliptical orbit is just outside that of Titan. Measuring 350 km by 235 km by 200 km, this heavily cratered object includes both angular and rounded parts. This moon is very dark, and, although its mass has not been determined, it is presumed to be composed of ice and rock. In addition to craters as large as 120 km, the surface displays scarps that may be part of a connected, arcuate feature nearly 300 km long that could be the remnant of a crater. Because of its irregular shape, heavily cratered surface, and a possible huge impact scar, Hyperion is considered by most investigators to be the remnant of a much larger object that was shattered by collisional or impact processes.

9.8 The ring system

Saturn's rings occur in sets (**Figs. 9.1** and **9.2**) and are composed mostly of water-ice, which ranges in size from tiny dust specks to 10 m boulders. The subtle ice colors include reddish tan and brown tones due to impurities, such as iron oxide, or to structural damage in the ice crystals caused by ultraviolet radiation from the Sun. Saturn's rings extend from about 53,000 km to nearly 360,000 km above Saturn, and the entire system is very thin (averaging 20 m thick) so that, when viewed from the side, it looks like a knife edge. Despite the enormous extent of the rings, if all of the ring material were accumulated, it would form a satellite less than 500 km across.

Ring elements are designated by letters that are based on the sequence of their discovery rather than on their position in relation to Saturn. The innermost (the D Ring) is barely visible and consists of very narrow ringlets composed of fine-granulated particles that appear to be "leaking" inward from the C Ring toward Saturn. The C Ring consists of light and dark bands of centimeter- to meter-sized particles but also includes some larger objects. This ring contains little or no dust, most of it evidently having been swept away to form the D Ring. The B Ring is the brightest of Saturn's rings and is about 2 km thick, making it thicker than any other ring. It is composed mostly of particles ~10 cm in diameter, although some may be as large as 10 m.

Transient, spokelike patterns appear superposed on the B Ring. Each "spoke" is 10,000 km long and 100–1,000 km wide. They radiate outward through the B Ring and appear and disappear with time. These are thought to be clouds of dust particles, perhaps levitated by electrostatic forces generated as the ring particles collide with each other.

The A Ring is the outermost of the large, bright rings and consists of several narrow ringlets, with the outermost ringlet having a sharp, well-defined edge. Atlas is a small satellite discovered during the *Voyager* mission and is in orbit just outside the A Ring. Referred to as a "shepherding" satellite, it appears to constrain material in the A Ring by gravitational forces. Separating the A Ring from the B Ring is the Cassini "division," named after the Italian astronomer Gian Domenico Cassini. First thought to be relatively free of particles, spacecraft images show that narrow rings are present within the division.

Pioneer data revealed the F Ring, which is ~700 km wide and orbits beyond the A Ring. It is bound by two "shepherding" satellites, Pandora and Prometheus, that appear to keep the ring particles constrained to a very narrow orbit. The arc-like G Ring is composed of ice grains just inside the orbit of Mimas while the wide, diffuse E Ring begins near Mimas and extends to Enceladus. The E ring is composed of fine ice grains, rocky material, and carbon dioxide originating from the Enceladus eruptions. The Phoebe Ring, discovered in 2009, resides just within the orbit of Phoebe, making it the outermost of Saturn's rings; it is thought to originate from impacts on Phoebe or to be otherwise derived from Phoebe.

Some planetary scientists suggest that Saturn's rings represent one or more satellites that were broken apart by tidal forces or through collisions. Alternatively, the rings could be composed of material that never accreted to form one or more satellites in the early stages of the Solar System's formation. In 2010, planetary scientist Robin Canup proposed a new theory of Saturn's ring formation based on *Cassini* data and elegant computational modeling. She suggested that early in the history of the Saturn system, there were large, differentiated moons composed of ices and rocky material, comparable to Titan. Tidal forces stripped the outer layers of ice-rich material from the outer shell of one such object, while the rocky part spiraled inward to Saturn. The icy material

spun outward to form the nearly pure water-ice ring particles seen today. Canup (2010) suggests that, as this system evolved, the mass decreased and "spawned" icy moons at the outer edge of the system, including Tethys.

9.9 Summary

Saturn, like Jupiter, is a giant planet composed mostly of gasses, likely surrounding a central rocky core. It, too, has complex cloud patterns driven by internal heating and leading to features such as the ephemeral Great White Spot. Of all the planets of the Solar System, Saturn has the greatest diversity of moons and ring systems. Its largest satellite, Titan, has long been held in fascination from a geologic perspective. Given its dense nitrogen atmosphere, the presence of methane, and a temperature regime in which all three states of methane (solid, gas, and liquid) could exist, following the *Voyager* results there was speculation that Titan's surface could be modified by extensive processes of gradation. This speculation was confirmed when the *Cassini Huygens* results showed extensive, integrated fluvial networks, many of which empty into pools of liquid methane. Vast dune-fields attest to the action of winds in the dense atmosphere, showing that aeolian processes are important, especially in the equatorial region of Titan. Although there is not yet direct evidence that gradation processes are currently active, the shrinking of lakes suggests that methane is cycling between the surface and the atmosphere as a function of Titan season.

Cassini images revealed the presence of active geysers on Enceladus, adding this moon to the small set of Solar System objects that demonstrate active internally driven processes. This discovery came as a surprise because Enceladus seemed too small (~504 km in diameter) to generate internal heat. The active geysers emanate from the so-called "tiger-stripe" terrain of the south pole region and can be traced more than 100 km above the surface. The geyser plumes consist mostly of water vapor and include other gasses and dust particles, some of which might "feed" into Saturn's E Ring. The presence of water and its internal energy make Enceladus a potential target for astrobiological exploration.

Saturn's other moons span a wide range of sizes, orbital characteristics, and geologic histories. While Mimas is in the same size class as Enceladus, its heavily cratered surface indicates little internal activity since its formation. The presence of canyon systems is attributed to adjustments of the ice-shell in response to the formation of the impact crater, Herschel. At 1,048 km in diameter, Tethys is about twice the size of Mimas and also has extensive canyon systems, most notably the 2,000 km long Ithaca Chasma. It, too, might have formed in response to an enormous impact (resulting in Odysseus crater).

Some of Saturn's moons show contrasting bright and dark hemispheres. For example, Iapetus has a very dark leading hemisphere and a bright trailing hemisphere, with a gradational boundary between the two. The difference is commonly attributed to the formation of a dark lag deposit of carbonaceous material left on the surface after sublimation of an icy matrix. Additional dark material might be implanted from dust derived from the neighboring moon, Phoebe.

Galileo discovered Saturn's hallmark ring system in the early 1600s. Subsequent telescopic observations and spacecraft data show that the system is highly complex, featuring interactions with many of Saturn's moons. Rings consist mostly of water-ice grains that range in size from dust to boulders. The various ring elements are designated by letters that indicate their sequence of discovery, not their distance from Saturn. For example, the A Ring is the outermost large, bright ring, while the F Ring was discovered through *Pioneer* observations and is bounded by two small moons that seem to keep the ring particles within a very narrow band. Current models suggest that most of the rings were generated by the tidal breakup of one or more satellites.

Assignments

1. Construct one or more graphs of Saturn's moons to assess potential relationships among satellite density, diameter, orbital distance from Saturn, retrograde versus prograde orbit, and geologic characteristics, and then discuss the results.

2. Compare and contrast the tectonic features seen on Mimas and Tethys.

3. Rhea and Iapetus have about the same diameter, but their densities are different. Examine images of these two moons and identify a crater that is of equal size on each. Compare their morphologies and explain any differences between the two as related to differences in gravity between Rhea and Iapetus.

4. Some of Saturn's moons have distinctive differences in albedo from one part of the satellite to other parts of the surface. Explain how these differences might originate.

5. Discuss briefly the relationships between the satellites of Saturn and the various rings.

6. Examine images of Titan's dunes and discuss the differences and similarities with dunes seen on Venus and Mars in terms of sizes, morphologies, and inferred directions of formative winds.

CHAPTER 10

The Uranus and Neptune systems

10.1 Introduction

Sir William Herschel was an amateur astronomer in Bath, England, who built his own telescopes to view stars. In March 1781 he observed a point of light that did not behave like a normal star. At first, he thought that it was a comet, but, with further observations, the object was shown to be the seventh planet – the first to be discovered in historical times. Although initially he suggested that it be named *Georgium Sidus* (George's star) after the reigning King George III, it was subsequently named Uranus after the Roman god of the heavens.

Following the discovery of Uranus, Neptune's existence was predicted before it was actually seen. Observations of the orbit of Uranus and the application of the laws of physics suggested that some object ought to exist beyond Uranus to account for its motions. On the basis of independent calculations by the British mathematician John Couch Adams and the French mathematician Le Verrier, a German astronomer, Johann Gottfried Galle, used a telescope at the Berlin Observatory to search for the proposed object. On the very first night of his search, September 23, 1846, Galle found the suspected planet within one degree of the predicted position.

Herschel continued viewing Uranus and in 1787 discovered the two largest satellites, later named Titania and Oberon. Subsequently, in 1851, Umbriel and Ariel were found, but it was not until nearly 100 years later that the fifth large moon of Uranus was discovered by Gerard Kuiper and named Miranda. The discovery of Neptune's large moons followed a similar path to the discovery of those of Uranus. Only 17 days after the discovery of Neptune in 1846, its largest moon, Triton, was found by William Lassell. Again, it was more than a century until the other large moon, Nereid, was found in 1949, again by Kuiper.

Most information on the geology of the uranian and neptunian satellites comes from the *Voyager 1* and *2*

flybys in 1986 and 1989, respectively, the trajectories of which were optimized for coverage of the moons. There were, however, some critical aspects of the flybys that had to be taken into account. At Uranus' distance of 3 billion kilometers from the Sun, and an even greater distance for Neptune, the amount of light illuminating the satellites is extremely low, making it difficult for *Voyager*'s vidicon imaging system to take pictures. Some camera exposures had to be as long as 96 seconds, but, because the spacecraft was traveling so fast, the long exposures would have resulted in image "smear" (imagine trying to snap a picture of an oncoming car through the window of your car with a camera exposure time of nearly two minutes). Fortunately, the talented engineers at JPL devised an **image motion compensation** routine in which the spacecraft was rotated while staying locked in position on the target, thus enabling high-quality images to be taken despite the long exposure times.

10.2 Uranus and Neptune

While Jupiter, Saturn, Uranus, and Neptune are referred to as giant planets because they are so large, there are significant differences between the sets Jupiter–Saturn and Uranus–Neptune with regard to their interiors. As noted in **Chapters 8** and **9**, Jupiter and Saturn are composed primarily of hydrogen in various forms (including metallic hydrogen) and helium, while Uranus and Neptune include substantial amounts of water in both frozen and liquid states, along with ammonia, methane, and proportionally large rocky cores. Such a configuration allows electrical dynamo conditions, enabling the generation of magnetic fields.

Uranus takes 84 Earth years to orbit the Sun once and is unique among the major planets in our Solar System in that its axis of rotation is tilted about 82°, making it nearly horizontal with the ecliptic plane (**Fig. 10.1**). Although no

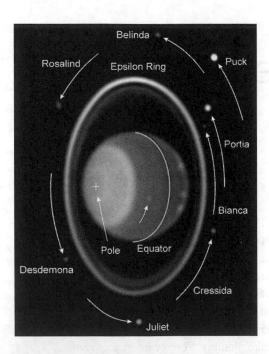

Figure 10.1. Uranus was the first planet to be discovered in historic times. As shown in this *Hubble Space Telescope* image, Uranus lies "on its side" with the spin axis inclined about 82° to the ecliptic plane. Haze, cloud patterns, and the tracks of faint rings and small satellites form a bull's eye pattern around the south pole of the planet. Eight of Uranus' small moons are shown here (NASA *Hubble* PIA01278).

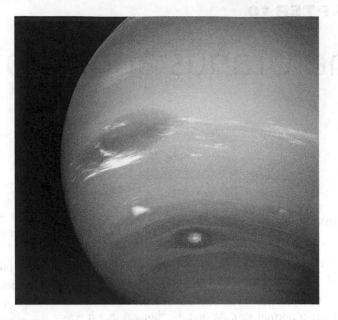

Figure 10.2. Neptune's Great Dark Spot is shown in this *Voyager 2* image, along with several other features in the upper atmosphere (NASA *Voyager 2* PIA02219).

definitive answer has been offered to explain this orientation, it is commonly thought that an enormous impact might have knocked Uranus on its side. At visible wavelengths, both Uranus and Neptune appear rather bland, with only hints of any structure in the upper atmosphere. Uranus is the coldest major planet in the Solar System, generating far less heat from the interior than Neptune. In natural color, Uranus' south polar region appears brownish-orange, while Neptune has an overall bluish cast. In the infrared, however, features such as small clouds, bands, and other structures are revealed. For example, *Hubble Space Telescope* views of both planets show more detail of these features than indicated in *Voyager* images. While atmospheric features of both planets are much more subdued than those of Jupiter and Saturn, Neptune shows much more activity than Uranus, perhaps driven by the greater amount of heat released from its deep interior. For example, *Voyager* images show a feature resembling Jupiter's Great Red Spot, leading to its name, the Great Dark Spot (**Fig. 10.2**), which probably is driven by thermal up-welling in the atmosphere. Tracking of other cloud features in sequential

Voyager images indicated wind speeds up to 580 m/s in Neptune's atmosphere.

Earth-based and spacecraft data show that all of the giant planets possess ring systems. Uranus' rings, first discovered from telescopic observations in 1977, are now seen to include at least 13 sets, some of which are broad and diffuse, others of which are narrow and have sharp boundaries. Particles in the rings range in size from microns to meters in diameter and are very dark, similar to soot. Uranus' rings could represent one or more moons that were smashed apart by collisions. Neptune has at least six ring systems, all of which are very faint and are named after individuals who contributed to Neptune science in the 1800s. The outermost main ring, named Adams Ring, consists of five arc-segments, which are thought to be held in place by Galetea, a small moon.

10.3 Uranian moons

Uranus has at least 27 moons, but those of geologic interest are (from Uranus outward) Miranda, Ariel, Umbria, Titania, and Oberon (**Fig. 10.3**), named after characters from Shakespeare's *A Midsummer Night's Dream* and *The Tempest* and Pope's *The Rape of the Lock*. All of the moons are in synchronous rotation, and their densities of 1.5 to 1.7 g/cm^3 are higher than

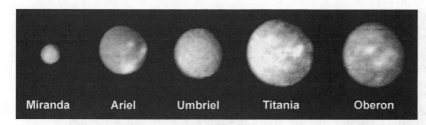

Figure 10.3. A composite view of the large moons of Uranus. Miranda, the innermost of the large moons, shows the most complex geology, while the outermost moon, Oberon, preserves the early bombardment history in the outer Solar System (NASA *Voyager* PIA01975).

those of Saturn's satellites, suggesting a higher proportion of rocky material. Although surface compositions are predominantly water and carbon dioxide ices (Grundy *et al.*, 2006), their rather low albedos suggest the presence of some dark material. On the other hand, methane-ice exposed to photons from the Sun releases hydrogen and leaves a higher proportion of carbon with time, as reported by Cruikshank *et al.* (1983) from laboratory experiments.

All of the satellites show impact scars, evidence of tectonic deformation, suggestions of resurfacing, and forms of gradation. There appears to be a relationship between the degree of geologic activity and the satellite position in orbit around Uranus, with those closest to the planet having been modified by extensive deformation, while the outer moons preserve more of the record of heavy cratering.

The orientation of the Uranus spin axis with respect to its orbit around the Sun means that the orbits of the major satellites result in extreme seasonal cycles in which each pole alternates in being in complete darkness for about 42 years and is then in complete sunlight for the next 42 years. Thus, volatiles in the north pole in its summer would migrate to the south pole's extreme darkness, and then the cycle would reverse in the next 42 years.

Miranda. Although this innermost moon is the smallest of Uranus' major satellites, it displays the most complex geology (**Fig. 10.4**). Miranda's surface is composed mostly of water-ice. Its geology includes both old, heavily cratered terrain and younger complex terrains marked by ridges and grooves, huge scarps, ovoid "racetrack" patterns (coronae), and banded zones, all reflecting intense tectonic deformation (**Fig. 10.5**). Many ideas have been proposed to explain Miranda's surface features, including disruption by diapirs, crustal extension, and one or more huge impacts. The grabens and other faults (**Fig. 10.6**) are thought to represent exterior features related to tidal heating generated from inferred orbital eccentricities and early

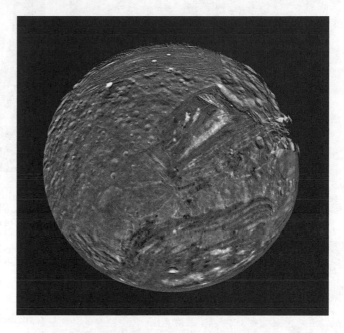

Figure 10.4. Although Miranda is only 474 km in diameter, parts of its surface display grabens reflecting extensive tectonic deformation; other parts of the surface are heavily cratered and are relatively unmodified by internal processes (NASA *Voyager 2* PIA01490).

interactions with the moon Ariel. Such heating might have led to diapirs, which could account for Miranda's unusual tectonics. Internal heating also could have generated cryovolcanic resurfacing to form the younger, smooth plains seen in some parts of the satellite.

Ariel is in synchronous rotation with Uranus and is immersed in its magnetosphere, which could account for the darkening on the trailing hemisphere, much like for Jupiter's moon Europa. Sputtering of ice by ions and possible implantation of non-ice materials has led to darkening of the surface in comparison with the leading (lighter) hemisphere. As with Miranda, Ariel's surface displays ancient heavily cratered terrain and evidence of tectonic deformation (**Fig. 10.7**), although not nearly as extensive as that on Miranda. Ariel's density suggests that it is composed of about half water and half rocky material;

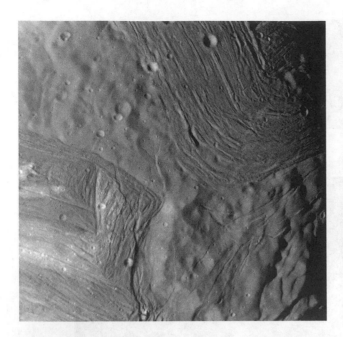

Figure 10.5. Part of the feature informally called the "chevron" (left side) and sets of ridges and grooves that cut across older cratered terrain; the area shown is 220 km across (NASA *Voyager 2* PIA00038).

Figure 10.7. A mosaic of four *Voyager 2* images showing the 1,159 km in diameter Ariel; as with Miranda, this moon shows evidence of crustal deformation in the form of grabens, faults, and smooth areas that might have been resurfaced by cryovolcanism; the prominent graben toward the bottom of the mosaic is named Korrigan Chasma (NASA *Voyager 2* PIA01534).

Figure 10.6. Miranda's surface has been cut by faults, some of which form scarps as high as 5 km, as seen in this limb view (NASA *Voyager 2* PIA00140).

it is thought to be a differentiated object. Assessments of the crater size–frequency distributions show an absence of large craters, leading to speculation that Ariel might have experienced one or more episodes of extensive

resurfacing early in its history. As with Miranda, internal heating might have been generated by tidal interactions with other objects resulting from its orbital geometry. Some smooth plains visible today show that resurfacing has occurred more recently, which would reflect internal heating. Fractures and narrow, straight valleys, which are likely to be grabens, indicate extensional tectonism.

Umbriel is a heavily cratered object and is the darkest of the large moons of Uranus (**Fig. 10.8**). Its relatively high density suggests a water-ice composition plus about 40% non-ice components, such as rock and carbonaceous materials. Locked in synchronous rotation, Umbriel is completely immersed in the magnetosphere generated by Uranus, leading to intense bombardment on its trailing hemisphere, as for Ariel. In addition to water-ice, remote sensing data indicate the presence of carbon dioxide on the surface, mostly on the trailing hemisphere. *Voyager 2* images are of rather low resolution and cover about 40% of the surface. There are hints of structures such as scarps, grabens, and hexagon-shaped features, but the image resolution is inadequate for confirmation. Craters are common and range in size up to the 208 km crater

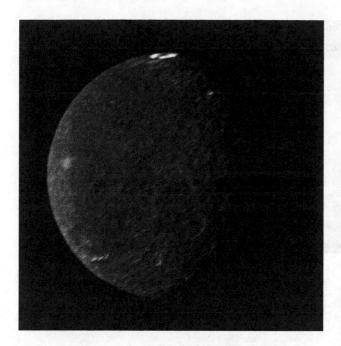

Figure 10.8. The 1,170 km in diameter Umbriel is very dark, suggesting the presence of non-ice materials on the surface. Although it is heavily cratered, the relative absence of bright ejecta rays suggests some process of surface darkening. The prominent crater toward the top of the image on the terminator has a bright central peak; the bright ring near the equator also could be the remnant of an impact (NASA *Voyager 2* PIA00040).

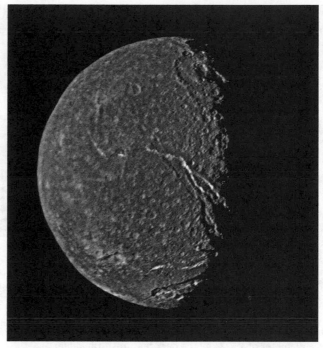

Figure 10.9. A composite of two *Voyager* images for Titania, the largest moon of Uranus at 1,578 km in diameter, showing abundant impact craters and tectonic deformation of the ice-rich crust (NASA *Voyager* PIA00039).

Wokolo in the southern hemisphere. Many of the craters have central peaks, some of which are rather bright, which could indicate exposures of fresh ice. The most prominent feature is a bright ring within crater Wunda, which is 131 km in diameter.

Titania is the largest of the uranian satellites (**Fig. 10.9**). It is thought to be differentiated, and its high density (1.71 g/cm^3) suggests about equal amounts of water-ice/water and rocky material; Titania might even have liquid water at the interface with the rocky interior if ammonia or salts are present to lower the melting temperature of the ice. As with the other uranian moons, Titania is in synchronous rotation and its trailing hemisphere is influenced by the plasma of the magnetosphere, causing preferential darkening. The surface shows abundant impact craters as large as 326 km (crater Gertrude), many of which have central peaks. Some of the impact craters are relatively shallow, which is thought to result from relaxation of the ice. Relatively smooth plains are present, reflecting one or more episodes of resurfacing (possibly from cryovolcanism). Tectonism is represented by huge canyon systems, including Messina Chasma, which is nearly 1,500 km long.

The chasmata and the smaller grabens indicate tectonic extension and are the youngest features seen on the 40% of the surface imaged by *Voyager 2*.

Oberon is the second largest moon of Uranus and is the most heavily cratered (**Fig. 10.10**). Like Titania, it is thought to be composed of nearly equal amounts of water-ice and rocky materials and might also have liquid water at the interface with the rocky interior. It, too, is in synchronous rotation with Uranus, but it is not completely immersed in the magnetosphere. Thus, unlike Titania, Oberon's trailing hemisphere is not darkened but instead is slighter brighter. Moreover, the leading hemisphere surface tends to be somewhat red, which is probably due to implantation of material derived from uranian irregular (i.e., captured) satellites, as suggested by Bonnie Buratti of the Jet Propulsion Laboratory and Joel Mosher (Buratti and Mosher, 1991). Crater Hamlet is the largest confirmed impact seen in *Voyager* images at 206 km in diameter. However, NASA scientist Jeff Moore and colleagues (Moore *et al.*, 2004) have suggested that a topographic feature seen on Oberon's limb could be the central peak of an even larger impact structure. Many of the numerous impact craters have bright rays of fresh ice and the floors

Figure 10.11. A mosaic of *Voyager* images for Triton, Neptune's largest moon. The so-called cantaloupe terrain of Monad Regio is on the top half of the image, while the smooth plains of Uhlanga Regio are on the bottom (NASA *Voyager* PIA00317).

Figure 10.10. A *Voyager* image of Oberon, Uranus' second largest moon, with a diameter of 1,522 km; the dark feature at the center of the image is the floor of Hamlet, an impact crater 206 km in diameter that exhibits bright ejecta rays (NASA *Voyager* PIA00034).

of some craters are dark, which could represent penetration into a dark substrate. In addition to craters, Oberon's surface is marked by chasmata and scarps that reflect crustal extension. The largest feature seen is Mommur Chasma, which is 537 km long and is found in the equatorial region.

10.4 Neptunian moons

Neptune has 13 known satellites, six of which were discovered in *Voyager* data and three of which were discovered more recently from ground-based telescopic observations. The largest, Titan, is in the same size class as Earth's Moon and is in a retrograde orbit. Outward from Titan, six of the moons are irregular satellites with highly inclined orbits and are thought to be captured objects. Inward from Titan, the other six moons are in prograde orbits and might have formed directly in association with Neptune. Many of these latter moons are within the ring-arcs and might be responsible for contributing material to the rings.

Triton has a density of 2.1 g/cm^3, the highest of all the outer planet satellites, suggesting that it is about one-third ice and two-thirds rocky material. Most planetologists

consider Triton to be an object gravitationally captured by Neptune. Since Triton is a retrograde satellite, its orbit is slowly decaying, or being pulled inward, toward Neptune. No doubt, in the process of its capture, Triton would have experienced enormous tidal stresses, leading to internal heating and likely differentiation. Given its overall density, it is thought to have a rocky core surrounded by a water mantle, parts of which could be liquid, and a surface crust of ices composed of methane and nitrogen, plus frosts of water.

Voyager images reveal a fascinating set of terrains on Triton, the most prominent being an older rugged unit, informally called *cantaloupe terrain* due to its fanciful resemblance to a melon skin (**Fig. 10.11**), and younger plains that are relatively smooth. The cantaloupe terrain includes irregular pits and bumps that could reflect extrusion and deformation of ice onto the surface. Linear features include ridges and grooves (**Fig. 10.12**) reflecting deformation of the crust and possibly additional extrusion of ice onto the surface. This terrain is pock-marked by impact craters, most of which are found on the leading hemisphere of Triton and have about the same size–frequency distribution as that of lunar maria. The lack of larger impact craters and the absence of heavily cratered units suggest that even the oldest terrain on Triton is geologically young.

Smooth plains cut across the rugged cantaloupe terrain and nearly lack impact craters, indicating youth. Many of the smooth plains have irregular-shaped

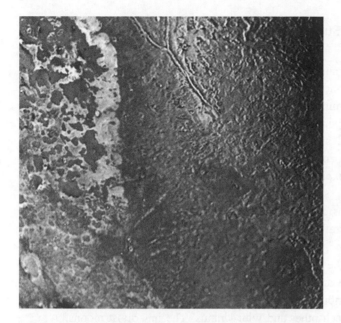

Figure 10.12. Detail of the boundary between the "cantaloupe" terrain, characterized by ridges, grooves, and irregular domes, and the smooth plain with its mottled surface (NASA *Voyager* JPL P-34694).

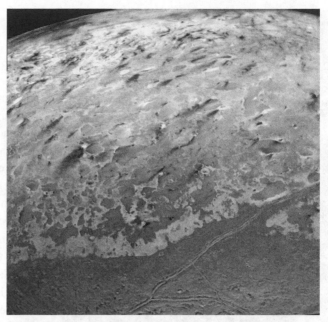

Figure 10.14. The dark wind streaks seen in the smooth plains are deposits left by explosively erupting plumes of material caught by upper-altitude winds and carried downwind toward the lower left; these streaks are as long as 100 km (NASA *Voyager* PIA34714).

Figure 10.13. Large caldera-like depressions are found in Triton's smooth plains and are considered to be sites of cryovolcanic eruptions of liquids from the subsurface; area shown here is about 290 km across (NASA *Voyager* JPL-34692).

depressions resembling volcanic calderas (**Fig. 10.13**), and many are considered to be sites where liquid water, nitrogen, and methane erupted from the subsurface to form local lakes.

Triton is the fifth object in the Solar System known to have active eruptions; several active geysers were seen during the *Voyager* flyby. The geysers spewed material as high as 8 km above the surface, where it was caught by upper-altitude winds and carried downwind some 100 km, leaving dark deposits on the surface (**Fig. 10.14**). Analysis of images revealed many more such dark surface streaks, suggesting that the eruptions are common. Although many ideas have been proposed to explain the process of geyser activity, the current concept is that they are a type of solid state "greenhouse," which is similar to the explanation for the "spider" terrain on Mars (in fact, the idea was first proposed for Titan and then modified for application to Mars). Nitrogen-ice, a common component on Triton's surface, is transparent in its pure form. Thus, solar energy penetrates through the nitrogen-ice and is trapped. Only a 4 K rise in temperature is needed to volatilize the nitrogen-ice at its base, and the resulting build-up of gas pressure can reach sufficient levels to rupture the overlying ice to release the gas explosively. The dark material could be silicates or carbonaceous material carried upward from deposits along the interface of the ice and the substrate.

Figure 10.15. An image of Proteus, the 416 km in diameter moon of Neptune, showing an impact scar named Pharos that is about 230 km across (NASA *Voyager* PIA00062).

Although Triton's tenuous atmosphere is extremely thin (exerting a surface pressure of less than 0.01 mbar), it is sufficient to generate winds. Analysis of Titan's cloud patterns and the dark plumes left by the geysers suggest that the prevailing winds at 1–3 km above the surface are toward the east, while higher-altitude (~8 km) winds are toward the west.

Proteus, Neptune's second largest moon at about 416 km in diameter, was discovered in *Voyager* images (**Fig. 10.15**) and is one of the darkest objects seen in the Solar System. It resides just outside Neptune's ring system and orbits very rapidly, taking only 1.1 Earth days to make one complete orbit. Despite its relatively large size, the overall shape of Proteus is irregular and the surface is rugged with topographic relief of 20 km. Geophysical models suggest that it is about the largest satellite that can retain an irregular shape without collapsing gravitationally into a sphere. Crater-like features are seen, including one structure more than 200 km across.

Nereid is the third largest satellite of the Neptune system and has a diameter of about 340 km. It, too, was imaged by *Voyager* but not in resolution sufficient to characterize its surface. Nereid is in a highly inclined orbit and takes 359 days for one complete orbit around Neptune.

10.5 Summary

The giant planets Uranus and Neptune have extensive atmospheres surrounding liquid and frozen water, along with ammonia, methane, and other volatiles, all surrounding large rocky cores. Both planets have ring systems and myriad satellites. Unlike the giant planets Jupiter and Saturn and their systems of moons and rings that have been explored by orbiters (*Galileo* and *Cassini*, respectively), spacecraft data for the Uranus and Neptune systems come only from the brief flybys of the *Voyager* spacecraft. Thus, our knowledge of this part of the Solar System is much more limited.

Of the five large moons of Uranus, Miranda, Umbriel, Titania, and Oberon are in synchronous rotation and appear to be influenced by tidal heating through their interactions with each other and with Uranus. The innermost moons, such as Miranda, show evidence of the greatest amount of internal heating, while the outermost, Oberon, preserves the greatest amount of impact cratering. All but Oberon display darker trailing hemispheres in comparison with the leading hemispheres. This is attributed to their immersion in the Uranus magnetosphere, in which radiation and implantation of ions occur preferentially on the trailing hemispheres because the speeds of the satellite orbits are less than the speeds within the magnetosphere.

About half of Neptune's known moons are in retrograde orbits, suggesting that they are captured objects. The largest of these, Triton, is about the size of Earth's Moon and exhibits active geysers, thought to be driven by a "greenhouse" effect in which solar energy penetrates ice on the surface and heats volatiles; pressure of the expanded gasses eventually exceeds the strength of the ice, venting explosively to the surface, along with entrained dark (carbonaceous) material. Above the surface the geyser plumes are caught by winds in Triton's tenuous atmosphere and are carried down wind, leaving dark streak deposits on the surface. Cryovolcanism is suggested on Triton, as evidenced by caldera-like depressions with smooth floors.

Many planetary scientists consider Triton to be an analog for what will be seen when *New Horizons* reaches Pluto and returns images of this dwarf planet and its large moon, Charon. All three objects occupy the same general zone in the Solar System (in fact, Pluto's orbit carries it inside the orbit of Neptune at times), and all three have similar densities and surface spectral properties.

Assignments

1. Compare and constrast geysers on Enceladus with those on Triton.

2. Compile a table for the chasmata on Titania, Oberon, and Ariel. Note chasmata sizes, geographic locations (trailing hemisphere, polar region, equatorial region, etc.), and potential cross-cutting relations with other features such as fractures.

3. Go on the web and summarize a recent scientific paper that describes the darkening of ices in the environments of the outer planet satellites.

4. Compare and contrast the surface features on Miranda and Triton and identify the geologic processes responsible for the formation of the features observed.

5. Compare and contrast the large moons of Jupiter and Uranus with regard to the degree of internal activity and their orbital distance from their parent planet.

CHAPTER 11

Planetary geoscience future

11.1 Introduction

Planetary geosciences are advanced primarily through new data and are stimulated by physical and computational modeling, theoretical studies, and field studies of terrestrial analogs in support of planetary data analysis. As shown in **Fig. 1.11**, missions are currently in flight for Mercury, Venus, the Moon, Mars, Jupiter, Saturn, and Pluto, as well as for a host of asteroids and comets. Particularly noteworthy is the *New Horizons* mission, which will give us our first close-up views of Pluto in 2015. Spacecraft for these missions carry sophisticated scientific payloads, including imaging systems that will provide additional coverage, higher resolution, or first ever views of planetary objects of geoscience interest.

Most planetary missions are flown by NASA, some in partnership with the European Space Agency, which also flies planetary missions independently of NASA. In addition, the space agencies of Japan, India, and China are becoming increasingly important, especially in lunar exploration and the eventual return of humans to the Moon. (After a hiatus of many years, in 2012 Russia attempted to resume planetary exploration with a mission to one of the moons of Mars, Phobos, but that mission failed shortly after launch.) Although NASA had planned lunar exploration by humans early in the twenty-first century, such plans have been deferred because of the economic climate. In its place, considerations are being given to sending humans to one or more asteroids because of the lower costs (it is easier to return to Earth from these low-gravity bodies), the high scientific potential of asteroids, and the need to assess asteroids as hazards to Earth.

Much of the motivation for planetary exploration is the search for life beyond the Earth. This involves searching for signs of past or present life and characterizing environments conducive to life as we know it. Such exploration raises the issue of "planetary protection."

11.2 Planetary protection

Concern over contamination of planetary objects began with deep space exploration in the early 1960s and is termed **planetary protection**. Planetary contamination includes both **forward contamination**, in which terrestrial organisms from Earth might be carried to other objects by spacecraft, and **backward contamination**, in which potential non-Earthly organisms are brought to Earth. International protocols through the United Nations were established for both conditions and most space-faring nations and agencies have agreed to follow the procedures given in the protocols. Through international agreements for planetary protection, five categories of missions are defined.

Category I (lowest level) missions include those to the Sun, Mercury or other objects that are not of direct interest for prebiotic chemistry or the origin and evolution of life.

Category II missions include those to objects such as the Moon, Venus, Jupiter or other targets that are of interest for prebiotic chemistry and the origin and evolution of life but for which there is an insignificant probability of contamination by Earth life.

Category III missions are flybys or orbiters to objects such as Mars and Europa that could be hosts for life and for which there is a possibility of contamination by Earth life.

Category IV missions are landers or probes to objects such as Mars and Europa that could be hosts for life and for which there is a possibility of contamination by Earth life.

Category V (highest level) missions are sample returns to Earth from locations that have the potential to support life.

To avoid forward contamination, spacecraft and spacecraft components, including the scientific instruments, must be sterilized to certain levels. Sterilization can be achieved by methods that include heating to specified temperatures for minimum times or exposure to radiation. After sterilization, the surfaces are swabbed, the swabs are placed in media to encourage growth of potential cultures, and each swab is analyzed. It is nearly impossible to reach 100% sterilization and thus maximum allowable levels are defined by mission category in the protocol before the spacecraft and components are certified for flight. If there is a failure to achieve the acceptable levels, the process is repeated until the acceptable levels are reached. As you might imagine, the process for sterilization and certification for flight is expensive and adds significantly to the cost of a mission.

Another consequence of sterilization is the impact on sensitive components, such as scientific instruments, many of which cannot survive heating or exposure to radiation. In these cases, components must be housed in "vaults" that would contain any Earthly organisms, even in the event of a catastrophic failure such as an explosion. Again, such an approach increases the cost and the complexity of planetary missions.

One could ask, why go through this at all? From a scientific perspective, recall that one of the primary motivations for planetary exploration is the search for life beyond Earth. If forward contamination were to occur, the problem would be that we might "find ourselves" on some future mission, thus calling into question forever the existence of extraterrestrial life on the contaminated planet. Moreover, should life already be present on the planet, there is the possibility that forward contamination could cause irreparable harm through introduced diseases, mutations, or even extinction.

Backward contamination is the stuff of countless science fiction stories but is also of great concern in planetary exploration. For example, in the early *Apollo* program, the astronauts and the lunar samples returned from the Moon were placed in quarantine to determine whether any lunar organisms had caught a ride back to Earth. Cultures were grown and other tests were carried out, all with negative results, and follow-on *Apollo* missions abandoned the procedure. Subsequent studies have shown that the quarantine procedures during *Apollo* were rather flawed and had a great many "leaky" paths that could have led to problems if any lunar organisms had been brought to Earth. Interestingly, pieces of the unmanned *Surveyor 3* spacecraft (**Fig. 4.6**) were returned to Earth by the *Apollo 12* astronauts

and were found to contain bacteria that presumably had been carried to the Moon by forward contamination. This meant that the organisms had survived in the harsh lunar environment for some 31 months, although some investigators suggested that the bacteria were from Earthly contamination after return from the Moon.

In more recent times, extraterrestrial samples have been returned to Earth from deep space, including those of the *Stardust* and *Hayabusa* missions. These have all been placed in laboratories under controlled conditions and analyzed. Currently, plans are being formulated by NASA and the ESA for a facility to receive the anticipated samples to be returned from Mars.

11.3 Missions in flight and anticipated for launch

The previous chapters on planetary systems outline the current missions that are in flight. For example, missions at the Moon (**Chapter 4**) include the *Lunar Reconnaissance Orbiter* (*LRO*). This mission was designed primarily to collect data to support future robotic and human exploration activities. After completion of these goals in 2010, the *LRO* continued operations to meet a host of scientific objectives, many of which are directly contributing to lunar geology. Similarly, the *ARTEMIS* mission was originally launched to collect data on lunar solar interactions. In 2011, the spacecraft was reposititioned to collect geophysical information for the Moon to complement the Discovery-class *GRAIL* mission, led by MIT PI Maria Zuber, which was launched in 2011. *GRAIL* consists of two orbiters collecting detailed information on the nature of the lunar interior. These missions were followed by *LADEE* (*Lunar Atmosphere and Dust Environment Explorer*) in 2012, which was designed to collect data on the mysterious "glow" that has been seen for decades near the lunar surface and is possibly caused by light scattering from a nebulous atmosphere and dust raised by electrostatic processes. China's launch of *Chang'e 2* in 2010 enabled acquisition of high-resolution images of the Moon in support of future missions. China's plans call for the launch of a lunar rover, possibly in 2012, to be followed by a second rover with the return of lunar samples to Earth in about 2017.

MESSENGER was placed into orbit around Mercury in 2010 and will complete its prime mission in 2012. Data from this Discovery-class mission are revolutionizing our understanding of this innermost planet of the Solar System, as discussed in **Chapter 5**. If all goes well, the mission is likely to be extended as long as the spacecraft and instruments are healthy and new data can be obtained by the project. Study of Mercury will continue with the launch of *BepiColombo* in 2014. This joint ESA–JAXA mission will place two spacecraft in orbit around Mercury in 2020 to obtain more information on the geology, interior, magnetosphere, and tenuous atmosphere.

As the most Earth-like planet in our Solar System and because it is a close neighbor, Mars is a focus in planetary exploration (**Chapter 7**). NASA's *Mars Science Laboratory* (*MSL*), launched in 2011, carries a rover named *Curiosity* (**Fig. 11.1**), which is the size of a small car, with a scientific payload designed to advance astrobiology exploration (Grotzinger, 2009). The primary objective is to assess the biological potential of a key target site to determine the nature of organic compounds and the chemical building blocks of life in the search for signs of past and present life (Grotzinger, 2009). In addition, the general geology of the site is being studied, including the rock record, to assess long-term atmospheric evolution and cycling of water and carbon dioxide between the surface and the atmosphere. The rover is powered by a radioisotope thermoelectric generator and will operate for at least one Mars year (686 Earth days). The landing site for the *MSL* is on the floor of Gale crater (**Fig. 11.2**) in the equatorial region of Mars. The crater is about 154 km in diameter and has a huge central peak that shows layering suggestive of repeated episodes of sedimentary deposition.

As part of NASA's Discovery Program, the *Mars Atmosphere and Volatile Evolution Mission* (*MAVEN*), led by PI Bruce Jakosky, will be launched in 2013 to be

Figure 11.1. A family portrait of NASA's Mars rovers, showing progression in size from the Mars Pathfinder rover, *Sojourner* (center), that operated on Mars in 1997, to the Mars Exploration Rovers (left side), two of which began operations in 2004, to the car-size *Mars Science Laboratory* (right) (NASA-Jet Propulsion Laboratory).

Figure 11.2. Gale impact crater in Mars' equatorial region is the landing site for the *Mars Science Laboratory*. The area of operations is shown by the ellipse in this THEMIS image. Gale crater is about 154 km in diameter and displays complex sedimentary deposits on its floor, some of which appear to have been emplaced by fluvial channels cut into the crater rim. (NASA PIA14290).

placed in Mars orbit in the fall of 2014. The mission will focus on Mars' upper atmosphere and interactions with the Sun to determine the loss of volatiles such as carbon dioxide and water. The results will provide unprecedented insight into the history of the atmosphere and climate, especially as related to the planet's habitability. The scientific payload includes instruments to characterize the solar wind, Mars' ionosphere, and the atmospheric composition. The *MAVEN* spacecraft is scheduled to operate for one Earth year, or about one-half of a Mars year.

The more distant future of Mars exploration is less certain. NASA and the ESA have recently been studying projects leading to the eventual return of samples from Mars to Earth. The return of martian samples from a site of known geologic context has been a high priority in planetary exploration for many decades. But such a mission is not so easy, and will not be accomplished in a single step. Rather, the first step would be to send a large rover, launched possibly in 2018. In addition to conducting *in situ* experiments focused on astrobiology and geology, this mission would "cache" (store) samples for return to Earth in the 2020s using a system yet to be designed. Current plans involve lift-off of the samples from Mars, with a small capsule being placed in orbit around Mars. This would later be captured by a separate spacecraft for return to Earth.

Exploration of the outer Solar System continues, but not at the same pace as for Mars and the inner Solar System. Currently in flight are *Cassini* (**Chapter 9**), scheduled to continue orbiting Saturn until September 2017, *New Horizons*, which will execute a flyby of Pluto in 2015, and *Juno*, which was launched in 2011 and is scheduled to begin orbiting Jupiter in 2016 (**Fig. 11.3**). During *Juno*'s one-year mission, it will complete 33 orbits and then de-orbit to plunge into Jupiter's dense atmosphere, ending the mission. *Juno*'s scientific objectives are to determine the amount of water in the atmosphere, measure the composition, temperature, and cloud motions in the atmosphere, and gain insight into Jupiter's magnetosphere and gravity fields. Collectively, the anticipated data will enable new perspectives on the origin and evolution of the giant planets. While not providing direct data on the geology of the satellites, the new data will be relevant to the overall evolution of the outer planets and their satellites.

Comets and asteroids are important objects for understanding the origin and evolution of the Solar System.

Figure 11.3. An artist's rendition of the *Juno* spacecraft in operation around Jupiter in 2016.

Many of the missions to these objects described in **Chapter 1** continue to return new data, including information relevant to planetary geology. For example, *Dawn* returned close-up images of asteroid Vesta (**Fig. 11.4**) and will be programmed to leave this object to reach asteroid Ceres in 2015. In addition to these missions, the ESA's *Rosetta* spacecraft was launched in 2004 and is scheduled to rendezvous with Comet 67P/Churyumov-Gerasimenko in 2014 to begin a detailed analysis of this 4 km object. The spacecraft will orbit the comet as it approaches the Sun to collect data on processes in and on the icy nucleus as it is warmed by the Sun. A small lander will be released from the orbiter to make *in situ* measurements of the comet's composition and physical properties, as well as to image its surface.

In 2016, the *OSIRIS-REx (Origins–Spectral Interpretation–Resource Identification–Security–Regolith Explorer)* spacecraft will be launched to reach the near-Earth asteroid 1999RQ36 in 2020. In this mission, led by planetary scientist Dante Lauretta, after six months of intensive study and surface mapping, the spacecraft will approach a site on the asteroid selected by the science team and extend a robotic arm to collect a small sample. The samples will be returned to Earth in 2023 and will enable comparisons of asteroid materials with cometary material that was collected from Comet Wild 2 and returned to Earth by the *Stardust* spacecraft.

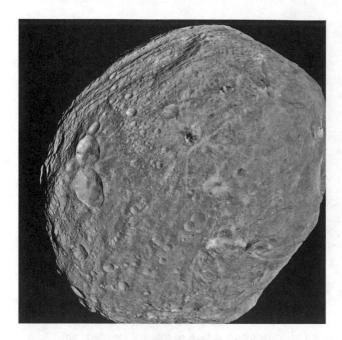

Figure 11.4. This image of asteroid Vesta was taken by the *Dawn* spacecraft in 2011 shortly after orbiting the asteroid. Vesta is about 530 km in diameter and displays a complex geologic history; differences in albedo among the impact crater ejecta deposits suggest a heterogeneous composition (NASA PIA14317).

11.4 Extended missions

When spacecraft projects are approved, they have specific scientific goals and objectives for the **primary mission**, which defines the period of operation. Typically, as long as the spacecraft and all or most of the payload is functioning, additional funds can be requested for an **extended mission** to continue operations. Often, more than one extension can be obtained, so long as there is scientific justification. Part of the rationale is that most of the costs of spaceflights are for building the spacecraft with its payload and for the launch rocket; the cost for an extended mission is only a small fraction of these initial expenses. Nonetheless, with numerous missions in flight, the total cost for extensions can be significant, with the sum often totaling that for a completely new mission. Consequently, NASA has established a policy in which all requests for extended missions are evaluated together by a panel of senior scientists to determine the priorities of the requests and to assess the cost–benefit ratios. Thus, what once had been fairly routine extensions are now evaluated in light of the overall program of Solar System exploration.

11.5 Summary

As outlined in **Chapter 1**, NASA generally follows the recommendations of the National Research Council (NRC, 2011) for establishing the priorities in developing and flying missions. These recommendations place a high value on the scientist-led missions such as those proposed through the Discovery and New Frontiers programs. For example, Discovery missions currently inflight or soon to be launched include *GRAIL* to the Moon, *Dawn* to the asteroid Vesta, and *MAVEN* to Mars. The lower-cost Discovery missions are flown every 24–36 months and are completely open as to destination through competition. In contrast, the higher-cost New Frontiers missions are flown less frequently and are subject to the priorities identified by the NRC. For the decade 2013–2023, priority missions for the New Frontiers program include sample return from a comet, sample return from the South Pole–Aiken basin on the Moon, a Saturn atmospheric probe, a Venus lander, an Io observer, or a lunar network. However, only one or two missions are likely to be selected from this set for flight, by competition among proposals.

The rate and complexity of missions are very much a function of their costs and the availability of funds. When the "politics" of space is also considered, the result is a great deal of uncertainty regarding which missions might be sent for future exploration. For example, the large-cost, or "flagship," missions of NASA have Mars sample return to Earth as the highest priority, followed by a mission to explore Jupiter's moon Europa. However, with the poor state of the economy, it is not clear when these missions might be approved and eventually flown.

Despite these uncertainties, there is an incredible wealth of data already returned from missions recently flown or anticipated from spacecraft that are in flight. These data are waiting to be "mined" for their scientific treasures by current and future planetary scientists, and it is likely that geosciences will continue to be a hallmark in understanding our Solar System.

Assignments

1. Since the inception of the Space Age, there has been debate on the role of humans in the exploration of the Solar System. Outline the pros and cons of robotic versus human exploration. Visit various websites dealing with this topic for background.

2. Sample return to Earth from Mars is a high priority. From your perspective, what type of sample should be of highest priority for science? Explain the requirements for such a sample with respect to the type of site that should be selected on Mars to meet your priorities.

3. In times of limited resources, there is debate regarding whether funds should be devoted to flying new missions or to analyzing data already returned to Earth. Discuss which you would favor, giving the justifications for your answer.

APPENDICES

Appendix 1.1 Some key websites for planetary science

Website name	Comments
European Space Agency http://www.esa.int/	General website; look for specific missions by name
Exploring the Planets http://www.nasm.si.edu/research/ceps/etp/	National Air and Space Museum Solar System tour
GRIN http://grin.hq.nasa.gov/	Great Images in NASA
JMARS http://jmars.asu.edu	Enables rapid searches and provides analytical tools for planetary data
JPL NASA Solar System http://www.jpl.nasa.gov/solar_system/	JPL's Solar System Tour
Lunar and Planetary Science Institute http://www.lpi.usra.edu	General planetary and related topics
Mars Society http://www.marssociety.org/	To further the goal of the exploration and settlement of the Red Planet
NASA http://www.nasa.gov/	General website; look for specific missions by name
NASA Astrobiology Institute http://nai.nasa.gov	Astrobiology topics
NASA-JPL Solar System Simulator http://space.jpl.nasa.gov/	Simulates positions of entire Solar System or members at any time from any vantage point (including various spacecraft)
NASA Remote Sensing Tutorial http://rst.gsfc.nasa.gov/	Extensive tutorial on remote sensing
NASA's Solar System Exploration http://solarsystem.nasa.gov/planets/profile.cfm?Object=SolarSys	Solar System Tour from NASA
National Space Society http://www.nss.org/	An independent, educational, grassroots nonprofit organization dedicated to the creation of a spacefaring civilization
NSPIRES http://nspires.nasaprs.com/external/	NASA Research Opportunities
NSSDC Photo Gallery http://nssdc.gsfc.nasa.gov/photo_gallery/	Most commonly requested NSSDC images

Website name	Comments
NSSDC Planetary Image Catalog http://nssdc.gsfc.nasa.gov/imgcat/	The National Space Science Data Center serves as the permanent archive for NASA space science mission data
Planetary Geomorphology http://www.psi.edu/pqwq/images/index.html	Images selected for geomorphology
Planetary Images http://pages.preferred.com/~tedstryk/	Rarely seen planetary images from Soviet and US missions
Planetary Image Atlas http://pds-imaging.jpl.nasa.gov/Atlas/	JPL image atlas arranged per mission
Planetary Photojournal http://photojournal.jpl.nasa.gov/index.html	Planetary Image Archive
The Planetary Society http://www.planetary.org/home/	The world's largest space-interest group
USGS Astrogeology Program – Browse the Solar System http://astrogeology.usgs.gov/Projects/BrowseTheSolarSystem/	USGS Solar System Tour
USGS Astrogeology Science Center http://planetarynames.wr.usgs.gov/	Gazetteer of Planetary Nomenclature
Welcome to the Planets http://pds.jpl.nasa.gov/planets/index.htm	Planetary Data System

Appendix 1.2 NASA regional planetary image facilities

United States locations

Arizona State University
Space Photography Laboratory
School of Earth and Space Exploration
Box 871404
Arizona State University
Tempe AZ 85287-1404
Director: David Williams
Data Manager: Daniel Ball
Phone: 480-965-7029
Fax: 480-965-8102
E-mail: RPIF@asu.edu

Brown University
Northeast Regional Planetary Data Center
Department of Geological Sciences
Brown University
Box 1846
Providence RI 02912-1846
Director: Peter H. Schultz
Data Manager: Peter Neivert
Phone: 401-863-3243
Fax: 401-863-3978
E-mail: Peter_Neivert@brown.edu

Cornell University
Spacecraft Planetary Imaging Facility
Center for Radiophysics and Space Research
317 Space Sciences Building
Cornell University
Ithaca NY 14853-6801
Director: Joseph Veverka
Data Manager: Rick Kline
Phone: 607-255-3833
Fax: 607-255-9002
E-mail: kline@astro.cornell.edu

Jet Propulsion Laboratory
Regional Planetary Image Facility
Mail Stop 202-100
Jet Propulsion Laboratory
4800 Oak Grove Drive
Pasadena CA 91109
Director: Timothy J. Parker

Data Manager: Debbie Martin
Phone: 818-354-3343
Fax: 818-354-0740
E-mail: jpl_rpif@jpl.nasa.gov

Lunar and Planetary Institute

Center for Information and Research
    Services
Lunar and Planetary Institute
3600 Bay Area Boulevard
Houston TX 77058-1113
Director: Paul Spudis
Data Manager: Mary Ann Hager
Phone: 281-486-2182
Fax: 281-486-2186
E-mail: cirs2@lpi.usra.edu

National Air and Space Museum

Center for Earth and Planetary Studies
National Air and Space Museum, MRC 315
PO Box 37012
Washington DC 20013-7012
Director: Thomas Watters
Data Manager: Rosemary Aiello
Phone: 202-633-2480
Fax: 202-786-2566
E-mail: steinatr@si.edu

US Geological Survey-Flagstaff

Regional Planetary Image Facility
Branch of Astrogeology
United States Geological Survey
2255 North Gemini Drive
Flagstaff AZ 86001
Director: Justin Hagerty
Data Manager: David Portree
Phone: 928-556-7264
Fax: 928-556-7090
E-mail: RPIF-flag@usgs.gov

University of Arizona

Space Imagery Center
Lunar and Planetary Laboratory
University of Arizona
1629 E. University Boulevard
Tucson AZ 85721-0092
Director: Shane Byrne
Data Manager: Maria Schuchardt
Phone: 520-621-4861
Fax: 520-621-4933
E-mail: mariams@LPL.arizona.edu

University of Hawai'i at Manoa

Pacific Regional Planetary Data Center
Hawai'i Institute of Geophysics and
    Planetology
School of Ocean and Earth Science and
    Technology
2525 Correa Road
Honolulu HI 96822
Director: B. Ray Hawke
Data Manager: Chris A. Peterson
Phone: 808-956-3131
Fax: 808-956-6322
E-mail: prpdc@higp.hawaii.edu

International locations

INAF/Istituto di Astrofisica Spaziale e Fisica Cosmica (IASF)

Southern Europe Regional Planetary Image Facility
C.N.R. Area ricerca di Roma Tor Vergata
Istituto di Astrofisica Spaziale e Fisica Cosmica (IASF)
Via del Fosso de Cavaliere, 100
00133 Roma, Italy
Director: Pietro Ubertini
Data Manager: Livia Giancomini
Phone: 39-6-49934449
Fax: 39-6-49934182
E-mail: livia.giacomini@ifsi-roma.inaf.it

Deutsches Zentrum für Luft und Raumfahrt e.V.

German Aerospace Center
Institute of Planetary Research
Regional Planetary Image Facility
Rutherfordstr. 2 12489 Berlin, Germany
Director: Ralf Jaumann
Data Manager: Susanne Pieth
Phone: 49-30-67055-333
Fax: 49-30-67055-372
E-mail: rpif@dlr.de

Institute of Space and Astronautical Sciences

Regional Planetary Image Facility
Institute of Space and Astronautical Sciences
Division of Planetary Science
3-1-1 Yoshinodai
Sagamihara-Shi, Kanagawa 229-8510, Japan
Director: Akio Fujimura
Data Manager: Satoshi Tanaka
Phone: 81-427-59-8198
Fax: 81-427-59-8516
E-mail: tanaka@planeta.sci.isas.jaxa.jp

Ben-Gurion University of the Negev
Department of Geography and Environmental
 Development
P.O. Box 653
Beer-Sheva 84105, Israel
Director: Dan Blumberg
Data Manager: Shira Amir
Phone: 972-8-647-7939
Fax: 972-8-647-2821
E-mail: blumberg@bgu.ac.il

University College London
Regional Planetary Image Facility
Department of Earth Sciences
University College London
Gower Street
London WC1E 6BT, UK
Director: Jan-Peter Muller
Data Manager: Peter Grindrod
Phone: 44-20-7679-2134
Fax: 44-20-7679-7614
E-mail: p.grindrod@ucl.ac.uk

University of New Brunswick
Planetary and Space Science
 Centre
Department of Geology
University of New Brunswick
P.O. Box 4400
Fredericton NB, Canada E3B 5A3
Director: John Spray
Data Manager: Beverly Elliott
Phone: 506-453-3560
Fax: 506-453-5055
E-mail: passc@unb.ca

University of Oulu
Nordic Regional Planetary Image
 Facility
Department of Physical Sciences
Astronomy Division
University of Oulu
FIN-90014 University of Oulu,
 Finland
Director: Jouko Raitala
Data Manager: Veli-Petri Kostama
Phone: 358-(0)8-553-1946
Fax: 358-(0)8-553-1934
E-mail: petri.kostama@oulu.fi

Université de Paris-Sud
Photothèque Planétaire d'Orsay
Département des Sciences de la Terre
Bâtiment 509
F-91405 Orsay, France
Director: Chiara Marmo
Phone: 33 (0) 1 69 15 61 49
Data Manager: Laurent Daumas
Phone: 33 (0) 1 69 15 61 51
Fax: 33 (0) 1 69 15 48 63
E-mail: datamanager@geol.u-psud.fr

Appendix 2.1 Common planetary imaging systems

Although charge-coupled devices (CCDs) are the primary detectors for current imaging systems, cameras on previous missions used a variety of detectors, including traditional film and video systems. Because the data from the previous missions still provide a wealth of planetary information, these systems are described below.

A2.1.1 Film systems

The highest-resolution planetary images were produced on photographic film that was returned to Earth. Thus far in planetary exploration, this has been achieved only from some of the Soviet *Zond* spacecraft and the manned *Apollo* missions to the Moon. Of the unmanned NASA lunar and planetary missions, only the *Lunar Orbiter* (**Fig. 2.20**) used film sensors, but the film was not returned to Earth; it was developed on board and then the images were transferred to Earth as an electronic signal. Although there was overlap of the frames, providing stereoscopic models, the nature of the electronic transfer and reconstruction introduced artifacts in the stereo models that were never fully resolved. Nonetheless, the five *Lunar Orbiters* returned the first near-global imaging data set for the Moon.

Returned film from the *Apollo* missions enabled analyses without the complexities of electronic transformation and provided reliable stereoscopic models useful for photogrammetry. Hand-held cameras using 70 mm film (color and black and white) were used both from orbit and on the ground. Some of these images were used to make panoramic views of the terrain from overlapping frames.

The *Apollo 14* mission carried a sophisticated mapping camera, but a malfunction early in the mission resulted in there being very little usable data. *Apollos 15, 16,* and *17* all carried two high-quality camera systems, a panoramic camera for detailed geologic studies and a metric (mapping) camera that enabled topographic data to be derived. The "pan" camera, using a lens of focal length 610 mm, was highly sophisticated and obtained stereoscopic views of the surface with resolutions as good as 1 m from an orbital altitude of 100 km. The camera rotated in a direction across the path of the orbiting spacecraft, giving a panoramic view.

The *Apollo* metric image system consisted of two cameras, one pointing downward and one pointing away from the Moon. Both used 76 mm lenses with the upward-pointing camera being used to fix precisely the geometric position of the spacecraft in relation to the star field and with the Moon's surface being photographed by the other camera. Although the spatial resolution of metric frames is only about 20 m, the precision of location enabled the production of high-quality maps.

The *Lunar Orbiter* and *Apollo* hand-held camera images, and the *Apollo 15–17* panoramic and metric camera images, are being digitized under the direction of Mark Robinson of Arizona State University, and are available through NASA websites.

A2.1.2 Vidicon systems

The imaging system most widely used in Solar System exploration in the 1970s and 1980s involved **vidicon** cameras. These systems employed a small electron gun and a photoconductor. The image was optically focused onto the photoconductor so that a beam of electrons from the gun was transformed into a current that varied with the intensity of the light reflected from the scene. This current was in either analog (flown on early systems, such as the lunar *Ranger* landers) or digital format and was either stored on board the spacecraft or transmitted directly to Earth. The signals were transferred into pixels and DN levels for computer processing.

A2.1.3 Facsimile systems

Facsimile cameras scan across the scene by rotating a slit across the scene as a panorama. Facsimile cameras were used on the two martian *Viking* lander spacecraft and on the Soviet *Venera* landers on Venus. The returned data were in pixel formats with DN levels.

REFERENCES

Antonenko, I., Head, J. W., Mustard, J. F., and Hawke, B. R. (1995). Criteria for the detection of lunar cryptomaria. *Earth, Moon, and Planets*, **69**, 141–172.

Arvidson, R. E., Ruff, S. W., Morris, R. V. *et al.* (2008). Spirit Mars Rover Mission to the Columbia Hills, Gusev Crater: mission overview and selected results from the Cumberland Ridge to Home Plate. *J. Geophys. Res.*, **113**(E12), E12S33.

Bagnold, R. A. (1941). *The Physics of Blown Sand and Desert Dunes*. London: Methuen and Co.

Barlow, N. G., Costard, F. M., Craddock, R. A. *et al.* (2000). Standardizing the nomenclature of Martian impact crater ejecta morphologies. *J. Geophys. Res.*, **105**, 26,733–26,738.

Bibring, J.-P., Langevin, Y., Mustard, J. F. *et al.* (2006). Global mineralogical and aqueous Mars history derived from OMEGA/Mars Express data. *Science*, **312**, 400–404.

Brownlee, D., Tsou, P., Aléon, J. *et al.* (2006). Comet 81P/Wild2 under a microscope. *Science*, **314**, 1,711–1,716.

Buratti, B. J. and Mosher, J. A. (1991). Comparative global albedo and color maps of the Uranian satellites. *Icarus*, **90**, 1–13.

Canup, R. M. (2010). Origin of Saturn's rings and inner moons by mass removal from a lost Titan-sized satellite. *Nature*, **468**, 943–946.

Carr, M. H. (2006). *The Surface of Mars*. Cambridge: Cambridge University Press.

Carr, M. H. and Head III, J. W. (2010). Geologic history of Mars. *Earth Planet. Sci. Lett.*, **294**, 185–203.

Christensen, P. R., Bandfield, J. L., Hamilton, V. E. *et al.* (2001). Mars Global Surveyor Thermal Emission Spectrometer experiment: investigation description and surface science results. *J. Geophys. Res.*, **106**, 23,823–23,871.

Clow, G. D. and Carr, M. H. (1980). Stability of sulfur slopes on Io. *Icarus*, **44**, 268–279.

Collins, G. and Nimmo, F. (2009). Chaotic terrain on Europa. In *Europa*, ed. R. Pappalardo, W. McKinnon, and K. Khurana. Tucson, AZ: University of Arizona Press, pp. 259–281.

Cruikshank, D., Bell, J. F., Gaffey, M. J. *et al.* (1983). The dark side of Iapetus. *Icarus*, **53**, 90–104.

Davies, A. (2007). *Volcanism on Io: A Comparison with Earth*. Cambridge: Cambridge University Press.

Davies, M. E., Dwornik, S. E., Gault, D. E., and Strom, R. G. (1976). *Atlas of Mercury*. NASA Special Publication 423.

Dence, M. R. (1972). The nature and significance of terrestrial impact structures. In *Proceedings of the 24th International Geological Congress*, Section 15, pp. 77–89.

Donahue, T. M., Hoffman, J. H., Hodges, R. R., and Watson, A. J. (1982). Venus was wet: a measurement of the ratio of D to H. *Science*, **216**, 630–633.

Esposito, L. W. (1984). Sulfur dioxide: episodic injection shows evidence for active Venus volcanism. *Science*, **223**, 1,072–1,074.

Figueredo, P. H., Greeley, R., Neuer, S., Irwin, L., and Schulze-Makuch, D. (2003). Locating potential biosignatures on Europa from surface geology observations. *Astrobiology*, **3**, 851–861.

French, B. M. (1998). *Traces of Catastrophe: A Handbook of Shock-Metamorphic Effects in Terrestrial Meteorite Impact Structures*. Houston, TX: Lunar and Planetary Science Institute.

Frey, H. (2011). Previously unknown large impact basins on the Moon: implications for lunar stratigraphy. In *Recent Advances and Current Research Issues in Lunar Stratigraphy*. Boulder, CO: Geological Society of America, pp. 53–75.

Gaddis, L. R., Staid, M. I., Tyburczy, J. A., Hawke, B. R., and Petro, N. E. (2003). Compositional analyses of lunar pyroclastic deposits. *Icarus*, **161**, 262–280.

Garcia-Ruiz, J. M., Hyde, S. T., Carnerup, A. M. *et al.* (2003). Self-assembled silica–carbonate structures and detection of ancient microfossils. *Science*, **302**, 1,194–1,197.

Gault, D. E., Quaide, W. L., and Oberbeck, V. R. (1968). Impact cratering mechanisms and structures. In *Shock Metamorphism of Natural Materials*, ed. B. M. French and N. M. Short. Baltimore, MD: Mono Book Corp., pp. 87–99.

Gault, D. E., Guest, J. E., Murray, J. B., Dzurisin, D., and Malin, M. C. (1975). Some comparisons of impact craters on Mercury and the Moon. *J. Geophys. Res.*, **80**, 2,444–2,460.

Golombek, M. P. and Phillips, R. J. (2010). Mars tectonics. In *Planetary Tectonics*, ed. T. Watters and P. Schultz. Cambridge: Cambridge University Press, pp. 183–232.

Greeley, R. and Batson, R. (2001). *The Compact NASA Atlas of the Solar System*. Cambridge: Cambridge University Press.

Greeley, R. and Iversen, J. D. (1985). *Wind as a Geological Process: Earth, Mars, Venus, and Titan*. Cambridge: Cambridge University Press.

Greeley, R. and Spudis, P. D. (1981). Volcanism on Mars. *Rev. Geophys. Space Phys.*, **19**, 13–41.

Greeley, R., Fink, J. H., Gault, D. E., and Guest, J. E. (1982). Experimental simulation of impact cratering on icy satellites. In *Satellites of Jupiter*, ed. D. Morrison. Tucson, AZ: University of Arizona Press, pp. 340–378.

Greenberg, R., Hoppa, G. V., Tufts, B. R. *et al.* (1999). Chaos on Europa. *Icarus*, **141**, 263–286.

Grotzinger, J. P. (2009). Mars exploration, comparative planetary history and the promise of Mars Science Laboratory. *Nature Geosci.*, **2**, 1–3.

Grundy, W. M., Young, L. A., Spencer, J. R. *et al.* (2006). Distributions of H_2O and CO_2 ices on Ariel, Umbriel, Titania, and Oberon from IRTF/SpeX observations. *Icarus*, **184**, 543–555.

Head, J. W. and Wilson, L. (1986). Volcanic processes and landforms on Venus: theory, predictions, and observations. *J. Geophys. Res.*, **91**, 9,407–9,446.

Head, J. W., Crumpler, L. S., Aubele, J. C., Guest, J. E., and Saunders, R. S. (1992). Venus volcanism: classification of volcanic features and structures, associations, and global distribution from Magellan data. *J. Geophys. Res.*, **97**, 13,153–13,198.

Head, J. W., Murchie, S. L., Prockter, L. M. *et al.* (2008). Volcanism on Mercury: evidence from the first MESSENGER flyby. *Science*, **321**, 69–72.

Head, J. W., Marchant, D. R., Dickson, J. L., Kress, A. M., and Bakerm, D. M. (2010). Northern mid-latitude glaciation in the Amazonian period of Mars: criteria for the recognition of debris-covered glacier and valley glacier land system deposits. *Earth Planet. Sci. Lett.*, **294**, 306–320.

Heiken, G. H., Vaniman, D. T., and French, B. M., eds. (1991). *Lunar Sourcebook: A User's Guide to the Moon*. Cambridge: Cambridge University Press.

Hiesinger, H., Head III, J. W., Wolf, U, Jaumann, R., and Neukum, G. (2011). Ages and stratigraphy of lunar mare basalts: a synthesis. In *Recent Advances and Current Research Issues in Lunar Stratigraphy*. Boulder, CO: Geological Society of America pp. 1–51.

Howard, K. A. (1967). Drainage analysis in geological interpretation: a summation. *Am. Assoc. Petrol. Geol. Bull.*, **51**, 2,246–2,259.

Jaumann, R., Kirk, R. L., Lorenz, R. D. *et al.* (2009a). Geology and surface processes on Titan. In *Titan*, ed. H. Brown, J-P. Lebreton, and J. H. Waite. Dordrecht: Springer, pp. 75–140.

Jaumann, R., Clark, R. N., Nimmo, F. *et al.* (2009b). Icy satellites: geologic evolution and surface processes. In *Saturn from Cassini–Huygens*, ed. M. Dougherty, L. Esposito, and S. Krimigis. Dordrecht: Springer, pp. 637–681.

Kivelson, M. G., Bagenal, F., Kurth, W. S. *et al.* (2004). Magnetospheric interactions with satellites. In *Jupiter – The Planets, Satellites, and Magnetosphere*, ed. F. Bagenal, T. Dowling, and W. McKinnon. Cambridge: Cambridge University Press, pp. 513–536.

Lunine, J. I. (1990). Evolution of the atmosphere and surface of Titan. In *Formation of Stars and Planets, and the Evolution of the Solar System. Proceedings of the 24th ESLAB Symposium, Friedrichshafen*. Noordwijk: ESA, pp. 159–165.

McCauley, J. F. (1977). Orientale and Caloris. *Phys. Earth Planet. Inter.*, **15**, 220–250.

McCauley, J. F., Smith, B. A., and Soderblom, L. A. (1979). Erosional scarps on Io. *Nature*, **280**, 736–738.

McKay, D. S., Heiken, G., Basu, A. *et al.* (1991). The lunar regolith. In *The Lunar Sourcebook A User's Guide to the Moon*, ed. G. H. Heiken, D. T. Vaniman, and B. M. French. Cambridge: Cambridge University Press, pp. 285–356.

McKee, E. D., ed. (1979). *A Study of Global Sand Seas*. Reston, VA: US Geological Survey.

McKinnon, W. B. (1999). Convective instability in Europa's floating ice shell. *Geophys. Res. Lett.*, **26**, 951–954.

McKinnon, W. B., Zahnle, K. J., Ivanov, B. A., and Melosh, H. J. (1997). Cratering on Venus: models and observations. In *Venus II – Geology, Geophysics, Atmosphere and Solar Wind Environment*, ed. S. W. Bougher, D. M. Hunten, and R. J. Philips Tucson, AZ: University of Arizona Press, pp. 969–1,014.

Melosh, H. J. (1984). Impact ejection, spallation and the origin of meteorites. *Icarus*, **59**, 234–260.

Melosh, H. J. (1989). *Impact Cratering: A Geologic Process*. New York, NY: Oxford Univerisity Press.

Michael, G. G. and Neukum, G. (2010). Planetary surface dating from crater-size-frequency distribution measurements: partial resurfacing events and statistical age uncertainty. *Planet. Space Sci. Lett.*, **294**, 223–229.

Moore, J. M., Schenk, P. M., Bruesch, L. S., Asphaug, E., and McKinnon, W. B. (2004). Large impact features on middle-sized icy satellites. *Icarus*, **171**, 421–443.

Murchie, S., Watters, T. R., Robinson, M. S. *et al.* (2008). Geology of the Caloris basin, Mercury: a view from MESSENGER. *Science*, **321**, 73–76.

Namiki, N., Iwata, T., Matsumoto, K. *et al.* (2009). Farside gravity field of the Moon from four-way Doppler measurements of SELENE (Kaguya). *Science*, **323**, 900–905.

Neukum, G., Ivanov, B. A., and Hartman, W. K. (2001). Cratering records in the inner solar system in relation to the lunar reference system. *Space Sci. Rev.*, **96**, 55–86.

Nimmo, F., Spencer, J. R., Pappalardo, R. T., and Mullen, M. E. (2007). Shear heating as the origin of the plumes and heat flux on Enceladus. *Nature*, **447**, 289–291.

NRC (2011). *Vision and Voyages for Planetary Science in the Decade 2013–2022*. Washington, DC: National Academies Press.

Oberbeck, V. R. (1975). The role of ballistic erosion and sedimentation in lunar stratigraphy. *Rev. Geophys. Space Phys.*, **13**, 337–362.

Oberbeck, V. R. and Quaide, W. L. (1967). Estimated thickness of a fragmental surface layer of Oceanus Procellarum. *J. Geophys. Res.*, **72**, 4,697–4,704.

Pappalardo, R. T., Collins, G. C., Head, J. W. *et al.* (2004). Geology of Ganymede. In *Jupiter – The Planets, Satellites, and Magnetosphere*, ed. F. Bagenal, T. Dowling, and W. McKinnon. Cambridge: Cambridge University Press, pp. 363–396.

Phillips, R. J., Davis, B. J., Tanaka, K. L. *et al.* (2011). Massive CO_2 ice deposits sequestered in the south polar layered deposits of Mars. *Science*, **332**, 838–841.

Porco, C. C., Thomas, P. C., Weiss, J. W., and Richardson, D. C. (2007). Saturn's small inner satellites: clues to their origins. *Science*, **318**, 1,602–1,607.

Prockter, L. and Patterson, G. W. (2009). Morphology and evolution of Europa's ridges and bands. In *Europa*, ed. R. Pappalardo, W. McKinnon, and K. Khurana. Tucson, AZ: University of Arizona Press, pp. 237–258.

Roatsch, T, Jaumann, R., Stephan, K., and Thomas, P. C. (2009). Cartographic mapping of the icy satellites using ISS and VIMS data. In *Saturn from Cassini-Huygens*, ed. M. Dougherty, L. Esposito, and S. Krimigis. Dordrecht: Springer, pp. 763–781.

Schenk, P. M. (1995). The geology of Callisto. *J. Geophys. Res.*, **100**, 19,023–19,040.

Schenk, P. M., Chapman, C. R., Zahnle, K., and Moore, J. M. (2004). Ages and interiors: the cratering record of the Galilean satellites. In *Jupiter – The Planets, Satellites, and Magnetosphere*, ed. F. Bagenal, T. Dowling, and W. McKinnon. Cambridge: Cambridge University Press, pp. 427–456.

Schmincke, H.-U. (2004). *Volcanism*. Berlin: Springer-Verlag.

Schultz, P. H. and Gault, D. E. (1975). Seismic effects from major basin formation on the Moon and Mercury. *Moon*, **12**, 159–177.

Sharpton, V. L. (1994). Evidence from Magellan for unexpectedly deep complex craters on Venus. In *Large Meteorite Impacts and Planetary Evolution*, ed. B. O. Dressler, R. A. F. Grieve, and V. L. Sharpton. Boulder, CO: Geological Society of America.

Smrekar, S. E., Elkins-Tanton, L., Leitner, J. J. *et al.* (2007). Tectonic and thermal evolution of Venus and the role of volatiles: implications for understanding the terrestrial planets. In *Exploring Venus as a Terrestrial Planet*, ed. L. W. Esposito, E. R. Stofan, and T. E. Cravens. Washington, DC: American Geophysical Union.

Solomon, S. and Head, J. W. (1982). Evolution of the Tharsis province of Mars: the importance of heterogeneous lithospheric thickness and volcanic construction. *J. Geophys. Res.*, **87**, 9,755–9,774.

Solomon, S. C., Smrekar, S. E., Bindschadler, D. L. *et al.* (1992). Venus tectonics: an overview of Magellan observations. *J. Geophys. Res.*, **97**, 13,199–13,255.

Solomon, S. C., McNutt, R. L., Watters, T. R. *et al.* (2008). Return to Mercury: a global perspective on MESSENGER's first Mercury flyby. *Science*, **321**, 59–65.

Sotin, C., Mitri, G., Rappaport, N., and Schubert, G. (2009). Titan's interior structure. In *Titan*, ed. H. Brown, J-P. Lebreton, and J. H. Waite. Dordrecht: Springer, pp. 61–73.

Spencer, J. R., Carlson, R. W., Becker, T. L., and Blue, J. S. (2004). Maps and spectra of Jupiter and the Galilean Satellites. In *Jupiter – The Planets, Satellites, and Magnetosphere*, ed. F. Bagenal, T. Dowling and W. McKinnon. Cambridge: Cambridge University Press, pp. 689–698.

Spencer, J. R., Barr, A. C., Esposito, L. W. *et al.* (2009). Enceladus: an active cyrovolcanic satellite, In *Saturn from Cassini-Huygens*, ed. M. Dougherty, L. Esposito, and S. Krimigis. Dordrecht: Springer, pp. 683–724.

Spudis, P. D. (1996). *The Once and Future Moon*. Washington, DC: Smithsonian Institution Press.

Spudis, P. D. and Guest, J. E. (1988). Stratigraphy and geologic history of Mercury. In *Mercury*, ed. F. Vilas, C. R. Chapman, and M. S. Matthews. Tucson, AZ: University of Arizona Press.

Spudis, P. D., McGovern, P. J., and Kiefer, W. S. (2011). Large shield volcanoes on the Moon. In *42nd Lunar and Planetary Science Conference*. Houston, TX: Lunar and Planetary Institute, abstract #1367.

Squyres, S. W. and the Athena Science Team (2003). Athena Mars rover science investigation. *J. Geophys. Res.*, **108**(E12), 8062, doi:10.1029/2003JE002121.

Stofan, E. R., Sharpton, V. L., Schubert, G. *et al.* (1992). Global distribution and characteristics of coronae and related features on Venus: implications for origin and relation to mantle processes. *J. Geophys. Res.*, **97**, 13,347–13,378.

Stoffler, D. and Ryder, G. (2001). Stratigraphy and isotope ages of lunar geologic units: chronological standard for the inner solar system. *Space Sci. Rev.*, **96**, 9–54.

Strom, R. G. (1979). Mercury: a post Mariner 10 assessment. *Space Sci. Rev.*, **24**, 3–70.

Strom, R. G. and Sprague, A. L. (2003). *Exploring Mercury: The Iron Planet*. Berlin: Springer-Verlag.

Thomas, P. C., Burns, J. A., Helfenstein, P. *et al.* (2007). Shapes of the saturnian icy satellites and their significance. *Icarus*, **190**, 573–584.

Treiman, A. H. (2007). Geochemistry of Venus' surface: current limitations as future opportunities. In *Exploring Venus as a Terrestrial Planet*, ed. L. W. Esposito, E. R. Stofan and T. E. Cravens. Washington, DC: American Geophysical Union.

Waters, T. R., Head, J. W., Solomon, S. C. *et al.* (2009). Evolution of the Rembrandt impact basin on Mercury. *Science*, **324**, 618–621.

Weber, R. C., Lin, P. Y., Garnero, E. J., Williams, Q., and Lognonne, P. (2011). Seismic detection of the lunar core. *Science*, **331**, 309–312.

Werner, S. C., Ivanov, B. A., and Neukum, G. (2009). Theoretical analysis of secondary cratering on Mars and an image-based study on the Cerberus Plains. *Icarus*, **200**, 406–417.

Whitford-Stark, J. L. (1982). Factors influencing the morphology of volcanic landforms: an Earth–Moon comparision. *Earth Sci. Rev.*, **18**, 109–168.

Wilhelms, D. E. (1980). Geologic map of lunar ringed impact basins. In *Papers Presented to Conference on Multi-Ring Basins*. Houston, TX: Lunar and Planetary Institute, pp. 115–117.

Wilhelms, D. E. and Davis, D. E. (1971). Two former faces of the Moon. *Icarus*, **15**, 368–372.

Williams, D. A. and Howell, R. R. (2007). Active volcanism: effusive eruptions. In *Io After Galileo: A New View of Jupiter's Volcanic Moon*, ed. R. M. C. Lopes and J. R. Spencer. Cambridge: Cambridge University Press, pp. 133–161.

Wilshire, H. G. and Howard, K. A. (1968). Structural patterns in central uplifts of crypto-explosive structures as typified by Sierra Madera. *Science*, **162**, 258–261.

FURTHER READING

Chapter 1

Anderson, R. S. and Anderson, S. P. (2010). *Geomorphology, the Mechanics and Chemistry of Landscapes*. Cambridge: Cambridge University Press.

Beatty, J. K., Petersen, C. C., and Chaikin, A., eds. (1999). *The New Solar System*. Cambridge: Sky Publishing and Cambridge University Press.

Lunine, J. I. (2005). *Astrobiology, a Multidisciplinary Approach*. San Francisco, CA: Addison Wesley.

Chapter 2

Greeley, R. and Batson, R. M., eds. (1990). *Planetary Mapping*. Cambridge: Cambridge University Press (see the chapter by Wilhelms on geologic mapping).

Sabins, F. F. (1997). *Remote Sensing: Principles and Interpretation*, 3rd edn. San Francisco, CA: W. H. Freeman and Company.

Chapter 3

Anderson, R. S. and Anderson, S. P. (2010). *Geomorphology, the Mechanics and Chemistry of Landscapes*. Cambridge: Cambridge University Press.

Basaltic Volcanism Study Project (1981). *Basaltic Volcanism on the Terrestrial Planets*. New York, NY: Pergamon Press.

Greeley, R. and Iversen, J. D. (1985). *Wind as a Geological Process: Earth, Mars, Venus, and Titan*. Cambridge: Cambridge University Press.

Melosh, H. J. (1989). *Impact Cratering, a Geologic Process*. New York, NY: Oxford University Press.

Melosh, H. J. (2011). *Planetary Surface Processes*. Cambridge: Cambridge University Press.

Schmincke, H.-U. (2004). *Volcanism*. Berlin: Springer-Verlag.

Watters, T. R. and Schultz, R. A. (2010). *Planetary Tectonics*. Cambridge: Cambridge University Press.

Chapter 4

Ambrose, W. A. and Williams, D. A., eds. (2011). *Recent Advances and Current Research Issues in Lunar Stratigraphy*. Boulder, CO: Geological Society of America

Heiken, G. H., Vaniman, D. T., and French, B. M. (1991). *Lunar Sourcebook, A User's Guide To The Moon*. Cambridge; Cambridge University Press.

Hiesinger, H. and Head, J. W. (2006). New views of lunar geoscience: an introduction and overview. *Rev. Mineral. Geochem.*, **60**, 1–81.

Jolliff, B. L., Wieczorek, M. A., Shearer, C. K., and Neal, C., eds. (2006). *New Views of the Moon*. Special issue of *Reviews in Mineralogy and Geochemistry*, **60**.

Spudis, P. D. (1996). *The Once and Future Moon*. Washington, DC: Smithsonian Institution.

Wilhelms, D. E. (1987). *The Geological History of the Moon*. Geological Survey, Professional Paper 1348, **302** pp., plus 24 plates.

Chapter 5

Papers in *Journal of Geophysical Research*, 1975, vol. 80, no. 17 on the *Mariner 10* results

Papers in *Science*, 2008, vol. 321, on the first flyby of Mercury by the *MESSENGER* spacecraft

Papers in *Science*, 2009, vol. 324, on the second flyby of Mercury by the *MESSENGER* spacecraft

Papers in *Icarus* special issue *Mercury after two MESSENGER Flybys*, vol. 209, September 2010.

Solomon, S. C. (2011). A new look at the planet Mercury, *Physics Today*, **64**, 50–55.

Strom, R. G. and Sprague, A. L. (2003). *Exploring Mercury: The Iron Planet*. Berlin: Springer-Verlag.

Vilas, F., Chapman, C., and Matthews, M., eds. (1988). *Mercury*. Tucson, AZ: University of Arizona Press.

Chapter 6

Basilevsky, A. T. and McGill, G. E. (2007). Surface evolution of Venus. In *Exploring Venus as a Terrestrial Planet*, ed. L. W. Esposito, E. R. Stofan and T. E. Cravens. Washington, DC: American Geophysical Union.

Bougher S. W., Hunten, D. M., and Phillips, R. J., eds. (1997). *Venus II – Geology, Geophysics, Atmosphere and Solar Wind Environment*. Tucson, AZ: University of Arizona Press.

Esposito, L. W., Stofan, E. R., and Cravens, T. E., eds. (2007). *Exploring Venus as a Terrestrial Planet*. Washington, DC: American Geophysical Union.

Grinspoon, D. H. (1997). *Venus Revealed: A New Look Below the Clouds of Our Mysterious Twin Planet*. Reading, MA: Addison-Wesley.

Journal of Geophysical Research, **97**, 13,063–13,689 and 15,921–16,382 (special Magellan issues Nos. E8 and E10).

Roth, L. and Wall, S. D. (1995). *The Face of Venus: The Magellan Radar Mapping Mission*. NASA Special publication 520.

Chapter 7

Bell, J., ed. (2008). *The Martian Surface: Composition, Mineralogy, and Physical Properties*. Cambridge: Cambridge University Press.

Boyce, J. M. (2002). *Mars*. Washington, DC: Smithsonian Institution Press.

Carr, M. H. (2006). *The Surface of Mars*. Cambridge: Cambridge University Press.

Kallenbach, R., J. Geiss, and W. K. Hartmann, eds. (2001). *Chronology and Evolution of Mars*. Dordrecht: Kluwer Academic Publishers.

Kieffer, H. H., B. M. Jakosky, C. W. Snyder, and M. S. Matthews, eds. (1992). *Mars*. Tucson, AZ: The University of Arizona Press.

Chapter 8

Bagenal, F., Dowling, T., and McKinnon, W., eds. (2004). *Jupiter – The Planet, Satellites, and Magnetosphere*. Cambridge: Cambridge University Press.

Greeley, R., Chyba, C. F., Head, J. W. *et al.* (2004). Geology of Europa. In *Jupiter – The Planets, Satellites, and Magnetosphere*, ed. F. Bagenal, T. Dowling, and W. McKinnon. Cambridge: Cambridge University Press, pp. 329–362.

Lopes, R. M. C. and Spencer, J. R., eds. (2007). *Io after Galileo: A New View of Jupiter's Volcanic Moon*. Chichester: Springer in association with Praxis Publishing.

Moore, J. M., Chapman, C. R., Bierhaus, E. B. *et al.* (2004). Callisto. In *Jupiter – The Planets, Satellites, and Magnetosphere*, ed. F. Bagenal, T. Dowling, and W. McKinnon. Cambridge: Cambridge University Press, pp. 397–426.

Pappalardo, R. T., McKinnon, W. B., and Khurana, K., eds. (2009). *Europa*. Tucson, AZ: University of Arizona Press.

Schenk, P. (2010). *Atlas of the Galilean Satellites*. Cambridge: Cambridge University Press.

Chapter 9

Brown, R. H., Lebreton, J-P., and Waite, J. H., eds. (2009). *Titan from Cassini–Huygens*. Dordrecht: Springer.

Dougherty, M. K., Esposito, L. W., and Krimigis, S. M., eds. (2009), *Saturn from Cassini–Huygens*. Dordrecht: Springer.

Greenberg, R. and Brahic, A., eds. (1984). *Planetary Rings*. Tucson, AZ: University of Arizona Press.

Chapter 10

Bergstrahl, J. T., Miner, E. D., and Shapley Matthews, M., eds. (1991). *Uranus*. Tucson, AZ: University of Arizona Press.

Burns, J. A. and Shapley Matthews, M., eds. (1986). *Satellites*. Tucson, AZ: University of Arizona Press.

Cruikshank, D. P., ed. (1995). *Neptune and Triton*. Tucson, AZ: University of Arizona Press.

Hunt, G. E. and Moore, P. (1994). *Atlas of Neptune*. Cambridge: Cambridge University Press.

INDEX

Page numbers in **bold type** indicate where a term is defined; page numbers in *italics* refer to images or diagrams.